Advanced Series in Agricultural Sciences 22

Co-ordinating Editor: B. Yaron, Bet-Dagan

Editors: H. Van Keulen, Wageningen
L.D. Van Vleck, Ithaca
F. Tardieu, Montpellier

Advanced Series in Agricultural Sciences

K.K. Tanji   B. Yaron   (Eds.)

Management of Water Use in Agriculture

With 53 Figures and 41 Tables

Springer-Verlag
Berlin Heidelberg New York
London Paris Tokyo
Hong Kong Barcelona
Budapest

Prof. Dr. KENNETH K. TANJI
University of California
Dept. of Land, Air and Water Resources
Veihmeyer Hall
Davis, CA 95616
USA

Prof. Dr. BRUNO YARON
Agricultural Research Organization
Institute of Soils and Water
The Volcani Center
Bet Dagan 50250
Israel

ISBN 3-540-57309-7 Springer-Verlag Berlin Heidelberg New York
ISBN 0-387-57309-7 Springer-Verlag New York Berlin Heidelberg

Library of Congress Cataloging-in-Publication Data. Management of water use in agriculture / edited by K.K. Tanji and B. Yaron.   p.   cm. – (Advanced series in agricultural sciences:
v. 22) Includes bibliographical references and index. ISBN 3-540-57309-7. – ISBN 0-387-57309-7.
1. Irrigation efficiency. 2. Water-supply, Agricultural. I. Tanji, Kenneth K. II. Yaron, B. (Bruno).
1929–   . III. Series: Advanced series in agricultural sciences; 22 S619. E34M36   1994   333.9′ 13
– dc20   93-43691.

Typesetting: Macmillan India Ltd., Bangalore-25

SPIN: 10038576   31/3130/SPS – 5 4 3 2 1 0 – Printed on acid-free paper

Preface

As the world population increases, there is increasing competition for water quantity as well as water quality. Since agricultural water use is typically large in comparison to other water uses such as municipal and industrial, it is likely that some shift in water from agriculture to other uses will occur. Moreover, with increasing concern for the environment, additional water is expected to be transferred from agriculture to fish and wildlife, recreational uses, and other ecological and esthetic values. Therefore, it is essential that wise management and judicial use of water for crop and animal production be promulgated.

This book on *Management of Water Use in Agriculture* addresses the above dilemma of increasing competition for freshwater resources, and the need for more efficient use of water in agriculture. Our goal was to focus on water management in agriculture from a wide range of perspectives and present them in one volume for the reader to assess Water Resources and Water Quality (Part I). Water Conservation and Technology (Part II), Problem Water Uses and Treatment (Part III), and Policy and Management Evaluation (Part IV).

Each of the 13 chapters in this book addresses a specific aspect in the management of water use in agriculture. Part I contains overviews on global water resources and agricultural water use as well as water quality considerations for crop and animal production. Part II consists of five chapters on water conservation and technology, including crop irrigation, irrigation application systems, drainage and shallow groundwater management, runoff irrigation, and efficient methods of water use in rainfed agriculture. Part III comprises three chapters on the use and treatment of problem waters, including irrigation with saline waters and treated sewage effluents as well as treatment and disposal of unusable drainage waters. Part IV contains three chapters on the economics of nonuniform water infiltration, irrigation under drought conditions, and management and policy issues in irrigation of agricultural crops.

These 13 chapters are, however, by no means a comprehensive and exhaustive treatise on that specific aspect. But what the chapters do present are contemporary appraisals on important issues and problems on several topics in irrigated agriculture, use of marginal quality waters for irrigation, rainfed agriculture, and agricultural water policy matters. Collectively, the chapters offer valuable insights to the many facets in managing water use in agriculture under increasing constraints on available quantities and qualities of water.

The editors of this book have brought together a select group of 24 contributing authors having expertise in different and numerous disciplines for a scientific exchange of views which is indispensable to the understanding of water use in agriculture. Each of the chapters was peer-reviewed by two or more colleagues to whom we are indebted for their critical reviews.

The editors believe that *Management of Water Use in Agriculture* will be of interest and value to a wide range of readers. This book presents an up-to-date assessment of recent developments and a research knowledge base for those involved in the planning, managing, and evaluating water use in agriculture.

The editors acknowledge that contributions from our colleagues are of high quality. They certainly met our goals and we thank them for their efforts. We apologize for the unexpected delays in getting this book published, not an unusual problem for multi-authored contributions.

Davis, USA   K. K. Tanji
Bet Dagan, Israel   B. Yaron
March 1994

Contents

Part IV   Policy and Management Evaluations

11   Economics of Non-Uniform Water Infiltration in
       Irrigated Fields
       E. Feinerman. With 6 Figures   249

12   Irrigation Management Under Drought Conditions
       N.K. Whittlesey, J. Hamilton, D. Bernardo, and
       R. Adams. With 1 Figure   269

Contributors

Adams, R.

Department of Agricultural
Economics
Washington State University
Pullman, WA 99162-6210
USA

Asano, T.

Department of Civil and
Environmental Engineering
University of California
Davis, CA 95616-8628, USA

Ben-Asher, J.

Blaustein Institute for Desert
Research Ben Gurion
University of the Negev
Sede Boqer Campus 84993
Israel

Berliner, P.R.

Blaustein Institute for Desert
Research, Ben Gurion
University of the Negev
Sede Boqer Campus 84993
Israel

Bernando, D.

Department of Agricultural
Economics
Washington State University
Pullman, WA 99162-6210
USA

Clemmens, A.J.

US Water Conservation
Laboratory, USDA-ARS
Phoenix, AZ 85040, USA

Dedrick, A.R.

US Water Conservation
Laboratory, USDA-ARS
Phoenix, AZ 85040, USA

Dinar, A.   Department of Economics
                                 Ben Gurion University of Negev
                                 PO Box 653
                                 Beer Sheva 84105, Israel

Enos, C.A.   Department of Land, Air
                                 and Water Resources
                                 University of California
                                 Davis, CA 95616-8628, USA

Feinerman, E.   Department of Agricultural
                                 Economics, Faculty of
                                 Agriculture, The Hebrew
                                 University of Jerusalem
                                 PO Box 12
                                 Rehovot 76100, Israel

Frenkel, H.   Institute of Soils and Water
                                 Agricultural Research Organization
                                 The Volcani Center
                                 Bet Dagan 50250, Israel

Grattan, S.   Department of Land, Air
                                 and Water Resources
                                 University of California
                                 Davis, CA 95616-8628, USA

Hadas, A.   Institute of Soil and Water
                                 Agricultural Research Organization
                                 The Volcani Center
                                 Bet Dagan 50250, Israel

Hamilton, J.   Department of Agricultural
                                 Economics
                                 Washington State University
                                 Pullman, WA 99162-6210
                                 USA

Lee, E.W.   Consulting Engineer
                                 32 Thurles Place
                                 Almeda, CA 94501, USA

Letey, J.   Department of Soil and
                                 Environmental Sciences
                                 University of California
                                 Riverside, CA 92521, USA

Meiri, A.   Institute of Soils and Water
                                   Agricultural Research Organization
                                   The Volcani Center
                                   Bet Dagan 50250, Israel

Murugaboopathi, C.   Department of Biological
                                   and Agricultural Engineering
                                   North Carolina State University
                                   Raleigh, NC 27695-7625, USA

Plaut, Z.   Institute of Soils and Water
                                   Agricultural Research Organization
                                   The Volcani Center
                                   Bet Dagan 50250, Israel

Rawitz, E.   Department of Soil and
                                   Water Sciences, Faculty of
                                   Agriculture, The Hebrew
                                   University of Jerusalem
                                   Rehovot 76100, Israel

Skaggs, R.W.   Department of Biological
                                   and Agricultural Engineering
                                   North Carolina State University
                                   Raleigh, NC 27695-7625, USA

Tanji, K.K.   Department of Land,
                                   Air and Water Resources
                                   University of California
                                   Davis, CA 95616-8628, USA

Whittlesey, N.K.   Department of Agricultural
                                   Economics
                                   Washington State University
                                   Pullman, WA 99164-6210
                                   USA

Yaron, B.   Institute of Soils and Water
                                   Agricultural Research Organization
                                   The Volcani Center
                                   Bet Dagan 50250, Israel

Part I Water Resources and Water Quality

1 Global Water Resources and Agricultural Use

K.K. Tanji and C.A. Enos

1.1 Introduction

With the world's population growing from 1.6 billion to more than 5 billion
in the twentieth century, agricultural use of water has progressively increased to meet the growing demands for food and fiber. About 17%, or 250
million ha of the world's cropland is irrigated today (Postel 1989). In some
countries, more than half of the domestic food production relies on irrigated
agriculture. Development of new irrigation projects in the past two decades
have markedly slowed down. The factors contributing to this slowdown
include: (1) a general economic demise, with fewer investments made in
agriculture and water development; (2) increase in competition for scarce
water resources from other sectors of society; (3) rising costs of developing
new supplies; and (4) potentially detrimental off-site environmental and
ecological effects from irrigated agriculture, such as wildlife toxicity from the
discharge of agricultural drainage waters into the Kesterson National
Wildlife Refuge in California's San Joaquin Valley (NRC 1989), and the
desiccation of the Aral Sea and the accompanying ecological, environmental, and public health impacts in the Commonwealth of Independent
State's central Asian plains (Micklin 1991).

This book, therefore, is a timely contribution, assessing the management
of water use in agriculture, because water can be stretched only so far. This
introductory chapter presents an overview of global water resources, agricultural water uses, anticipated impacts on future agricultural water use, and
strategies for the future.

1.2 Global and Continental Water Resources

1.2.1 World Water Cycle and Budget

The biosphere, which may be defined as "that part of the earth in which life
exists" (Hutchinson 1970), is sustained through its interaction with three
other spheres – the atmosphere, the lithosphere, and the hydrosphere

Adv. Series in Agricultural Sciences, Vol. 22
K.K. Tanji/B. Yaron (Eds.)
© Springer-Verlag Berlin Heidelberg 1994

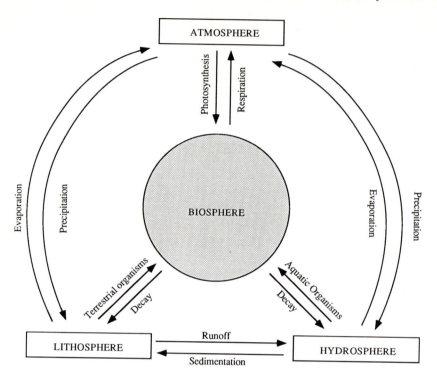

Fig. 1.1. The role of the hydrosphere in sustaining the biosphere

(Fig. 1.1). These three spheres allow the cyclic transfer of biologically important materials to and from the biosphere. The hydrosphere is that part of the earth where water and/or water vapor is held. It extends as far as 5 km under the earth's crust and 11 km above the earth's surface. The hydrosphere is essential to the biosphere as its source of water for, without water, life as we know it cannot exist.

The hydrosphere contains approximately 1.4 billion km^3 of water in its three phases: liquid, ice, and vapor (van der Leeden et al. 1990). However, most of that water is saline. Approximately 97.2% is present as seawater in the oceans (Fig. 1.2). In addition, much of the fresh water is frozen. Approximately 2.15% is in the form of polar ice caps and glaciers. Consequently, only 0.65 % (8.5 million km^3) of the total volume of water is in the form of fresh water. Of that fresh water, approximately 0.16% is in the form of water vapor in the atmosphere. The remaining water, 0.49% of the earth's total water, is present as groundwater or in lakes, rivers and streams, and is the water available for use by humans. However, not all of these supplies of water are economically manageable or usable.

Water is continually moving from one reservoir to another in what is called the hydrologic cycle (Fig. 1.3). The primary driving force for the water

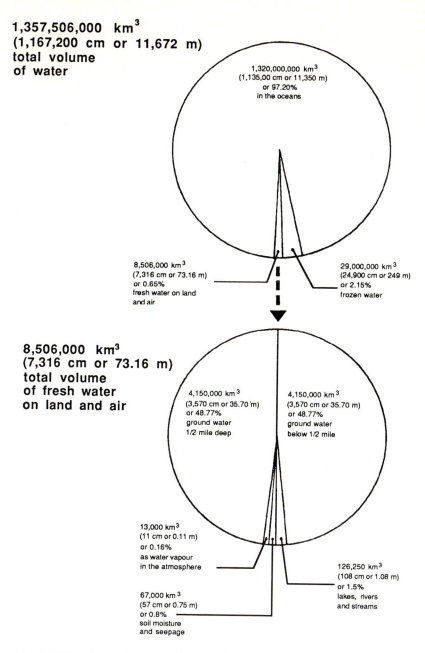

1,357,506,000 km³
(1,167,200 cm or 11,672 m)
total volume
of water

1,320,000,000 km³
(1,135,00 cm or 11,350 m)
or 97.20%
in the oceans

8,506,000 km³
(7,316 cm or 73.16 m)
or 0.65%
fresh water on land
and air

29,000,000 km³
(24,900 cm or 249 m)
or 2.15%
frozen water

8,506,000 km³
(7,316 cm or 73.16 m)
total volume
of fresh water
on land and air

4,150,000 km³
(3,570 cm or 35.70 m)
or 48.77%
ground water
1/2 mile deep

4,150,000 km³
(3,570 cm or 35.70 m)
or 48.77%
ground water
below 1/2 mile

13,000 km³
(11 cm or 0.11 m)
or 0.16%
as water vapour
in the atmosphere

126,250 km³
(108 cm or 1.08 m)
or 1.5%
lakes, rivers
and streams

67,000 km³
(57 cm or 0.75 m)
or 0.8%
soil moisture
and seepage

Fig. 1.2. World water inventory. (van der Leeden et al. 1990)

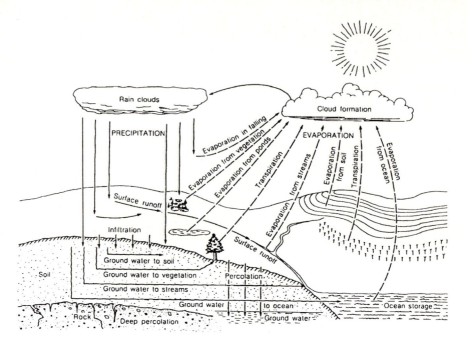

Fig. 1.3. A descriptive representation of the hydrologic cycle. (van der Leeden et al. 1990)

cycle is solar energy. Water enters the atmosphere through evaporation and transpiration, where it remains for an average of 10 days before returning to the earth as rain or snow (Penman 1970). The oceans receive the majority of the precipitation, a mean of 107 cm/yr (Berner and Berner 1987). The continents receive significantly less precipitation, 74 cm/yr, which is approximately two-thirds of the volume that the oceans receive (Berner and Berner 1987). Part of the precipitation on land runs off into rivers or accumulates in lakes. Other parts flow underground to be discharged to lakes and rivers at a later time.

To achieve a mass balance on water, global evaporation must balance global precipitation. However, the mean evaporation rate (117 cm/yr) over the oceans is generally greater than precipitation (107 cm/yr) (Berner and Berner 1987; Fig. 1.4). On the other hand, precipitation on land (74 cm/yr) generally exceeds evaporation (49 cm/yr). The difference in each case is made up by water transported from the oceans to the land in the form of water vapor, or by water returned to the oceans as river runoff. Thus, the global precipitation/evaporation rate is 97 cm/yr.

Although on a global scale evaporation equals precipitation, there is considerable variation in evaporation and precipitation from one region to another. Figure 1.5 illustrates the mean annual precipitation for different regions of the world. Net precipitation, which equals precipitation minus

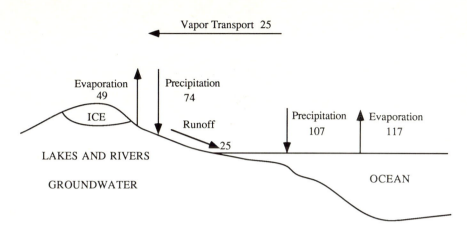

Fig. 1.4. Fluxes of water in the hydrologic cycle. Values are in cm/yr. (Berner and Berner 1987)

evaporation, is positive if precipitation exceeds evaporation for the region (Fig. 1.6). Net precipitation is greatest near the equator and at 35° to 60° N and S. Net evaporation occurs when evaporation exceeds precipitation, i.e., when net precipitation is negative. In order for evaporation to occur, there must be a source of heat, low moisture in the air, and a source of water to evaporate. Evaporation is highest near the equator from 30° N to 30° S. The greatest negative net precipitation occurs over subtropical regions such as the Gulf Stream. In the subtropical regions, high evaporation rates result in higher regional ocean salinities due to the removal of pure water during evaporation and the subsequent concentration of the remaining dissolved mineral salts.

1.2.2 Continental Water Resources

The water resources of each continent vary, depending on precipitation, evaporation, runoff, stream flow, and groundwater storage. The continent with the greatest total precipitation is Asia (excluding Europe) (32 690 km^3/yr) (van der Leeden et al. 1990). However, the continent with the greatest precipitation per square kilometer is South America, which receives 1658 mm of mean annual precipitation, whereas Asia only receives 726 mm per year. The continent with the lowest mean annual precipitation per square kilometer is North America, which receives only 670 mm/yr. Evaporation is another major factor in continental water resources. The evaporation rate is greatest in South America, which loses 1065 mm of water per

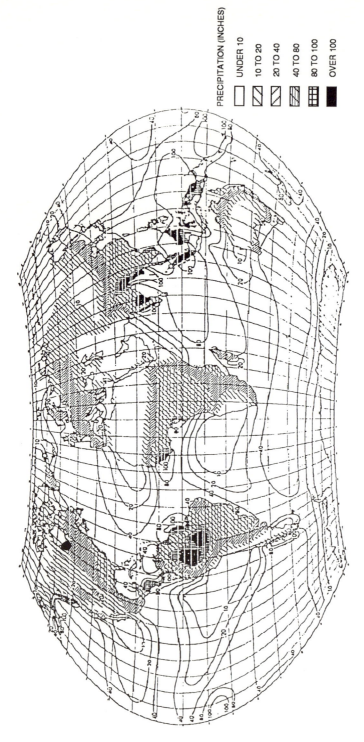

Fig. 1.5. General pattern of annual world precipitation. (van der Leeden et al. 1990)

Fig. 1.6. Net precipitation (precipitation minus evaporation) as a function of latitude. (Berner and Berner 1987)

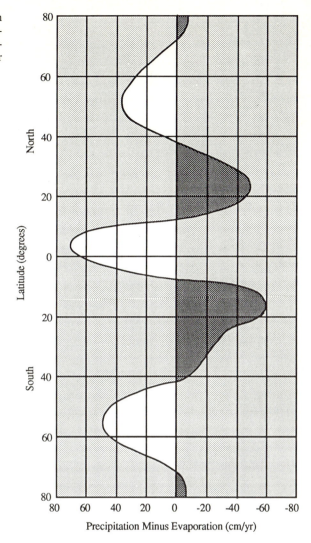

year. North America has the lowest evaporation rate, 383 mm/yr. Not surprisingly, river runoff is directly related to precipitation and evaporation. Total river runoff for a continent is equal to the net precipitation. The greatest river runoff is found in South America (583 mm/yr), which receives the highest precipitation per square kilometer. The least river runoff occurs in Africa (139 mm/yr). This is the result of a low precipitation rate (686 mm/yr) and a high evaporation rate (547 mm/yr). North America, which has the lowest precipitation rate (670 mm/yr), has a median river runoff of 287 mm due to its low evaporation rate (383 mm/yr). Another continental water resource is stream flow. The greatest annual stream flow

Table 1.1. Estimated groundwater storage by continent (van der Leeden et al. 1990)

Continent	Total storage (Millions of km^3)	Storage per km^2 (km^3/km^2)
Europe	1.6	0.16
Asia	7.8	0.17
Africa	5.5	0.18
North America	4.3	0.21
South America	3	0.17
Australia	1.2	0.14

per area of land is in Oceania (New Zealand, Fiji, and New Guinea), which has 1610 mm of stream flow per year (Shiklamanov 1990). The greatest total stream flow occurs in Asia, which has 14 410 km^3 of flow per year. The lowest stream flow per land area occurs in Australia and Tasmania. Their combined annual stream flow is 45 mm/yr. The lowest total stream flow also occurs in Australia and Tasmania, which have a total flow of 348 km^3/yr.

Groundwater storage is another important continental water resource. Table 1.1 lists the estimated groundwater storage by continent (van der Leeden et al. 1990). The largest total estimated groundwater storage is in Asia (7.8 million km^3). The largest groundwater storage per square kilometer is in North America, with 0.21 million km^3/km^2 of groundwater. Asia's groundwater storage is only 0.17 million km^3/km^2. Australia is the continent with the least total groundwater storage, as well as the least groundwater per land area (1.2 and 0.14 million km^3/km^2, respectively).

1.2.3 Global Water Uses

Water is used by everyone every day. Water is a necessary component in society's agricultural, industrial, and domestic sectors. In the face of a rapidly growing population, it is not surprising that the majority of the world's water is used for agriculture. In 1990, approximately 70% of the total water usage was for agriculture, while 21% was used for industry (Shiklamanov 1990). The remainder was for municipal water supply and other uses.

The percent of water used by each sector varies greatly from country to country. Egypt uses an average of 98% of its annual water consumption for agriculture (van der Leeden et al. 1990). The rest of its water is used for public supply and industry. On the other hand, Malta uses an average of 100% of its annual water consumption for public supply. No water is used in agriculture. The biggest consumer of water for industry is Finland, which uses approximately 85% of its water for industry.

The USA is one of the greatest consumers of water, using approximately 1986 m^3 of water per person per year (van der Leeden et al. 1990). This results in a total consumption of 472 000 km^3/yr. In sharp contrast, Malta's annual per capita use is only 60 m^3, and its total consumption is only 0.023 km^3.

1.3 Agricultural Uses of Water

1.3.1 United States Agricultural Water Uses

The majority of the cropland in the USA is rainfed. In 1982, only 13.4% of the nation's total cropland was irrigated (van der Leeden et al. 1990). However, irrigated crops accounted for 31.8% of the total value of agricultural crops (van der Leeden et al. 1990). In 1985, agriculture accounted for 42% of all fresh water diversions in the USA. This amounts to approximately 533 750 million l/day (Moore et al. 1987). Of that water, 97% was for irrigation and 3% for livestock. Corn, hay and wheat, the three crops grown on the highest number of hectares accounted for 3.9, 3.4 and 1.9 million ha of cropland, respectively, in 1982 (van der Leeden et al. 1990).

Large regional differences occur in the amount of land irrigated. The majority of irrigation occurs in the western states, with California leading all the states in the total amount of farm products sold from irrigated farms. Irrigation makes it possible to grow crops in arid areas of the western states, where the rainfall alone is too low to support agriculture. In the eastern, more humid states, irrigation is used mainly to prevent crop losses from drought and frost, to improve productivity, and to increase the yield and quality of fruit, vegetables, and specialty crops (Moore et al. 1987).

Approximately two-thirds of the water used in irrigation is derived from surface water and the remaining third is derived from groundwater (Moore et al. 1987). The majority of the surface water use is in the west, where public development of surface-water supply systems has made possible the long-distance transportation and distribution of water. The Plains states use mostly groundwater from private development of ground water supplies.

Livestock production is the agricultural water use with the greatest economic return for the water used. Water use for livestock amounts to approximately 3% of the total agricultural water consumption but accounts for more than 50% of agricultural cash receipts (Moore et al. 1987). In 1985, approximately 17 million l/day were used for livestock, including aquaculture (van der Leeden et al. 1990). Livestock water uses include drinking water for livestock, evaporation from stock-watering ponds, and related sanitation and waste disposal practices.

1.3.2 United States Irrigation Budget

The USA is one of the top consumers of water for irrigation. In 1986, it had the fourth largest land area under irrigation, surpassed only by China, India, and what was then known as the USSR (van der Leeden et al. 1990). Unfortunately, irrigation is a comparatively inefficient process. Less than half of the water diverted for irrigation is actually consumed by the crops. Significant amounts of water are lost in the transport and application of water. Figure 1.7 illustrates the irrigation water budget of the USA. Of gross water diversions, 78% actually makes it to the farm, while the other 22% is lost during delivery and by spills. Once on the farm, another 37% is lost. Only 41% of the gross water diversion makes it to the crops for consumptive use. About 46% of the gross diversion is returned to the system, while 1.3% is lost through evaporation and seepage.

1.3.3 Global Agricultural Water Uses

The percent of water used for agriculture varies greatly from country to country. For example, Oman uses an average of 98% of its annual water consumption for agriculture, while Malta does not use any (van der Leeden et al. 1990). Agricultural water uses include irrigation of food and fiber crops; livestock watering; irrigation for agroforestry and biomass; and aquaculture.

In 1981, the total land area used for agriculture in the world was approximately 4570 million ha (van der Leeden et al. 1990). The three largest agricultural regions are Subsaharan Africa, Latin America, and the

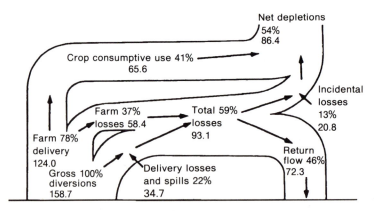

Fig. 1.7. Irrigation water budget of the United States. (Percent of diversion; billions gallons per day). (van der Leeden et al. 1990)

Commonwealth of Independent States. These three regions make up almost half of the total agricultural land area. Approximately 67% of the agricultural land is used to support animal agriculture, while 30% is used for annual crops and 2% is used for multi-year crops (Aspen Institute 1983).

Of the total land area used for agriculture in 1981 (4570 million ha), only 212 million ha, or approximately 5%, were irrigated (Aspen Institute 1983). Table 1.2 shows the total irrigated area for the world divided up into seven regions for 1972, 1982 and 1987. According to this table, the total irrigated area has increased steadily for every region, except North and Central America. The irrigated area in North and Central America increased between 1972 and 1982, from 21.8 million ha to 27.2 million ha, but decreased in 1987 to 25.7 million ha. In 1987, the total global irrigated area was approximately 227 million ha. The region with the largest irrigated area was Asia, with 142 million ha. The region with the least irrigated area was Oceania, with 2.1 million ha.

Food crops consist of cereals, oilseeds, and other crops. In 1981, North America had the greatest cereals and oilseeds production, 344.6 and 68.6 million metric tons, respectively (Aspen Institute 1983). The cereal crops produced in the greatest quantities were corn, with almost 200 million metric tons, followed by wheat, with 86.5 million metric tons. The oilseed crop produced in greatest quantity was soybeans, with a total production of 55.9 metric tons. Other crops include roots/tubers, fruits, coffee, sugar, and specialty crops.

Fiber crops include cotton, hemp, wool, and other similar products. The largest fiber producer in 1981 was South Asia, with 5 million metric tons (Aspen Institute 1983). The largest fiber crop produced was jute (2.5 million metric tons), followed by cotton lint (2.1 million metric tons).

The largest producer of livestock in 1981 was North America, with a total production of 26.9 million metric tons of meat (Aspen Institute 1983). The

Table 1.2. Irrigated area for each region of the world in 1972, 1982 and 1987 (United Nations 1988)

Continent	Irrigated area (1000 ha)		
	1972	1982	1987
Africa	9 125	10 319	11 058
North and Central America	21 838	27 161	25 740
South America	6 032	6 952	8 586
Asia	113 888	135 297	142 301
Europe	11 910	15 079	16 833
Oceania[a]	1 606	1 864	2 105
USSR	11 991	18 608	20 485
World	176 390	216 132	227 108

[a] Includes Australia, Fiji and New Zealand

majority of the meat was beef and veal (11.1 million metric tons), followed by pork (8.1 million metric tons). The 1981 livestock inventory for North America consisted of approximately 733 million animals, including 477 million chickens and 125 million cattle.

Another agricultural use of water is agroforestry which can be defined as "land use systems in which trees are grown in association with agricultural crops and/or pasture and livestock, either in a spatial arrangement or a time sequence, and in which there is both an economic and ecological interaction between the tree and non-tree components of the system" (Young 1988). Based on the nature of the components, agroforestry can be classified into three tree systems (Nair 1985): (1) agrisilviculture which involves crops and trees including shrubs/vines/trees, and trees; (2) silvopastoral involving pasture/animals, and trees; and (3) agrosilvopastoral involving crops, pasture/animals, and trees.

The function of agroforestry may be productive or protective. Productive functions include the output of timber, fuelwood, charcoal, fodder, fruits, nuts, oils, and thatching material. Protective functions include windbreaks, shade, improvement in soil fertility, and soil conservation.

Aquaculture is another agricultural water use for the culture of fish, mollusks, crustaceans, of seaweeds in fresh (warm or cold), brackish, or marine waters. Asia leads the world in aquaculture production. In 1980, Asia produced 84% (7.3 million metric tons) of the total global aquaculture production (8.7 million metric tons), including approximately 2.5 million metric tons of finfish, 2.6 million metric tons of mollusks, and 2.2 million metric tons of seaweed (Pillay 1982). In contrast, Africa produced only 0.05% (approximately 4500 metric tons) of the total aquaculture production in 1980.

1.4 Anticipated Impacts on Future Agricultural Water Use

1.4.1 Competition for Water Quantity and Quality from Other Sectors

A continuing increase in the population will result in an increasing demand for water. It is estimated that, from 1980 to 2000, global water use is expected to increase by 20% in North America, 32% in Africa, and 95% in South America (Shiklamanov 1990). By 2000, the global water demand will reach about 11.6% of global resources (Shiklamanov 1990). Some parts of the world will require 100% of the available water resources. For example, Egypt, a country of 55 million people, depends almost entirely on the Nile River for its water. It is likely that its water demands will exceed its supplies within a decade or so.

As the population increases and cities grow, competition for water between cities and farms is expected to increase. Municipal world water demand is expected to increase from 6% to 8.5% between 1990 and 2000 (Shiklamanov 1990). Some countries are already facing a substantial loss in agricultural water supplies to municipal use. In China, dozens of cities face acute shortages. Farmers near Beijing could lose 30 to 40% of their current water supply to municipal uses by the next decade. In the USA, water is being transferred from agriculture by cities that are willing to pay a premium to ensure sufficient water for future growth. In some instances, farmers can make more money by selling their water to cities than by using it to irrigate crops. Another loss of agricultural water to cities is through water ranching. In states where it is difficult to buy water rights, cities purchase agricultural land as "water farms".

The increase in domestic water demands will inevitably be accompanied by an increase in industrial water demands. The world water demand for industry is expected to increase from 21% in 1990 to 25% in 2000 (Shiklamanov 1990). The value of water in crop production is often less than in other activities, such as industry. In China, a given amount of water in industry will generate 60 times the economic value of the same amount in agriculture. Therefore, it is likely that as water supplies decrease, water use will be shifted from the agricultural sector to the more economically profitable industrial sector.

A growing concern for the environment has created another source of competition for water supplies. As more attention is being given to instream values, such as fish and wildlife habitat, recreational uses, and other ecological and esthetic values, demands for increased instream water are being heard. In the past, water was diverted for offstream uses without regard to these instream values. However, in the last few years, all the western states, except New Mexico, have taken action to protect instream water values. Some state agencies now have the right to buy water from private holders to boost stream flow levels. The public trust doctrine can also be used to protect instream values. This doctrine, which states that governments can hold certain rights in trust for people and can take steps to protect these rights from private interests, could have far-reaching effects, since it allows existing water rights to be revoked to prevent violation of a public trust. Recently, the public trust doctrine was invoked to protect California's Mono Lake, a scenic lake that was shrinking due to excessive water diversions to metropolitan southern California. Private-interest groups are also buying water for the protection of wetlands and waterfowl. Recently, the Nevada Waterfowl Association bought 43 175 m^3/yr of water from the Truckee-Carson Irrigation District for wetlands and waterfowl protection (Postel 1989).

Not only will competition for water quantity increase, but competition for water quality will also increase. As new water resources dwindle, so will the amount of water available to dilute wastewater discharges. In 1985, 48 000 km of rivers were severely affected by non-point pollution and an additional 140 000 km were moderately affected (van der Leeden et al. 1990).

Between 1972 and 1982, approximately 0.67 million ha of lakes were degraded (van der Leeden et al. 1990). With a reduction of good-quality fresh water, farmers will have to use lower-quality water for irrigation. Irrigation pollutants include salts, pesticides, nutrients, sediments, and trace elements. In arid regions, such as the western USA, the principal pollutants are salts. Sources of salinity include minerals dissolved in natural water or leached from soil and rock materials, salts added from previous uses of water, fertilizers, and other farm chemicals. The use of highly saline water results in poor crop yields and an inability to grow some salt-sensitive crops.

1.4.2 Increasing Regulation on Agricultural Waters

In 1972, the US Congress passed the Clean Water Act, the first law to establish a national strategy for reducing water pollution. The objective of the Act was "to restore and maintain the chemical, physical and biological integrity of the nation's water" (Olexa and Smart 1990). The act requires obtaining a permit before discharging from point-source, which is defined as "any discernable, confined, and discrete conveyance from which a pollutant may be discharged" (Olexa and Smart 1990). The act exempted non-point-source pollution which is defined as "diffuse pollution resulting from land runoff, precipitation, atmospheric deposition, drainage, or seepage, rather than a pollutant discharge from a specific, single location" (US General Accounting Office 1990). Non-point-source pollutants include agricultural and irrigation runoffs, urban runoff, and mining wastes. Although collected drainage waters are discharged as point sources, agricultural irrigation return flow was excluded from the Clean Water Act's definition of point-source pollution. In 1987, the government took a step towards curtailing non-point-source pollution by passing the Water Quality Act, which requires states to develop programs and activities that control non-point-sources. However, it leaves the primary responsibility of non-point-source pollution with the states, and the adoption of recommended management practices is voluntary.

The majority of the nation's remaining water-quality problems are due to non-point-source pollution. A large portion of the non-point-source pollution is agricultural. Seventy-six percent of the impaired acres of lake water are affected by non-point-source pollution (US Environmental Protection Agency 1989). Of that 76%, 64% is due to agriculture. According to the EPA (1989), 50 to 70% of assessed surface waters are adversely affected by agricultural non-point-source pollution stemming from soil erosion of cropland, overgrazing, and the application of pesticides and fertilizers.

As water-quality concerns heighten, regulations on agricultural water are expected to increase. In the future, it is likely that agricultural exemptions from the Clean Water Act will be revoked. Regulations similar to those that apply to point-source pollution will probably be applied to non-point-source pollution including that from agriculture. There will likely be uniform, national toxic-waste regulations to ensure water-quality standards. National registration and control of pesticides will most likely be enforced.

1.4.3 Ecological Impacts

In addition to the loss of agricultural water through increased population and regulation, there is the threat of climate change from increasing buildup of carbon dioxide and other "greenhouse gases" in the atmosphere. Carbon dioxide and other greenhouse gases allow the sun's energy to pass through the atmosphere to the earth, but trap the heat waves that are radiated back. A buildup of these gases could cause a global warming, which would alter temperature, precipitation, winds, humidity, and cloud cover.

Carbon dioxide is released naturally through respiration of oxygen-breathing organisms. However, increasing amounts of carbon dioxide are being emitted from anthropogenic sources, such as the burning of fossil fuels. Certain calculations predict an increase of 17 to 18% in carbon dioxide content by the end of the century and a doubling by 2030–2050 (Shiklamanov 1990). The increase in carbon dioxide could cause a rise in mean air temperatures by 1 °C (compared to 1960 temperatures) and 3 to 4 °C, respectively (Shiklamanov 1990). Such increases in temperature could cause increased variability in precipitation, with a 10% increase in some areas and a 10% decrease in others. A rise in temperature of 1 to 2 °C in combination with a 10% decrease in precipitation would result in a reduction of river runoff by 40 to 70% in arid regions.

These climate changes could have a profound effect on agriculture. The increased temperature would cause an increase in evaporation and transpiration, which would result in the need to irrigate cropland that was previously rainfed. A 3 °C increase in temperature could up irrigation needs by 15%. If this increase in temperature was accompanied by a 10% decrease in rainfall, irrigation needs would rise by 26% (Postel 1989). Globally, the climate change could force irrigation to cease on 5% of existing irrigated cropland. This would remove 13 million hectares from the global base (Postel 1989).

Not all of the possible scenarios are discouraging. It is possible that some areas could benefit from the change in climate. Some regions with harsh climates could find themselves in more favorable conditions, with milder winters, or more abundant rainfall, or both. Historically low rainfall areas, such as India, could receive additional rainfall, which would boost water supplies and allow for additional agricultural activity or increased productivity, or both. Studies on Japanese agriculture concluded that a climate change induced by a doubling of atmospheric carbon dioxide could result in an increase in rice yields of up to 25% (Ausubel 1991).

Deforestation and denudation of watersheds is another factor leading to water scarcity. Deforestation and denudation of vital watersheds upsets the natural water cycle. Without vegetation to trap the water, rainfall runs off as floods instead of percolating into the ground to recharge aquifers. This leaves seasonably high rainfall areas without water during the dry parts of the year. In India, tens of thousands of villages are living with water shortages due to deforestation and denudation of watersheds (Postel 1989).

Soil salination and the frequently accompanying problem of water-logging have plagued irrigated agriculture for 6000 years (Tanji 1990). Many ancient civilizations based on irrigated agriculture in arid climates have failed. Salination and water logging are not unique to ancient civilizations. For example, of the nearly 1 million hectares of irrigated cropland in the San Joaquin Valley's west side in California, about 38% is water logged and 59% salt-affected (NRC 1989). But, only about 21% of the entire west side has opportunities to discharge its saline subsurface drainage water for ocean disposal through the San Joaquin River. The 1983 discovery of selenium poisoning of waterfowl at the Kesterson National Wildlife Refuge from evapoconcentration of irrigation drainage has increased the complexity in managing irrigation-induced water-quality problems (NRC 1989). The 300 ppb selenium in the drain water impounded in the Kesterson Reservoir had biomagnified and bioconcentrated in the aquatic food chain, resulting in a reduced rate of reproduction and gruesome deformity and death among waterfowl. The US Department of Interior has implemented the National Irrigation Water Quality Program to ascertain if selenium and other naturally occurring trace elements in drainage are causing ecological damages at 26 sites in 15 western states. The potential of Kesterson-like wildlife toxicity occurring in other western states and elsewhere in the world where outcrops of geologic formations from the Cretaceous period exist is substantial.

A second example of an ecological crisis derived from irrigated agriculture is that of the Aral Sea in the Central Asian plains in the Commonwealth of Independent States (Micklin 1991). The Amu Daŕya and Syr Daŕya River systems formerly carried considerable water from their headwaters of the Pamir and Tyań Shaú Mountains across the deserts into the Aral Sea, formerly the fourth largest inland body of water. The flow of water in these rivers has been nearly depleted by withdrawals to establish new irrigation projects in the 1950s and 1960s. In 1960, the level of the Aral Sea was about 53.4 m above sea level with a surface area of 68 000 km^2 and a volume of 1090 km^3. By 1989, the water level had dropped to 39.1 m, the surface area had shrunken to 40 400 km^2, and the volume had decreased to 370 km^3. The average salinity in the river increased from 10 000 mg/l in 1960 to 30 000 mg/l in 1989.

By early 1980s, all 24 native species of fish had disappeared, and the catch of commercial fish dropped from 44 000 metric tons in the 1950s to zero. The shrinkage of the Aral Sea resulted in heavy ecological damage to the plants and animals in the delta and shorelines. Exposure of huge tracts of dried-up sea bottom resulted in large quantities of salt deposits and salt/dust storms up to 400 km downwind, damaging natural ecosystems and agricultural lands. The reduction of river flow resulted in salinization and toxic pollution from irrigated lands to such a degree that drinking water supplies have been contaminated and public health has been severely impaired, including a rise in infant mortalities, intestinal diseases, kidney and liver ailments, esophageal cancer, and birth defects. The prognosis for the future of the Aral Sea and health of Central Asians remains a gloomy one (Micklin 1991).

1.5 Strategies for the Future

1.5.1 New Water Supplies

As fresh water supplies dwindle and new sources become more remote, alternative sources of agricultural water need to be found. Three promising sources are reuse of sewage effluent, desalinization of seawater, and use of saline water. Reuse of wastewater effluent has been increasing in popularity recently due to several factors: a need for more water; water pollution control regulations, which mandate expenditures for sewage treatment; legislative directives for more water conservation; escalating energy costs; and the economics of developing remote water sources (Heaton 1981). At least 16 countries in North Africa and the Middle East are currently reusing wastewater for irrigation. Other countries using wastewater for irrigation include South Africa, the Netherlands, Japan, and the USA. In 1978, 275 million m^3 of wastewater were used for agricultural irrigation in the USA (van der Leeden et al. 1990). It is estimated that, by 2000, 673 billion l/day of wastewater will be used. Israel, a country that uses 90% of its existing water sources, relies heavily on wastewater reuse. In 1986, 35% of total wastewater flow was reused for agriculture, irrigating 15 000 ha (Shuval 1987). In the USA, the majority of reclaimed water is used for irrigation of pasture, fodder, fiber, and seed crops, where health concerns are minimal and water of lower quality can be used.

Desalinization is another new source of agricultural irrigation water. Worldwide, approximately 4000 desalinization plants produce a total of 12.9 billion l/day (Abelson 1991). About 60% of the 12.9 billion l/day are produced on the Arabian Peninsula. The principal desalinization process is multi-stage flash (MSF) distillation, in which saltwater is distilled at successively lower temperatures. The secondary process is reverse osmosis, where water is forced through membranes. The disadvantage of desalination is its high cost. The cost of producing potable water from seawater using multistage flash distillation is about $4.00 per 3800 l (Abelson 1991). For domestic use, that is acceptable, but for the farmer it is dismayingly expensive. The cost of reclaiming moderately polluted water by reverse osmosis is more reasonable – approximately $0.50 per 3800 l (Abelson 1991). Progress is being made to reduce the cost of desalinization. Cost-reducing strategies include alternate energy, such as solar and nuclear energy; a dual system of seawater desalinization and electrical generation; improvements in technology; and better management (Dabbagh and Al-Sagabi 1989).

Finally, another promising source of agriculture irrigation water is saline water. In certain situations, saline waters can be used effectively for irrigation. Moderately saline waters may be used for the irrigation of salt-tolerant crops, such as cotton, wheat, barley, sugar beet, alfalfa, rye grass, and wheat grass. Another technique is to use saline water to irrigate crops when they are at a stage of growth that can tolerate the saline water, and to use the normal irrigation water at other stages (Rhoades 1984). Another possibility is the development of new strains of plants that are more salt-tolerant and could be irrigated with saline water. Research indicates that wheat may be bred for greater salt tolerance (Postel 1989). Currently, one promising salt-tolerant substitute is Salicora, a succulent that can be irrigated with seawater and possibly used as a substitute for fodder in dry areas. It can contribute up to 10% of a fodder mix for cattle, sheep, and other livestock (Postel 1989).

1.5.2 Water Conservation

As water becomes more scarce, it becomes increasingly more important to conserve the available water. There are several off-farm and on-farm measures to conserve water (Hoffman et al. 1990; Tanji 1990). The off-farm measures include improvements in water delivery to the farms that are flexible, reliable, and include stream flow controls. Canal seepage is a major problem that could be reduced by a number of canal lining practices. The on-farm measures involve applying water more uniformly and efficiently. Irrigation scheduling would aid in determining the time and amount of water application. Surface irrigation systems, such as furrow, border, and basin methods, may be improved. Furrow irrigation could be substantially improved by using shorter runs, modifying set times, and using tailwater recovery systems. Land leveling, especially the use of laser leveling, can substantially improve the application efficiency of surface systems. Using pressurized systems, such as high and low volume sprinklers, as well as surface and subsurface drip/trickle systems give more precise controls on water application. Reuse of irrigation drain water until it is no longer usable would contribute toward water conservation. Shallow groundwater existing within reach of root-water extraction and of suitable quality could be used to satisfy a portion of the crop evapotranspirational needs.

1.5.3 Better Use of Rainwater

Better use of rainwater could considerably increase the amount of available water in areas with low rainfall. Rainfall collection, runoff concentration, and rainwater harvesting are all used to describe "the collection and concentration of rain falling on natural slopes, cleared and compacted slopes or sealed catchments to irrigate crops or supply water to livestock or to meet domestic needs" (Barrow 1987). Rainfall collection could have considerable potential where rainfall is low. It could allow farmers to grow reasonably secure, reasonably good crops where it is presently too dry. Where rainfall is more plentiful, it could increase the diversity of the crops. Better crop management techniques could also result in better use of the available rainwater. By improving soil permeability and reducing runoff during the rainy season, soil moisture from rainfall can be greatly increased.

1.5.4 Institutional and Policy Changes

Institutional and policy changes could help to reduce much of the water waste. Substantial overuse and mismanagement of water comes from the failure to price it adequately. In the domestic sector, underpricing results in the overuse of water in the household. In the agriculture sector, under-charging results in insufficient funds to maintain canals and other irrigation works adequately. In Third World countries, government revenues from irrigation average no more than 10 to 20% of the full delivery cost (Postel 1989). For example, fees paid by Pakistani farmers cover only 13% of the government's costs (Postel 1989). In California, farmers who receive water from the federal Central Valley's Project have repaid only 5% ($50 million) of project costs ($931 million) over the last 40 years (Postel 1989). Low water costs explain why many farmers do not invest in water-conserving improvements. Pricing strategies to increase the efficiency of agricultural water use include incentive payments to return unused irrigation water to the common supply; tiered water rates; market incentives to conserve water for resale at a profit; and implementation of drainage fees for irrigation runoff water (Thomas and Leighton-Schwartz 1990).

Pricing is not the only available strategy for inducing agricultural water conservation. Several policy changes at federal, state, and local levels could help to improve conservation: water duties that prescribe a maximum amount of water allowed per crop; water conservation planning enforced by water agencies; orders by the government to take steps to eliminate wasteful or unreasonable practices; the use of best management practices made mandatory by the government; and elimination of government water supply subsidies (Thomas and Leighton-Schwartz 1990). When the price of water increases or governments increasingly regulate water use, people will begin to use resources more carefully.

1.6 Summary and Conclusions

Water is a finite resource that is essential for all life forms. Most of the water in the hydrosphere is salty and much of the fresh water is frozen. Moreover, water is not distributed uniformly in space or time. The supply of fresh water resources, both surface and groundwaters, is becoming scarce in many countries and continents. There is increasing competition by various sectors of society for not only water quantity, but also water quality. One of the largest users of our fresh water supply is agriculture.

Agriculture's use of water has progressively increased in the twentieth century to meet the growing demands of food and fiber, as world population increased from 1.6 billion to more than 5 billion. But, in the past two decades, new irrigation projects have markedly slowed down, due in part to economic constraints and environmental concerns. Off-site effects, such as wildlife toxicity at the Kesterson National Wildlife Refuge in California and the salinization of Aral Sea in the Commonwealth of Independent State's central Asian plains, with its devastating ecological, agricultural and public health effects, have forced us to re-examine the practice of irrigated agriculture.

Kesterson and Aral Sea have resulted in critical agricultural and ecological crisis. There will be increasing governmental regulations imposed on the amounts of water diverted for agriculture, as well as on the discharge of agricultural drainage into sensitive environments.

As water becomes scarce, it becomes increasingly important to conserve available water. A number of off-farm and on-farm measures may need to be imposed to use water more efficiently. In some areas, changes in water policy and institutional arrangements or structure may be required.

Because water can be stretched only so far, agriculture will be increasingly challenged to use water more beneficially and efficiently.

References

Abelson PH (1991) Desalination of brackish and marine waters. Science 251(4999):1289
Ahlgren RM (1989) Potential for water reuse in conjunction with desalination systems. Desalination 75(1-3):315–328
Al-Khafaji AA, Howarth DA (1989) Wastewater reclaimation and reuse in Aspen Institute (ed) (1983) Europe, Middle East and North Africa. Desalination 75(1-3):289–314
Aspan Institute (ed) 1983)
Ausubel JH (1991) A second look at the impacts of climate change. Am Sci 79(3):210–221
Ayers RS, Westcot DW (1985) Water quality for agriculture. FAO irrigation and drainage Pap 29 rev 1. FAO, Rome
Bajwa RS, Crosswhite WM, Hostetler JE (1978) Agricultural irrigation and water supply. US Dep Agric, Econ Res Serv, Washington DC
Barrow C (1987) Water resources and agricultural development in the tropics. John Wiley & Sons, New York
Berner EK, Berner RA (1987) The global water cycle. Prentice-Hall, New York
Biswas AK (1987) Irrigation in Africa. ICID Bull 36(2):1–11
Brown EE (1983) World fish farming: cultivation and economics. 2nd edn. AVI, Westport, CT
Bunting AH (1987) Irrigation in Africa's agricultural future. ICID Bull 36(2):12–23
Bushnak AA (1990) Water supply challenge in the Gulf region. Desalination 78(1-3):133–145

Ciuff CB (1989) Water harvesting systems in arid lands. Desalination 72(1-3):149–159
Dabbagh TA, Al-Sagabi A (1989) The increasing demand for desalination. Desalination 73(1-3):3–26
Davis T (ed) (1986) Development of rainfed agriculture under arid and semiarid conditions. Proc 6th Agric Sector Symp, Jan 6–10, 1986, World Bank
Deevey ES Jr (1970) Mineral cycles. Sci Am 223(3):81–92
Dooge JCI (1984) The waters of the Earth. Hydrol Sci J 29(2):149–176
Heaton RD (1981) Worldwide aspects of municipal wastewater reclamation and reuse. In: D'Itri FM, Martinez JA, Lambarri MA (eds) Municipal wastewater in agriculture. Academic Press, New York, pp 43–74
Hoffman GJ, Howell TA, Soloman KH (1990) Management of farm irrigation systems. Am Soc Agric Eng Monogr, St Joseph, MO
Huisman EA (ed) (1986) Aquaculture research in the Africa region. Proc Afr Sem Aquacult, 7–11 October, 1985, Kisumu, Kenya
Hutchinson GE (1970) The biosphere. Sci Am 223(3):1–11
Jordan WR (ed) (1987) Water and water policy in world food supplies. Proc Conf, May 26–30, 1985, Texas A&M Univ, College Stn, TX
King HR, Ibrahim KH (1988) Village level aquaculture development in Africa In: Proc Commonw Consultative Worksh Village level aquaculture development in Africa, 14–20 Feb, 1985, Freetown, Sierra Leone
Michael G (1987) Managed aquatic ecosystems. Ecosystems of the world 29. Elsevier, Amsterdam
Micklin PP (1991) The water management crisis in Soviet Central Asia. Carl Beck Pap, Univ Pittsburgh, PA
Middlebrookes EJ (1982) Wastewater reuse. Ann Arbor Science, Ann Arbor MI
Moore M, Crosswhite W, Hostetler J (1987) Agricultural water use in the United States 1950–85. US Geol Surv Water Supply Pap 2350
Nair PKR (1985) Classification of agroforestry systems. Agrofor Syst 3:97–128
National Research Council–NRC (1989) Irrigation-induced water quality problems. National Academy Press, Washington, DC
Olexa MT, Smart GC (1990) Laws governing use and impact of agricultural chemicals: an overview. Univ Press, Gainsville, FL
Oswald WJ (1989) Use of wastewater effluent in agriculture. Desalination 72(1-3):67–80
Penman HL (1970) The water cycle. Sci Am 223(3):37–45
Pillay TVR (1982) State of aquaculture, 1981. In: Bilio M, Rosenthal H, Sindermann CJ (eds) Realism of aquaculture: achievements, constraints, perspectives. Rev Pap World Conf Aquaculture, 21–25 Sept 1981, Venice, It
Postel S (1989) Water for agriculture: fancing the limits. World Watch Institute. World Watch Pap 93. World Watch Inst, Washington, DC
Rhoades JD (1984) Use of saline water for irrigation. Cal Agric 38(10):42–43
Shiklamanov IA (1990) Global water resources. Natl Resour 26(3):34–43
Shuval HI (1987) The development of water reuse in Israel. Ambio 16(4):186–190
Tanji KK (1990) Agricultural salinity assessment and management. Am Soc Civil Eng Manual 71, New York
Thomas GA, Leighton-Schwartz M (1990) Legal and institutional structures for managing agricultural drainage in the San Joaquin Valley. San Joaquin Valley Drainage Program, Sept 30
United Nations (ed) (1988) FAO production yearbook 1988. Statistic series 88. FAO, Rome
US Committee on Irrigation and Drainage (ed) (1988) How can irrigated agriculture exist with toxic waste regulations. Denver, CO
US Environmental Protection Agency (ed) (1989) Non-point sources: agenda for the future. Washington, DC
US General Accounting Office (ed) (1990) Water pollution: greater EPA leadership needed to reduce non-point source pollution. Washington, DC

Van der Leeden F, Troise FL, Todd DK (1990) The water encyclopedia, 2nd edn. Lewis, Michigan

World agriculture, review and prospects into the 1990s. Proc 1st Annu Aspen Inst Forum Food, water, and climate, Sept 1983, Wye Plantation, Winrock

Young A (1988) Agroforestry and its potential to contribute to land development in the tropics. J Biogeogr 15:19–30

2 Water Suitability for Agriculture

B. Yaron and H. Frenkel

2.1 Introduction

Both crop and animal production are water-dependent. The quality and quantity of water used in agriculture are two interrelated properties controlling the production capacity of the land and the quality of the environment and the water resources.

Global water resources are unequally distributed. Areas characterized by water scarcity are generally characterized by water resources of poor quality as well. Water quality in an area with large water reserves can also be negatively affected by human activity.

When is water considered suitable for agricultural use? When it has no osmotic or specific toxic effects on crop or animal production, when it contains no solutes affecting the chemical and hydraulic properties of the soil, and when it does not cause the deterioration of surface or groundwater.

Management of agricultural production should be designed as a function of the quality of water to be used, since water quality is the main parameter to be considered in the selection of an economically efficient and ecologically safe agricultural system.

A large body of interdisciplinary and multidisciplinary research on water suitability for agricultural uses has been carried out over the years, but it is not within the scope of the present chapter to review the existing knowledge. Comprehensive books and reviews dealing with these topics have been published in the last decade, among which we would cite Bresler et al. (1982), Shalhevet and Shainberg (1984), Yaron et al. (1984a), Ayers and Westcot (1985), Frenkel and Meiri (1985), Pallas (1985), Moore (1990), Tanji (1990).

2.2 Water Quality – Origin Dependence

Water quality is defined by its relation to specific use. The chemical and microbiological composition of any water resource is a result of the resource's interaction with its natural surroundings and of the exogenous chemicals entering the water system.

Adv. Series in Agricultural Sciences, Vol. 22
K.K. Tanji/B. Yaron (Eds.)
© Springer-Verlag Berlin Heidelberg 1994

2.2.1 Natural Origin

Water can be considered an aggressive solvent that interacts with geologic media exposed on the landscape (Falkenmark and Allard 1991). The quality of a water resource is controlled primarily by the geochemical processes of the water system (Angino 1983). The chemical composition of water is affected by reactions occurring at the earth's surface and leading to a condition of equilibrium or by those characterized by slow reaction rate (Fig. 2.1). The reactions leading to equilibrium are reversible and include: (1) dissolution of crystalline material involving various aqueous species and complexes; (2) reversible dissolution and deposition reactions – essentially the many types of hydrolysis and dissociation; and (3) ionic reactions involving changes in oxidation state and a reversible dissolution – deposition reaction (Hem 1970). The reactions that are affected by slow release require energy input or the involvement of organisms at certain stages leading to completion.

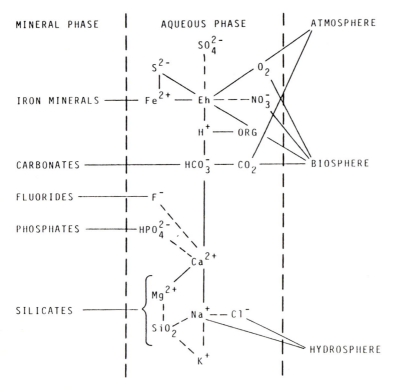

Fig. 2.1. Major components of natural fresh waters and their interaction with geological phases. (Falkenmark and Allard 1991)

The chemical composition of a water system at a given time is the end-product of all the reactions to which the water has been exposed in the hydrologic cycle. Deposition and dissolution are processes that continue as long as water is in contact with a soluble substance or until an equilibrium is reached. As a consequence, the chemical composition of a water system is in continuous chemical transition.

2.2.2 Man's Activities as the Origin

The major source of dissolved salt load of most natural water systems is the geological and soil environment. However, the quality of water resources can be drastically affected by such human activities as increased water supply, the addition of agrochemicals, the disposal of solid wastes, and accidental chemical spills.

An increase in water supply by irrigation or by an excessive groundwater overdraft can lead to changes in water quality and quantity. The use of saline drainage water or sewage effluents for irrigation can significantly change the water quality, not only through increased concentration and composition but also through the addition of "toxic" trace chemicals, organic soluble ligands, and microbial populations. Agrochemicals might constitute an additional source for new components of the water. Constituents such as nitrates, phosphates, heavy metals and pesticides may reach the water resources and affect their quality. Land-disposed solid waste and sludges constitute point sources of contamination of groundwater with nitrates, phosphates, heavy metals and toxic organic chemicals. A high organic percentage in the solid waste and sludges can affect the speciation of trace elements reaching the waters as well as the water microbial population. The application of manure on agricultural fields can affect the nitrate balance in soil and water. If the groundwater composition can be changed due to the transport of chemicals from the land surface to the unsaturated zone/groundwater interface, the composition of surface water can be changed by direct contamination or via runoff from agricultural lands. Finally, the chemical composition of water resources can be altered by accidental addition to the soil-water system of toxic organic chemicals in high concentration from a point source.

2.3 Quality Parameters Relevant to Crop Production

The assessment of water suitability for crop production involves a consideration of its salinity, acidity-alkalinity, and specific ionic and toxic organic effects.

2.3.1 Salinity

The most important water quality parameter for salinity is total salt concentration, which is usually expressed as the total dissolved solids (TDS) or the electrical conductivity (EC). Both measures are correlated with plant growth. The effect of salinity on plant growth is due mainly to total soluble solids (TSS). An osmotic effect is stressful to most crops, since it depresses the external water potential by making water less readily available. If the environmental osmotic potential becomes lower than that of the plant cells, crops will suffer an osmotic desiccation. This will result in a reduction in the plant growth rate with an enlargement and the synthesis of metabolites and structural compounds. The salt stress effect on the plant growth can be explained by the increase in the energy required by the plant to acquire water from the soil and to make the biochemical adjustments necessary for growth under stress. Crop salt tolerance may be defined as the ability of the plant to survive and produce economic yields under adverse conditions caused by salinity. Maas and Hoffman (1977) express crop tolerance to salinity in terms of relative yield (Yr), threshold value (a), and percentage decrement value per unit increase of salinity in excess of the threshold (b), as follows:

$$Yr = 100 - b \, (ECe - a) \, ,$$

where ECe (in dS/m) is the EC of the soil saturation extract and Yr is the relative crop yield obtained under saline vs nonsaline at comparable conditions.

Maas (1984) presented in graphic form (Fig. 2.2) the tolerance to salinity of agricultural crops by plotting relative crop yield against the EC of the soil (ECe) and of the irrigation water (ECw) in dS/m. The agricultural crops were grouped in five classes corresponding to a salinity range of $< 1.3 - > 10.0$ dS/m (Table 2.1).

The tolerance limits from Table 2.1 are for primary applications to mature crops. During emergence and the early stages of growth, the tolerance limits are more restrictive. The tolerance to salinity varies from species to species and is affected by climatic conditions. Crops are generally more sensitive to salt stress in a hot and dry climate than in a cool and humid one. Saline water may cause less damage when applied through the soil surface (e.g., drip irrigation) than via sprinklers, when direct contact occurs between the water and the foliage. Soil fertility is an additional factor which may affect the crop's tolerance to salinity.

2.3.2 Acidity–Alkalinity

Acidity–alkalinity of water as measured by pH has little direct significance when the water is applied through the soil surface since, in general, it could

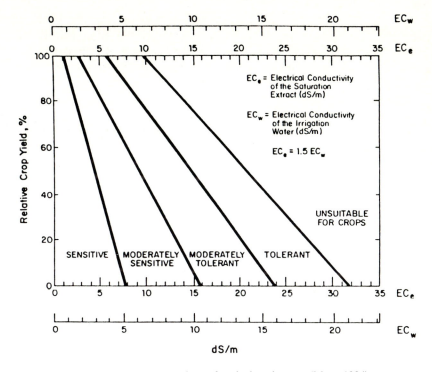

Fig. 2.2. Relative salt tolerance ratings of agricultural crops. (Maas 1984)

Table 2.1. Grouping of agricultural crops as a function of tolerance to salinity (Ayers and Westcot 1985)

Crop salinity tolerance rating	Salinity of soil saturation extract (ECe) at which loss begins
	dS/m
Sensitive	< 1.3
Moderately sensitive	1.3–3.0
Moderately sensitive	3.0–6.0
Moderately tolerant	6.0–10.0
Unsuitable[a]	< 10.0

[a] For most crops.

be buffered by the soil buffer system. However, the acidity–alkalinity might affect agricultural crops by foliar contact. In acid soils, however, when the buffer capacity of soils for acid water is limited, waters with a pH < 4.5 could enhance the solubilization of iron, aluminium or manganese in concentrations large enough to be toxic to crop growth. For instance, the rapid increase in the sulfate and nitrate contents in precipitation, due to atmospheric pollution, led during a 10-year period to the acidification of surface waters, from pH 5.5 to 4.5 (Jacks et al. 1984). Biological sources may also result in acidification of water. Water having a pH value in excess of 8.3 is highly alkaline and may contain a high concentration of sodium carbonates and bicarbonates. Values in the pH range of 5 to 8.5 should not present any problem.

2.3.3 Specific Ion Effect

Crop nutrition may be affected by an imbalance of common nutrient chemicals in water and soil solutions and can create an unfavorable environment for plant growth. Essential ions such as calcium, magnesium and potassium may deter growth if their total or relative concentration is out of balance. Läuchli and Epstein (1990), in their recent review on plant response to salinity, emphasize the environmental effect on crop nutrition. When sodium predominates over potassium, the plant's paramount nutritional requirement is potassium in adequate amounts. Saline conditions may inhibit the uptake of nitrates. Research results accumulated during the last 25 years have led to the conclusion that the ionic composition of water affects crop development, in contradiction to the long-accepted opinion that the total salt content of the solution governs the plant growth pattern.

Plants may be sensitive to the presence of moderate to very low concentrations of several specific ions in the irrigation water. Even a moderate concentration of electrolytes such as Na^+, Ca^{2+}, Cl^- and SO_4^{2-} can reduce growth and cause specific injury. Toxicity of sensitive crops occurs at a relatively low sodium or chloride concentration (> 3 meq/l). In one study, a phosphate toxicity was induced by a particular Ca^{2+}/Na^+ ratio of the irrigation waters (Grattan and Maas 1988). Boron in the water has a negative effect on plant growth, as it becomes toxic to most plants at a concentration above a few mg/l (Keren and Bingham 1984). Direct foliar contact of sensitive crops by overhead sprinkling with water containing boron can cause toxicity, which is not encountered with surface irrigation methods.

2.3.4 The Effect of Micropollutants

Trace elements and organic toxic chemicals are micropollutants which may be found in various water sources as a result of weathering processes or exogenous causes. Pratt and Suarez (1990) have summarized the existing knowledge. The limits of crop tolerance to trace elements (Table 2.2) are only indications and could vary as a function of environmental conditions and plant genetic resources. Page and Chang (1990) reported that plants differ in their ability to tolerate, absorb and accumulate trace elements. Table 2.3 (after Wolnick et al. 1983, 1985) shows the medium concentrations

Table 2.2. Recommended maximum concentration of selected trace elements in waters[a] (Pratt and Suarez 1990)

Element	Maximum concentration (mg/l)
Lead	5.00
Fluoride	1.00
Zinc	0.50
Manganese, copper, nickel	0.20
Chromium, vanadium	0.10
Selenium	0.02
Cadmium	0.01

[a] The recommended concentration varies as a function of environmental conditions and plant species.

Table 2.3. Medium concentrations of essential and non-essential trace elements in row crops grown in major producing areas in the USA (Wolnick et al. 1983, 1985)

Crop species	No. of observations	Trace element concentration (mg/kg, oven-dry weight)								
		Cu	Fe	Mn	Mo	Zn	Cd	Ni	Pb	Se
Lettuce	150	6.1	57	31	0.25	46	0.435	–	0.19	0.039
Spinach	104	8.5	200	81	0.22	43	0.80	1.1	0.53	–
Tomato	231	11.0	48	15	0.30	22	0.22	0.84	0.027	–
Wheat	280	4.9	36	43	0.43	29	0.036	–	0.02	0.19
Sweetcorn	268	1.8	18	7	0.16	25	0.008	0.26	0.009	0.014
Soybean	322	13.0	71	27	–	45	0.045	4.8	0.036	0.082
Rice	166	2.1	3	11	0.65	15	0.005	0.26	0.005	–
Carrot	207	4.7	27	12	0.098	20	0.16	0.41	0.055	–
Potato	297	4.4	20	7	0.19	15	0.14	–	0.025	0.013
Onion	228	3.6	13	9	0.14	16	0.09	0.32	0.038	–
Peanut	320	8.3	20	18	0.28	31	0.068	1.5	0.040	0.040

of selected trace elements in row crops grown in major producing areas of the USA.

Toxic organics can be assimilated by crops from a water source by direct foliar contact or through the roots via the soil environment. There are no tolerance limits for organo-toxic effects to plants, but carryover effects of herbicides from one crop to another are often cited in the literature.

2.4 Quality Parameters Relevant to Animal Production

Drinking water for animals should have more or less the same quality parameters as those for humans. Assessment of the suitability of water for animal production involves the determination of its electrolyte composition and concentration of toxic organic and inorganic micropollutants and microorganisms.

2.4.1 Chemical Parameters

Salinity is the most important characteristic for determining the suitability of a water source for livestock (Pallas 1985). Excessive intake of saline water may cause sickness and death. Adult sheep are fairly tolerant to salinity; cattle are less tolerant than sheep but more than pigs and poultry. Water quality standards for animals are defined as a function of environmental conditions. The Western Australian standards, for example (Table 2.4a), encompass a range of total salt concentrations from 2.8 to 12.8 g/l for different animals. A critical factor to be considered in defining the water quality for animals is daily water consumption. Salt poisoning symptoms, (e.g., abdominal pain, loss of appetite, diarrhea and excess urination) are produced by excessive daily consumption of saline water (Table 2.4b).

Not only total salt content but also ionic composition can affect water quality. Pallas (1985) gives the following tolerance limits for magnesium: lactating cows and horses, 0.25 g/l; dairy cattle, 0.40 g/l; and adult sheep, 0.50 g/l.

Table 2.5 summarizes water quality standards for animals in arid regions (Schoeller 1977), where total salt content and ionic composition are considered together with other water properties such as odor, color, turbidity and taste.

Animal production can also be affected by toxic organic and inorganic chemicals which may be found in water, such as micropollutants. The most stringent water quality standards which can be adopted for animal production are those proposed for drinking water by the US Environmental Protection Agency. These standards include both synthetic organic contaminants and inorganic contaminants. Selected tolerance limits for the above

Table 2.4a. Upper limits of total salt content of water for livestock (Pallas 1985)

Livestock	Total dissolved solid g/l
Poultry	2.8
Pigs	4.3
Horses	6.4
Cattle (dairy)	7.1
Cattle (beef)	10.0
Adult dry sheep	12.8

Table 2.4b. Total quantity of salts producing symptoms of salt poisoning (Pallas 1985)

Livestock	Total amount of salts g/day
Poultry	4–8
Pigs	100–200
Cattle	1800–3600
Sheep	100–200

Table 2.5. Water quality standards for animals in arid regions (Schoeller 1977)

Parameter	Suitability for permanent supply			
	Good	Fair	Moderate	Poor
Color	Colorless	Colorless		
Turbidity	Clear	Clear		
Odor	Odorless	Hardly perceptible	Slight	Slight
Taste at 20 °C	None	Perceptible	Pronounced	Unpleasant
Total dissolved solids (mg/l)	0–500	500–1000	1000–2000	2000–4000
EC (µs/cm)	0–800	800–1600	1600–3200	3200–6400
Na (mg/l)	0–115	115–230	230–460	460–920
Mg (mg/l)	0–30	30–60	60–120	60–120
$\left[\dfrac{Mg}{12}+\dfrac{Ca}{20}\right]$mEq/l	0–5	5–10	10–20	20–40
Cl (mg/l)	0–180	180–360	360–710	710–1420
SO_4 (mg/l)	0–150	150–290	290–580	580–1150

Table 2.6. Recommended limits of tolerance for microcontaminants in drinking water (according to the US Environmental Protection Agency 1986)[a]

Inorganic		Organic	
Chemical	Level (mg/l)	Chemical	Level (mg/l)
Arsenic	0.05	Aldicarb	0.009
Barium	1.5	Carbofuran	0.036
Cadmium	0.005	DBCP	0
Chromium	0.12	O-dichlorobenzene	0.62
Copper	1.3	2,4-D	0.07
Lead	0.020	EDB	0
Mercury	0.003	Ethylbenzene	0.68
Nitrate	10.0	Lindane	0.0002
Nitrite	1.0	Methoxychlor	0.4
Selenium	0.045	Monochlorobenzene	0.1
		Pentachlorophene	0.22
		Toluene	2.0
		Toxaphene	0
		Xylene	0.44

[a] Recommended by US Environmental Protection Agency (1986).

micropollutants are summarized in Table 2.6. The limits are very severe and in many countries are not standardized even for drinking water for human beings. The data in Table 2.6 could serve, however, as indices of the health hazard for animal production despite the fact that, for this section, lower standard values will be used.

2.4.2 Biological Parameters

Optimum animal production requires water which is bacteriologically clean as well as chemically satisfactory. Most organisms found in water are not pathogens, but even microorganisms which are not disease pathogens can cause pest and odor problems. Pathogens are difficult to identify, but coliform bacteria meet all criteria for an ideal, easily measurable indicator.

2.5 Water Quality–Soil Quality Relationship

Soil properties in all their facets are affected by interactions with incoming water. Changes in the quality of the water reaching the soil medium may lead to changes in the chemical, physical and biological properties of the soil.

2.5.1 Soils

Chemical properties of soils can be affected by their interaction with incoming water through the following processes: surface reactions, precipitation-dissolution, and oxidation-reduction. The chemical composition of the incoming water may also be changed during its interaction with the soil-solid phase by saturation phase speciation or hydrolytic processes.

   Surface reactions are controlled by the properties of the solid phase – mostly clay minerals and organic matter – and by the composition of the solute. Both clay minerals and organic matter exhibit a large specific surface area which, coupled with their permanent and pH-dependent charge, become the most reactive surface of the soil. The surface activity of the soil-solid phase is also affected by the presence of hydroxides of iron and aluminum and by the end-products of weathering.

2.5.1.1 Salts

All water reaching the soil is characterized by a specific salt content. The distribution of ions between the incoming water and the soil exchange phase is governed by the ion exchange processes controlled by the mass action principle. The Ca/Na exchange between soils and saline-sodic waters may be defined by a selectivity coefficient (known as Gapon exchange constant):

$$k_g = \frac{[\text{Na X}] [\text{Ca}^{2+}]^{1/2}}{[\text{Ca}_{1/2} \text{X}] [\text{Na}^+]},$$

where solution concentration is given in mol/l and the exchangeable ion concentration in Cmol_c/kg; k_g is the Gapon selectivity coefficient and assumed to be constant.

The US Salinity Laboratory (1954) considers that Mg behaves similarly to Ca in the adsorbed phase and that Ca, Mg and Na are the most common exchangeable cations in arid soils. As a result, they modified the Gapon equation as follows:

$$\frac{[Na\ X]}{[Ca\ X + Mg\ X]} = k_g \frac{[Na]}{[Ca + Mg]^{1/2}} = k_g\ SAR,$$

where SAR is the sodium adsorption ratio, defined as:

$$SAR = \frac{[Na]}{[Ca + Mg]^{1/2}}.$$

The above equation can be simplified to:

$$\frac{[Na]}{CEC - [Na\ X]} = k_g\ SAR = ESR,$$

where CEC is cation exchange capacity.

$$\text{Since } ESP = \frac{[Na\ X]}{CEC} \times 100,$$

where ESP is exchangeable sodium percentage, it follows that:

$$\frac{ESP}{100 - ESP} = K'_g\ SAR = ESR.$$

Recent evidence suggests that divalent ions are preferentially adsorbed in the interlayer spaces of the clay material and monovalent ions are adsorbed on the edges and planar surfaces (ion demixing).

2.5.1.2 Trace Elements

Alkali and alkaline earth materials, transition metals, nonmetal trace elements and heavy metals can be found in wastewaters reaching the land surface even after the standard purification procedure. Adsorption is a significant mechanism in distributing trace elements from the water to the soil-solid phase. Adsorption of trace elements can be nonspecific – involving simple electrostatic attractions, or specific – involving coordinate covalent bonding (Jenne 1976). A continuum of different types of sites exists in natural systems, ranging from extremely specific sites of a low concentration of trace elements to nonspecific cation exchange type sites at high concentration. In natural systems specific adsorption is usually the dominant process controlling the presence of trace metals and trace element oxyanions (Deverel and Fujii 1988). For example, boron – which is potentially toxic to crops – is adsorbed onto and released from the surface of soil particles. Soil solutions are buffered against rapid changes in B concentration. If the B in irrigation water is increased, it is adsorbed on the soil surface, resulting in a smaller increase in the B concentration than in the increase to irrigation water (Keren and Bingham 1984).

Trace metals compete with each other for adsorption on mineral and soil surfaces. Forbes et al. (1976) for example, found that trace mineral affinity for goethite is $Cu^{2+} > Pb^{2+} > Zn^{2+} > Co^{2+} > Cd^{2+}$.

2.5.1.3 Organic Micropollutants

Organic micropollutants in ground, drainage or surface waters may exhibit a particular charge or be apolar. The phase distribution of the organic micropollutants in the soil medium is controlled by the nature and properties of the soil colloids, the chemical and physicochemical characteristics of the organic molecules, and the nature of the soil environment (Yaron 1990). The soil constituents that control retention are soil organic matter, clays and amorphous minerals. Clay minerals have a negative charge that is balanced by exchangeable cations; this is important in the adsorption of charged and protonated organic molecules. Soil organic matter is generally considered responsible for the adsorption capacity of soils for non-ionic organic micropollutants, which is borne out by the relative constancy of the distribution coefficient values for a given compound in a variety of soils. Adsorption materials can predominate over clays and organic matter in certain soils and therefore should be considered mainly in dealing with the organic micropollutant adsorption on various sized particles.

The molecular properties of the adsorbate, and in particular electronic structure and water solubility, are of great importance in the adsorption reaction. The molecular structure determines the ionization potential of a compound. The ability of a molecule to ionize, which depends upon the electron distribution in the molecule and its electron mobility, is the primary reason that the soil interactions of many organic molecules are pH-dependent. For non-ionic compounds, aqueous solubility is the most important molecular parameter correlated to adsorption. In general, highly soluble non-ionic pesticides have low adsorption coefficients for soils, while the opposite holds true for chemicals of low solubility.

In addition to the properties of both soil components and organic micropollutants, the distribution of synthetic organic chemicals among the soil phases is influenced by the soil environment. The environment of a specific soil is determined not only by its intrinsic properties but also by external factors, mainly climatic conditions and agricultural practices. Among these factors, the most important are soil moisture content and temperature. The soil moisture content affects the adsorption process by modifying the accessibility of the adsorption sites and the surface properties of the adsorbent. For non-ionic organic molecules, competition with water for the adsorption sites and negative relationships between pesticide adsorption and soil moisture content have often been reported. As the adsorption processes are exothermic, changes in soil temperature can have a direct effect on the phase distribution of organic micropollutants in soils: adsorption usually increases as the temperature decreases.

The release of organic micropollutants in the soil-aqueous phase is usually expressed by desorption isotherms. In some cases the desorption isotherm is nearly identical to the adsorption isotherm. In many soils, however, hysteresis in the adsorption-desorption process occurs – the extent of which is related to the properties of the adsorbent and of the adsorbate, and to the environmental conditions. The existence of hysteresis is important for the transport of chemicals in soil.

Once an organic micropollutant reaches the soil, it undergoes transformation with an intensity that is controlled by the molecule's properties, the soil medium's characteristics and the ambient conditions. Biochemical and chemical degradation of pesticides are the main processes involved in this transformation, both occurring simultaneously with relative contribution difficult to define. In the root zone, degradation proceeds mainly via microbiological processes, which are more rapid than chemical processes. However, there is little biological activity below the root zone and the degradation therefore proceeds at a much slower rate in the deeper, unsaturated zone. It should be emphasized, however, that pesticide transformation is related to the appearance of metabolites with properties different from the parent material – sometimes more polar, more soluble in water and even more toxic – which may also reach the soil environment.

2.5.2 Soil Hydraulic Properties

Soil hydraulic properties can be affected by the composition and amount of electrolytes and suspended colloids present in the water-to-land interface.

2.5.2.1 Salinity–Alkalinity Effect

A high sodium level combined with a low electrolyte concentration can induce a decrease in soil permeability and infiltration capacity through swelling, dispersion of clays and slaking of aggregates. The conducting pores in soils are narrowed as a result of clay swelling (Quirk and Schofield 1955), the dimensions of the process being controlled by the soil mineralogy (Yaron and Thomas 1968). Soil permeability can be reduced by slaking and by dispersion of clays in the narrow necks of the soil pores (Felhendler et al. 1974; Frenkel et al. 1978), and can affect the soil's hydraulic conductivity due to its effect on pore size distribution.

An example of a combined salinity-SAR effect on infiltration rate is presented in Fig. 2.3. Ayers and Westcot (1985) emphasized that very low

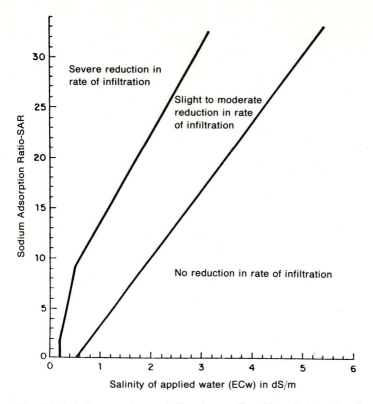

Fig. 2.3. Relative rate of water infiltration as affected by salinity and sodium adsorption ratio. (Ayers and Westcot 1985; in turn adapted from Rhoades 1977; Oster and Schroer 1979)

salinity water ($< EC = 0.2$ dS/m) invariably results in water infiltration problems, regardless of the relative SAR. Excessive sodium in irrigation water also promotes a decrease in the soil hydraulic conductivity, but only if sodium exceeds calcium by more than a ratio of approximately 3:1. Such a relatively high sodium content often leads to a severe water infiltration problem due to soil dispersion, which causes plugging and sealing of the surface. This process results from a lack of sufficient calcium to counter the dispersing effects of sodium. The prediction of the decrease of infiltration rate is improved by using an adjusted SAR, as calculated by Suarez (1981). The new term R_{Na} (adjusted sodium adsorption ratio) is defined as follows:

$$\text{adj } R_{Na} = \frac{Na}{\left[\dfrac{Ca_x + Mg}{2}\right]^{1/2}},$$

where:

Na = sodium in the irrigation water, reported in mEq/l;

Ca_x = a modified calcium value, reported in mEq/l. Ca_x represents Ca in the applied irrigation water but modified due to salinity of the applied water, its HCO_3/Ca ratio (HCO_3 and Ca in mEq/l) and the estimated partial pressure of CO_2 in the surface few millimeters of soil ($PCO_2 = 0.0007$ atm);

Mg = magnesium in the irrigation water reported in mEq/l.

Comparison of SAR and adj R_{Na} for various types of water from around the world shows that for most waters, the SAR calculation is within $\pm$ 10% of the value obtained after adjustment of the calcium concentration using the adjusted SAR equation (Ayers and Westcot 1985).

2.5.2.2 Effects of Suspended Particles

The presence of suspended particles in natural water resources, and especially in recycled water, can bring about a decrease in soil hydraulic conductivity due to the plugging of pores by colloidal materials. Since the colloidal suspended particles are generally rich in organic matter, they become a source of energy for the soil microbial populaltion enhancing the accumulation of biomass. The reduction of hydraulic conductivity occurs, for example, during prolonged infiltration of sewage effluent into field soils and sands (Rice 1974), and alleviated biological exudation (Avnimelech and Nave 1974) or physical clogging of pores (Kristianssen 1981). Vinten et al. (1983), Yaron et al. (1984b) and Metzger and Yaron (1987) quantitatively related the accumulation of suspended solids with the decrease in soil hydraulic conductivity as affected by the soil properties. Figure 2.4 shows, for example, the solid distribution, relative flow rate changes and suspended solids in the leachate of a silty loam and a coarse sand. Vinten et al. (1983) showed that the finer the soil texture, the greater the relative reduction in hydraulic conductance which occurs following irrigation with sewage effluent. Care should be exercised in using sewage effluent or surface water with a high content of suspended particles on soils which already have low hydraulic conductivity, because as irrigation proceeds the application rate may exceed the infiltration rate. This may result in ponding and increased risk of surface runoff, waste of water, and erosion if suitable management is not practiced. Surface sealing is a short-term problem, as the surface mat may be readily broken up during cultivation.

A rough calculation shows that with sewage effluent of the composition used by Vinten et al. (1983) and an annual irrigation of 1000 mm, 100 g/m² of solids are added to the soil per year. If 10% of this is accumulated in a 10-mm layer at the wetting front, the porosity of the soil at this depth will be reduced by approximately 0.2% per season. Thus, in the long run, this pan formation is feasible if little decomposition occurs at that depth, or if large

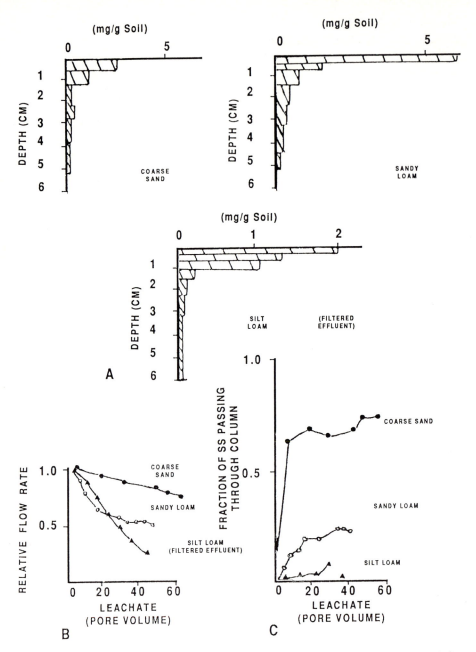

Fig. 2.4. A Deposited solids distribution in three soils leached with waste water (98 mg l[-1]): unaltered effluent: 38 mg l[-1] SS: filtered effluent. **B** The effect on the soil hydraulic conductivity and **C** suspended solids transport through the solis. (Vinten et al. 1983)

amounts of inorganic colloidal material are present in the irrigation water. In most cases the authors consider that the use of sewage effluent for irrigation is unlikely to pose major management problems due to a loss of soil permeability.

2.5.3 Soil Microflora

Soil microflora can also be affected by the presence in the incoming water of a high concentration of toxic organic chemicals which originated in crop protection practices and accidentally reached the land surface. The soil is a system biologically in equilibrium. Toxic organic chemicals reaching the soils through a polluted source of water may alter this equilibrium, modifying the microflora activity and consequently the soil fertility. Simon-Sylvester and Fournier (1979), in reviewing the effects of pesticides on microflora, point out the changes in the microbiological population and activity caused by synthetic organic compounds. The respiratory and enzymatic activity of soil bacteria, fungi, actinomycetes and algae can be affected. Toxic organic chemicals affect the biological cycles of the soil, e.g., the carbon, nitrogen, sulfur, phosphorus and manganese cycles.

The activity of the soil microflora is a result of regulative phenomena that make it possible to maintain the soil for given ecological data at a certain biological balance which generally prevents the excessive proliferation of certain organisms while protecting others. The presence of toxic organic chemicals in the water reaching the land surface can also affect these mechanisms which are of particular importance for the life and fertility of soils: the phytophotogenic action of some microorganisms and the antagonistic or synergistic relationships of these microorganisms with other species.

References

Angino EA (1983) Geochemistry and water quality. In: Thornton I (ed) Applied environmental geochemistry. Academic Press, New York, pp 172–199
Avnimelech Y, Nave Z (1974) Biological clogging of sand. Soil Sci 92:222–226
Ayers RS, Westcot DW (1985) Water quality for agriculture. FAO, Rome, M-56, pp 174
Bresler E, McNeal BL, Carter DL (1982) Saline and sodic soils. Springer, Berlin Heidelberg New York 237 pp
Deverel SJ, Fujii R (1988) Processes affecting the distribution of seleniuNew York, pp 49–79
Felhendler R, Shainberg I, Frenkel H (1974) Dispersion and hydraulic conductivity of soils in mixed solution. In: Trans 10th Int Congr Soil science, Moscow, vol 1. Kanku, Moscow, pp 103–112
Forbes EA, Posner AM, Quirk JD (1976) The specific adsorption of divalent Cd, Ca, Cu, Pb and Zu on goethite. J Soil Sci 27:154–166
Frenkel H, Geortze JO, Rhoades JD (1978) Effects of clay type and content, exchangeable sodium percentage and electrolyte concentration on clay dispersion and soil hydraulic conductivity. Soil Sci Soc J 42:3–8

Frenkel H, Meiri A (1985) Soil salinity. Van Nostrand Reinhold, New York, 440 pp

Grattan SR, Maas EV (1988) Effect of salinity of phosphate accumulation and injury in soybean. I. Influence of $CaCl_2NaCl$ ratio. Plant Soil 105:25–32

Hem JD (1970) Study and interpretation of the chemical characteristics of natural water. US Geol Surv, Washington, DC, Pap 1473, 84 pp

Jacks G, Knutsson G, Maxe L, Fylkner A (1984) Effect of acid rain on soil and ground water in Sweden. In: Yaron B, Dagan B, Goldshmid J (eds) Pollutants in porous media. Springer, Berlin Heidelberg New York, pp 94–113

Jenne EH (1976) Trace elements sorption by sediments and soils – sites and processes. In: Chapman W, Peterson K (eds) Symp Molybodenum in the environment, vol 2. Dekker, New York, pp 425–553

Keren R, Bingham FT (1984) Boron in water, soils and plants. Adv Soil Sci 1:228–275

Kristianssen R (1981) San filter trenches for purification of septic tank effluent. I. The clogging mechanism and soil physical environment. J Environ Qual 10:353–357

Läuchli A, Epstein E (1990) Plant response to saline and sodic conditions. In: Tanji KK (ed) Agricultural salinity assessment and management. ASCE, New York, pp 113–137

Maas EV (1984) Salt tolerance of plants. In: Christie BR (ed) Handbook of plant science in agriculture. CRC, Boca Raton

Maas EV, Hoffman GJ (1977) Crop salt tolerance: evaluation of existing data. J Irrig Drain Div ASCE 103 (IRZ/0 115–134)

Metzger L, Yaron B (1987) Influence of sludge organic matter on soil physical properties. Adv Soil Sci 7:141–163

Moore JW (1990) Inorganic contaminants of surface water. Springer, Berlin Heidelberg New York, 334 pp

Oster JD, Schroer FW (1979) Infiltration as influenced by irrigation water quality. Soil Sci Soc Am J 43:444–477

Page AL, Chang AC (1990) Deficiencies and toxicities of trace elements. In: Tanji KK (ed) Agricultural salinity assessment and management. ASCE, New York, pp 138–160

Pallas PH (1985) Water for animals. FAO, Rome, AGL/Misc/4/85, 76 pp

Pratt PF, Suarez DL (1990) Irrigation water quality assessments. In: Tanji KK (ed) Agricultural salinity assessment and management. ASCE, New York, pp 221–226

Quirk JP, Schofield R (1955) The effect of electrolyte concentration on soil permeability. J Soil Sci 6:163–178

Rhoades JD (1977) Potential for using saline agricultural drainage water for irrigation. Proc Conf Water Manag Irrig Drain, July 1977, ASCE, Reno, pp 85–116

Rice RC (1974) Soil clogging during infiltration of secondary effluent. J Water Pollut Control Fed 46:708–716

Schoeller HJ (1977) Geochemistry of ground water. Ground water studies, Suppl 3. UNESCO Tech Pap Hydrol 7. UNESCO, Paris

Shalhevet J, Shainberg I (eds) (1984) Soil salinity under irrigation. Springer, Berlin Heidelberg New York, 349 pp

Simon-Sylvester G, Fournier JC (1979) Effects of pesticides on the soil microflora. Adv Agron 31:1–91

Suarez DL (1981) Relation between pHc and SAR and an alternate method of estimating SAR of soil or drainage water. Soil Sci Soc Am J 45:469–475

Tanji KK (ed) (1990) Agricultural salinity assessment and management. Am Soc Civil Eng, New York, 619 pp

US Environmental Protection Agency (1986) Quality criteria for water. Washington DC, 440/5-86-001, 453

US Salinity Laboratory Staff (1954) Diagnosis and improvement of saline and alkali soils. US Dept Agric Handb 60:160 pp

Vinten AJA, Mingelgrin U, Yaron B (1983) The effect of suspended solids in wastewater on soil hydraulic conductivity. II. Vertical distribution of suspended solids. Soil Sci Soc Am J 47:408–412

Wolnick KA, Fricke FL, Capar SG, Braude GC, Meyer MW, Satzger RD, Bonnin E (1983) Elements in major raw agricultural crops in United States. J Agric Food Chem 31: 1240–1249

Wolnick KA, Fricke FL, Capar SG, Braude GC, Meyer MW, Satzger RD, Bonnin E (1985) Elements in major raw agricultural crops in United States. J Agric Food Chem 33:807–811

Yaron B (1990) Behavior of exogenous organic chemicals in soils as controlled by abiotic processes. In: Proc 14th Int Congr Soil Sci, Kyoto, Jpn 1:30–42

Yaron B, Thomas GW (1968) Soil hydraulic conductivity as affected by sodic water. Water Resour Res 4:418–427

Yaron B, Dagan G, Goldshmid J (eds) (1984a) Pollutants in porous media. Springer, Berlin Heidelberg New York, 296 pp

Yaron B, Vinten AJ, Fine P, Metzger L, Mingelgrin U (1984b) The effect of solid organic components of sewage on some properties of the unsaturated zone. In: Yaron B, Dagan B, Goldshmid J (eds) Pollutants in porous media. Springer, Berlin Heidelberg New York, pp 162–218

Part II   Water Conservation and Technology

3 Crop Irrigation

Z. Plaut and A. Meiri

3.1 Crop Water Status and Its Components

Many efforts have been made to determine, interpret and measure plant water status and various parameters have been used as indicators of plant water status. The most common one in recent years is the one based on thermodynamics, which can also serve as a basis for water application to crop by means of irrigation.

The chemical potential of water, which is a quantitative expression of its free energy as derived from thermodynamics, is influenced by three factors: concentration, pressure and gravity. All are expressed as the difference between the chemical potential at a given state and its value at a standard state. For practical purposes, another term defined as *water potential* is used which is the quotient of the chemical potential of water and its partial molal value ($18 \text{ cm}^3 \text{ mol}^{-1}$). Since the units of chemical potential are energy per mole, the water potential will be in units of pressure, such as kilo- and megapascals (KPa and MPa).

Water potential, which is symbolized by the Greek letter ψ, is a relative quantity compared to a reference standard state. This standard state, assigned as O Pa, is pure water at ambient pressure, sea level and at an identical temperature as that of the given state.

The concentration of water within plant cells can be changed into concentration of solutes, as the solutions are fairly dilute, and can be determined by their osmolality. Osmolality is a measure of the sum of concentration (more correctly the activities) of all solutes regardless of their molecular species (Nobel 1991). The term *osmotic potential*, which is symbolized by ψ_s, can thus be approximately calculated by:

$$\psi_s = - RT \sum C_j, \tag{3.1}$$

where R is the gas constant ($0.008314 \text{ L MPa mol}^{-1} \text{K}^{-1}$), T is absolute temperature (in K°) and $\sum C_j$ is the osmolality of the sum of all solutes in the solution.

The rigidity of the plant cell walls is responsible for the buildup of a hydrostatic pressure which is in excess of the ambient atmospheric pressure. This is known as *turgor potential*, symbolized by ψ_p, and is also in units of

Adv. Series in Agricultural Sciences, Vol. 22
K.K. Tanji/B. Yaron (Eds.)
© Springer-Verlag Berlin Heidelberg 1994

pressure. It is zero in open water bodies, but always positive in growing plant tissue.

Gravity causes water to move downward and it is thus of importance mainly for the flow of water to the top of tall trees. The effect of gravity on water potential depends on the height above a reference point, the acceleration constant (g) and the density of water.

Since in most cases the factor of gravity can be omitted, water potential will remain:

$$\psi = -\psi_s + \psi_p. \tag{3.2}$$

In a transpiring plant the absolute value of ψ_s will exceed that of ψ_p, yielding negative values for ψ.

The concept of water potential fulfills two main functions: (1) it governs the direction of water flow across cell membranes, and it is also the driving force for water movement from the soil into the root; and (2) it is a measure of *water status* of a plant, and as such of interest in the present chapter. Water potential is the most commonly measured parameter which is closely connected to plant functions. Thus a decrease in ψ under given conditions relative to ψ of well-watered plants can be correlated with yield and productivity (Hiler et al. 1972).

The component of ψ_s may also serve as parameter indicating plant water status. For instance, a reduction of ψ_s indicates an increase in osmotic solutes due to dehydration or increased solute content. Both can arise by plant water stress, but more likely is the dehydration. The other component of water potential, turgor potential (ψ_p) is probably a more meaningful parameter for plant water status than ψ_s (Hsiao and Bradford 1983). High ψ_p is a good indication of optimal water status which will enable high rates of physiological activities involved in growth. Values of ψ_p approaching zero indicate a substantial water stress. It should be noted, however, that this parameter in contrast to ψ and ψ_s cannot be directly determined, at least not under field conditions. Turgor potential is also influenced by the elastic properties of the cell wall. The more elastic the cell walls, the smaller is the change in ψ_p for a given change in cell volume. For this reason, ψ_p of young plant tissue will be less sensitive to water stress than mature tissue.

The sensitivity of various physiological processes to water deficit is however not equal. Cell expansion, for instance, is known to be sensitive and may be inhibited at approximately -0.4 to -0.8 MPa. This can probably be attributed to the need of a hydrostatic pressure to stretch cell walls, which can only be achieved when the differences between ψ_s and ψ are sizable. Photosynthetic CO_2 fixation is much less sensitive and was found to be reduced only at -1 to -2 MPa (Hsiao 1974). It is likely that apparent sensitivities to plant water status of the different plant processes involved in production are the result of additional parameters. At a significant but not extreme decrease in ψ, the decrease in solute concentration is minimal and changes in the hydration of macromolecules as enzymes is very small.

Metabolic activities, as carbon fixation processes, are thus expected to be of low sensitivity. At a similar decrease in ψ, the decrease in ψ_p may be sufficient to lead to a marked decrease in expansion growth.

The values of both ψ and ψ_s at given soil water status and atmospheric conditions will depend on the type of plant and its growing conditions. In many crop plants grown under mild climatic conditions and frequent water supply by rainfall and/or irrigation, ψ_s will be as low as -0.5 to -0.7 MPa and ψ will mostly be above -0.3 MPa. The level of ψ_s is probably set by the low concentration of dissolved ions, metabolites and other solutes in the cytoplasm. The high values of ψ are a result of high availability of soil water and a moderate transpiration demand. A temporary decrease in water availability or a sudden increase in transpiration demand may lead to a sharp response in plant water status, namely a decrease in ψ and ψ_p associated with small changes in ψ_s. Other crop plants will buildup high concentrations of solutes like ions, sugars, and amino and organic acids under conditions of soil water deficit, salinity and other external stress conditions. These may lead to ψ_s values as low as -2.5 to -3.5 MPa. Such levels of ψ_s may still permit the maintenance of positive turgor potential in spite of the decrease in ψ under those conditions. These plants will be much less sensitive to moderate changes in soil water availability and atmospheric conditions.

3.2 Evapotranspiration and Crop Water Requirement

The main practical aspects of irrigation are the determination of how much water to apply to a given crop and when to apply the water. The amount of water to be applied is based on the rate of evapotranspiration and thus requires an estimation of this rate. Evapotranspiration (ET) is comprised of evaporation from the soil surface and transpiration from the plant tissue. Monteith (1985) pointed out that the terms evaporation and transpiration are not identical. While evaporation refers to a change of phase from liquid to gas, transpiration refers to the flux of vapor from the plant to the atmosphere. He thus suggested use of the term total evaporation (TE). We shall, however, use the more common term evapotranpiration (ET) to express water loss from both sources, as the phase change from liquid to gas in the case of plants does not necessarily imply losses to the atmosphere.

The contribution of the two factors to total evapotransition can vary greatly throughout the growing season of an annual crop, or at different parts of the year in a perennial crop. As long as the leaf area index is low, evaporation from the soil surface contributes much more to ET than under full or nearly full coverage. Separate measurements of soil evaporation and transpiration, or partitioning of these two processes by models, were thus recommended (Ritchie and Johnson 1990). We must, however, keep in mind that the contribution of each component changes continuously due to the change in leaf area index (LAI) and in root distribution. It would therefore be easier, at least for practical purposes to determine crop water requirement related to total ET.

Evapotranspiration is the principal factor determining irrigation water requirements. Although the actual requirement may be influenced by additional factors, as an inability to apply water uniformity, a need for soil leaching and expected runoff losses of water, ET remains the main factor to be estimated. This can be done by either direct or indirect methods, and we shall only refer to those in general.

3.2.1 Direct Methods

Evapotranspiration is one of the variables in the soil water balance:

$$Q = P + I - ET - D - R , \tag{3.3}$$

in which Q is the actual soil water content, P is water added by precipitation, I is water added by irrigation, and D and R stand for water removed by drainage and runoff; all in units of mm. Rates of ET can be derived from this equation provided all other parameters are known. This could be acceptable for short periods of 1 to 2 weeks if there are no precipitations and irrigations and therefore no losses of water by drainage and runoff. In this case ET would be equivalent to $-\Delta Q$, which has, thus, to be determined. The classical method for obtaining soil water content in the effective rooting depth is by gravimetric sampling, which requires a high labor input and is subjected to extensive variability. Neutron probes or neutron scattering systems became more acceptable because labor requirements are somewhat less and the variability is reduced as moisture content is determined throughout the season at the same location. This method, however, needs calibration, which may also lead to serious errors. It should also be noted that there are difficulties in having reliable soil moisture determinations in the top 30 cm. Soil moisture content can also be measured by time-domain reflectometry. Although this device does not need much calibration and measures in-situ moisture content, it is not yet much in use for practical irrigation purposes.

Lysimeters, which are another method for direct ET determination, have the advantage that all the parameters of the hydraulic equation (3.3) are being considered. They can thus be used for any length of time and year round as all gains and losses of water are being accounted for. Lysimeters vary in the surface area being exposed and in depth, and these are of importance in order to evaluate reliable field ET values. The uniformity of the soil bulk density within the lysimeter, water profile, plant development as compared to the surrounding field and the size of this field, have also to be considered. The gravimetric and neutron-probe methods still have advantages over the lysimeter by having higher flexibility in location, possibility of multiple sites determination and much lower cost.

The velocity of water flow in the xylem of crop plants can directly be related to the state of transpiration. There are two methods which have been used for this purpose: the heat-balance technique (Sakuratani 1984) and the heat pulse technique (Cohen et al. 1987). Both methods do not account for direct evaporation from soil surface, which can probably be neglected at complete or nearly complete soil coverage. Otherwise, it has to be determined separately by other means, for instance by the micro-lysimeter technique (Boast and Robertson 1986). Another limitation of these methods is the need to get the method adapted to different crops than those few which were calibrated.

3.2.2 Indirect Methods

Numerous equations that require meteorological data have been proposed and several are commonly used to estimate ET for short period hourly to daily or for more extended periods. The estimation of ET from climate records is appealing because of its simplicity compared to the on-site direct ET measurements. A discussion on the use of indirect methods for evaluation of ET based on theoretical and empirical equations is given by Jensen (1974). If the surface from which evaporation takes place is at saturation, then the rate of water vapor transfer can be expressed as the latent heat transfer per unit area. This rate of water vapor tansfer from soil and plants was defined as Potential Evapotranspiration (PET). This value might be somewhat ambiguous, as claimed for instance by Brutsaert (1982), and may be replaced by the term Reference Crop Evapotranspiration (ETr), as proposed by Doorenbos and Pruitt (1977). This is the evapotranspiration of a known crop (a grass crop was used by some and alfalfa by others) grown under non-limiting water supply and is used for the calibration of the ET empirical equations. The amount of water lost from a crop and the soil under actual conditions is known as Actual Evapotranspiration (ETa). Actual evapotranspiration will thus, be:

$$ETa = Kc\ ETr\ ,\tag{3.4}$$

in which Kc is known as crop factor (coefficient). This crop factor will depend on crop species and cultivar, growth stage, soil water content, nutritional status and other environmental factors.

Another approach is to determine ETa directly rather than from ETr, using micrometeorological empirical methods. These are based on the separation of total solar energy flux on earth to soil heat flux, sensible heat flux and latent heat. Evapotranspiration estimated from micrometeorological methods can be based on surface aerodynamic properties, profile techniques, turbulence and a combination of these, and are outlined in some detail by Hatfield (1990). The main limitation of these methods is the need of extensive instrumentation and collection of accurate measurements. Their use for row-crops might be somewhat inadequate because of the usual lack in uniform canopy surface. These methods are most satisfactory in relatively large fields having a long fetch.

Empirical equations using standard meteorological data are also being used to estimate evapotranspiration. These equations are based on net or solar radiation fluxes, air temperature, wind factor and combinations of factors. A comprehensive review on these is givben by Doorenbos and Pruitt (1977) and will not be discussed.

Another indirect method used for the estimation of evapotranspiration is based on water evaporation rates from evaporation-pans. This method is very popular as it is easy to use and is low in cost. Crop coefficients estimating ET from pan evaporation data were suggested by many investigators at different locations and for a variety of crops (Doorenbos and Pruitt 1977; Iruthayaraj and Morachan 1978; Shalhevet et al. 1981). Evapotranspiration is determined using an appropriate crop factor and pan evaporation rate at the actual field site. It should, however, be kept in mind that the responses of pan evaporation and transpiration from a crop to climatic factors such as air temperature, humidity and wind velocity are not identical. Climatic changes throughout the growing season may thus result in varying crop factors. The use of a standard crop factor may thus lead to errors in the estimation of evapotranspiration.

3.3 Irrigation Timing

Irrigation timing is much more variable and flexible than the quantity of water to be applied and will depend on climatic factors, crop characteristics, soil properties and irrigation methods. Farm operation factors and water availability on the farm also play an important role in irrigation timing decision. In addition, salinity problems and needs of chemigation may also be of importance. The determination of irrigation timing (scheduling) may have to be based on more than one parameter.

3.3.1 Soil Moisture Content

The oldest and probably the most widely used parameter for irrigation timing is the allowable depletion of soil water. This determines what fraction of stored soil water is allowed to be depleted between irrigations and is thus used for the timing of upcoming irrigation. Provided soil moisture depletion is restricted to a given depth, namely the major part of the root zone, this parameter can be used quite successfully. A great number of investigations were conducted on different crops and in different regions in order to determine the allowable soil water depletion. Doorenbos and Kassam (1979) for instance suggested that a depletion of 50% of the available soil water within the root zone can be considered as a safe level for most crops grown in most soil types. Many field and orchard crops may deplete the available soil water up to 30–35% without a significant effect on their production, unless grown in a soil with a low water holding capacity. This would allow a less frequent scheduling of the irrigation in many soil types. A lower limit of 35% of the available soil water in the root zone volume was also recommended by Shalhevet et al. (1981).

Several soil and plant factors are involved in the relationship between allowable depletion of soil water and relative yield production. An increase in soil particle size (a more coarse texture), irrigation water or soil salinity or a restricted and dense root system will all tend to raise the lower limit of allowable water to be withdrawn. Uniformity of soil properties, sparse plant population and insensitive growth stages will tend to lower this limit.

3.3.2 Soil Water Tension

Tensiometers are probably the most common device to monitor soil water tension under field conditions for irrigation timing purposes. The range of soil water potential measurable with a tensiometer is only a small fraction of the water potential over which plants can extract water from the soil. However, as the rate of water uptake by plants and the damage to crops due to water deficit begins before soil water potential exceeds the tensiometer reading, this device can still be used. The principles and applications of tensiometry have been reviewed in detail by Cassel and Klute (1986).

Irrigation studies have been conducted on diverse crops in which irrigation timing was based on tensiometer readings. Many of these have shown that tensiometers can be used successfully for irrigation timing (for instance, Kaufman and Elfving 1972; Dubetz and Krogman 1973; Cassel et al. 1985). In spite of this, the use of tensiometers in practice has remained quite limited to some high value crops and/or where water is very scarce and expensive, and sometimes also under high water table conditions. One reason for restricted usage is probably the labor required for both the maintenance and the reading. Readings, however, can be automated by connecting the tensiometer to a pressure transducer and data logger, but this significantly increases the cost of the system. Another reason is the spatial variability of the soil and the non-uniform contact between tensiometer cup and soil which also causes variability. In addition, the tensiometer reading will be a function of the depth and distance from row and water application site, which requires standardization prior to use.

3.3.3 Evapotranspiration Rate

Evapotranspiration rates can be used as a parameter to determine the interval between irrigations. Once the availability of soil water is reduced the rate of transpiration and evapotranspiration will decrease and this change in rate can be used for the determination of the next irrigation. This requires less labor and is less subject to spatial variabilities than the determination of soil moisture content or tension. However, methods which determine ET or transpiration rates under limiting water availability are not always sufficiently reliable and are not sensitive enough to detect a depression of 5%. Such a small depression may already lead to a decrease in yield if this occurs during sensitive growth stages (Stegman 1983). Leaf water potential (ψ) is generally decreased and stomatal resistance increased under slight water stress, affecting sensitive physiological activities and potentially also yield.

Another approach is the use of a change in ratio between actual ET and maximal or potential ET as a parameter for irrigation scheduling. A critical ratio needed for irrigation timing will depend on a threshold level of available soil water, and this might be within a wide range (Morgan et al. 1980; Meyer and Green 1981). These weaknesses make the use of this parameter very limited.

3.3.4 Plant Parameters

The major plant parameters of potential use for irrigation timing are leaf water potential (ψ), leaf temperature, stomatal conductance, and growth rate of specific plant organs.

The determination of leaf water potential will most likely involve the use of the pressure chamber technique. Although the measurements obtained with the pressure chamber may not represent the actual leaf ψ, they may still be in good correlation to reflect plant water status. Other, and probably more reliable methods need more equipment and more constant environmental conditions and are therefore not very useful for this purpose in the field.

A diurnal pattern of leaf ψ values can be observed under most field conditions using the pressure chamber, even shortly after irrigation. Minimum values will be obtained during midday, when transpiration demand is at its peak, and maximal values shortly before sunrise, when full leaf water capacity is attained. Leaf ψ was found to decrease during midday and early afternoon as a result of decreased available soil water content. This is probably associated with a decrease in leaf osmotic potential (ψ_s) maintaining turgor potential close to or at the original value. Numerous studies were conducted on a wide assortment of species in order to determine threshold values of midday leaf ψ. Accordingly, water should be applied once ψ reaches a threshold value so that a further decline in ψ will be avoided. Such threshold values suggested were − 1.20 to − 1.25 MPa for corn (Stegman 1983), approximately − 1.5 MPa for wheat (Sojka et al. 1981), − 1.3 to − 1.5 MPa for sorghum (Musick 1976), and − 1.4 to − 1.5 MPa for cotton (Plaut 1983; Plaut et al. 1992). A fairly wide range of leaf ψ threshold values were outlined for some crops, depending on their developmental stage, e.g., in the case of sunflowers the value may decrease from − 1.0 to − 1.4 MPa throughout plant development (Stegman 1983).

The use of leaf temperature measurements for irrigation timing is based on a predicted rise in leaf temperature when leaf evaporative cooling is reduced due to a lower transpiration rate. Small portable infrared thermometers that measure radiation emitted from the entire canopy and soil surface within the instruments view are used for this purpose. An index can be developed using the difference between canopy and air temperature (Tc and Ta), related to vapor pressure deficit of the air (VPD), net radiation (Rn) and aerodynamics and canopy resistances (r_a and r_c). An equation to determine this crop-water stress index (CWSI) was outlined by Jackson (1982):

$$CWSI = \frac{Tc - Ta\ (Tc - Ta)_l}{(Tc - Ta)_u - (Tc - Ta)_l},$$ (3.5)

where $(Tc - Ta)_l$ and $(Tc - Ta)_u$ are at $r_c = 0$ and $r_c = \infty$, respectively, namely at a wet canopy acting as a free water surface and as a dry nontranspiring canopy. The CWSI was shown by Jackson (1982) to be:

$$CWSI = 1 - ET/ET_p.$$ (3.6)

Values of CWSI of 0.25 or higher represent already extreme conditions. A threshold value of CWSI is needed in order to set an irrigation timing and it has to be determined whether this is possible to obtain while preventing yield losses of different crops.

Leaf temperature changes due to water stress are mainly based on alterations in stomatal diffusion resistance during water stress. It might thus be possible to determine irrigation timing directly on the basis of stomatal resistance measurements. This was in fact suggested in the past and even used to some extent. However, spatial variability, effects of plant and leaf age, fluctuations in radiation intensity over the canopy both in time and space and other environmental factors are serious limitations. A high degree of standardization and many calibrations would be needed, which makes this method impractical.

As expansion growth is probably the most sensitive plant process to water stress, it may be adopted as a parameter for irrigation timing. Expansion growth was found to be stopped or greatly inhibited by water deficit, while transpiration continued anabated (Boyer 1970; Acevedo et al. 1971). The use of this parameter is probably more feasible in the case of cereals, in which the growing zone is at the leaf base and its elongation rate can easily be measured. This parameter can obviously be used only throughout the vegetative growth period. Moreover, it must be assured that no other factor, in addition to water deficit, limits growth. The leaf linear growth phase proceeds for only a few days, thus one would have to switch from one leaf to another, which introduces complications into the use of this parameter.

The use of stem elongation rates might thus be an easier parameter to determine irrigation timing, and can in fact be used for several crops.

3.4 Implication of Irrigation Method on Scheduling

The common irrigation methods differ in their irrigation rates, in the relative wetting area and in the flexibility of management, and as such have an influence on irrigation scheduling.

3.4.1 Surface Irrigation

This system needs a predetermined irrigation timetable for management purposes. Thus, in order to avoid soil water depletion to an unfavorable level, irrigation of most crops is scheduled at 50% of available water withdrawal out of the root zone depth. This may cause in some cases losses of water, if irrigation could be applied at a later date. The system is then operating just to refill the soil profile, unless leaching is required when there are salinity problems and the amount has to be increased. Applications of quantities smaller than 25–30 mm per irrigation are difficult with the surface irrigation system, because the infiltration rate controls the application depth. Smaller quantities given at higher intervals are thus not recommended.

3.4.2 Sprinkler Irrigation

This system can be controlled to deliver more uniform and precise applications, and is less dependent on infiltration rates. Variable amounts of water can be applied at each irrigation and the time can be set fairly easily according to crop needs and management demands. Sprinkling is therefore also used most favorably for germinating purposes when only small quantities of water are needed.

Moving sprinkler systems, either line sources or center pivot, are often used to apply small amounts of water. The quantity of water to be applied at each irrigation should be so that surface runoff is prevented. The irrigation frequency is then scheduled to satisfy crop ET requirements and applied when 50–60% (depending mainly on crop) of available stored water is depleted. Larger amounts of water per single irrigation will increase the interval between irrigations. This is more economical, as less water will be lost by direct evaporation and less machine hours will be needed. Each irrigation should apply sufficient water to fill the entire soil profile to the depth of the root zone with water.

3.4.3 Trickle and Drip Irrigation

These systems are typical for high frequency irrigation with small application depth and for partial watering at planting or shortly thereafter so that available soil water will never reach a threshold value, provided ET water is replaced. Most trickle systems are used for high value crops, in which water deficit should be avoided so that even short periods of water stress should be minimized.

Frequency of irrigation with a trickle or a drip system is high, and water is applied once in every 2–5 days. Allowable soil water depletion, soil variability and plant water stress are less important with this system due to the high irrigation frequency. The rate of irrigations are low and kept well below infiltration rates. The system has a high application efficiency as losses by runoff or deep percolation are avoided.

3.5 Effect of Crop Growth Stage on Irrigation Practices

In addition to climatic condition and soil storage capacity of available water, irrigation practice through the growing season will strongly be influenced by the crop growth stages. Growth of annual crops can be divided into three major stages with regard to irrigation practices: vegetative, flowering including fruit set, and fruit development. An irrigation may also be needed to ensure and enhance germination of seeds, which may be considered as an additional early and brief growth stage. Subsequent crop development and high yield production require uniform gemination and emergence. Sprinkler irrigation (especially moveable systems) are well suited because the amount of water added can be kept low during stand establishment.

3.5.1 Vegetative Growth

During the vegetative stage, the rate of ET is continuously increasing reaching a peak approximately at the time of flowering. For most crops a good moisture supply should be available to the plants at this growing stage. It is worth noting that the root system is still shallow, at least at the early part of the vegetative growth stage, so that irrigations should be more frequent and light as compared to later growth stages. In perennial crops or in crops which develop a deep root system fairly fast less frequent and heavier irrigations could be applied throughout the vegetative growth stage. Crops which are grown for economic use of their vegetative shoot organs, like lettuce, cabbage, alfalfa and grasses, will be harvested during the peak rates of ET and the quantities of water applied per irrigation should be kept at maximum up to the time of harvest. In crops which are grown for subsurface vegetative organs, like potatoes, onion, and beets, a decline in canopy and in leaf area index (LAI) takes place toward the end of their growing period. This implies that quantities of water per irrigation can be decreased but irrigation frequency should still be retained and available water stored in the soil should not be much reduced.

3.5.2 Flowering

Water requirement is maximal in most crops at flowering and early fruit development stage. LAI is probably the highest during this stage and water requirement is at its peak. Water stress should be minimized in order to avoid possible flower abortion and shedding of young fruit. The root system reaches in most crops its maximal stage of development shortly before flowering, so that the maximal water requirement during this period is partly offset by the depth of the root system. The deeper the root system the less frequent irrigation can be. Plants which are grown for their flowers, like cut flower and some vegetable crops, are also harvested at the peak of irrigation water requirement and it should not be decreased before the harvest.

In several crops excessive irrigation during the late vegetation and flowering stages may result in a reduction in fruiting. Cotton is an example of such a crop in which optimal water supply during this period will induce vegetative growth at the expense of boll formation (Kittock 1979; Guinn et al. 1981). This is believed to be due to utilization of assimilates for the growth of vegetative organs rather than development of bolls. A switch to mild water stress will favor growth of reproductive organs. Recommendations have been made to irrigate cotton at a lower frequency at late vegetation and flowering stages allowing the withdrawal of larger quantities of available soil water to induce slight water stress.

3.5.3 Fruit Development

The fruit development stage is associated in most crops with a decrease in ET rates even in those which are grown for fresh fruit. This is mostly due to the termination of vegetative growth and gradual senescence of the existing canopy. In non-determinant flower and fruit crops like fresh tomatoes, musk-melons, bell-pepper, strawberries and carnation, the mature foliage which is constantly senescing is replaced by new vegetative organs, generally smaller in size with decreasing LAI. It is, therefore, important that sufficient moisture is available during fruit development of fresh fleshy or juicy fruits. Since the transpiring leaves are a major sink for soil water, fruit size and as a result the marketable yield may be reduced even under mild water stress. However, in some cases, for example in processing tomatoes, total soluble solids in the juice, which is an important quality characteristic, can be increased by reducing quantity of water at each application during fruit growth (Z. Plaut, A. Meiri, and A. Grava pers. comm. 1993). Fresh grain fruit, such as green peas and fresh corn, also require high moisture availability during grain filling, otherwise they will not be firm and fully formed (Hansen et al. 1980). The quality of many of those fruits, therefore, can be controlled by irrigation management.

In dry grain crops the rate of ET decreases drastically during grain filling stage and during ripening, and transpiration almost ceases so that the plant is nearly dormant. There is thus no need to apply any irrigation later than the early grain filling stage, as the fairly deep roots will find sufficient available water to conclude the plant life cycle.

3.5.4 Irrigation at Different Growth Stages Under Limited Irrigation Water

Special attention should be given when or at what growth stages should irrigation be withdrawn if availability of irrigation water is limited. This refers mainly to the irrigation of grain and other field crops, since optimal irrigation practice for high value crops such as flowers, fruits and vegetable crops will probably be less altered.

Early studies came to the conclusion that stress sensitivity is generally greatest during floral through pollination stages. Grain yield was claimed to be less sensitive to stress during early vegetative stage or grain filling stage (Salter and Goode 1967). Irrigation frequency and quantities of water to be applied can thus be reduced during this growth stage if irrigation water is limiting.

These findings were verified and further interpreted by later studies (Stewart et al. 1975; Stegman 1982). Stress at an early growth stage may result in adjusting plant size and leaf area to low soil water availability and will thus decrease ET. Provided water is applied at flowering and part of it stored through grain filling, grain yield per unit water applied will be higher than irrigating at any other growth stage. Irrigation at vegetative stage only may result in insufficient water stored for grain filling so that yield per unit water applied will be lower. Irrigation at grain filling stage following dry early growth stages is not recommended because it will result in high transpiration rate late in the season, while plants might have shed part of their tillers, aborted flowers and lost assimilates during leaf drying.

The most favorable irrigation scheduling under limited availability or irrigation water will depend on the amount of water stored in the soil at planting, the size of the root system, the density of plant population and ET demands at different parts of the growing season. We have found, for instance, that irrigating corn during flowering and pollination stage, which was expected to give highest grain yield, resulted in lower yield as compared to the same quantity of irrigation water spread over the season up to grain filling stage (Z. Plaut and A. Grava pers. comm. 1993). It seems that small quantities of water during the vegetative growth stage prevented stress damages and retained much more assimilates within the vegetative organs, to be used later for grain filling.

3.6 Irrigation Management in Humid Areas

Irrigation in humid areas is very often economical even though annual rainfall may exceed annual ET. Feasibility of irrigation in humid areas will require that: (1) the distribution of crop water requirement throughout the growing season does not coincide with the distribution of rainfall; and (2) soil water storage within the root zone volume will not supply sufficient water to the crop during dry periods. The purpose of irrigation in humid areas will thus be to increase evapotranspiration but with minimal losses by runoff, drainage and leaching of fertilizers and pesticides. The ideal situation would thus be avoidance of water stress throughout the growing season, yet having no water losses. Scheduling under conditions of rainfall is, thus, the major goal of irrigation in humid areas. It is of importance to refill only part of the profile at each irrigation, leaving some storage to be filled by forthcoming rain.

One of the most common methods for irrigation scheduling is based on field soil water budget. The budget must be calculated at desirable intervals (between a day and a week), based on the effective rainfall and/or applied irrigation on the incoming side and on ET on the depletion side. Information is also needed on root depth, capacity of available water storage in the soil and its amount at the beginning of the season, and the allowable soil water depletion.

A major problem is the determination of effective rainfall as incoming water. As this is the portion of rainfall that contributes to ET, rain water that is lost by surface runoff or by subsurface drainage has to be neglected. Rainfall of high intensities or large amounts at a single storm may produce significant runoff and only part of the rain water can be considered as effective. Similarly, rainfall on a wet soil profile will be ineffective, as most of the water will be lost by drainage. Rainfall after the crop is at an advanced stage of development and has a low transpiration rate is of low benefit, unless it is stored for the next crop. Very light rains may not contribute much to the storage as most of the water will be directly evaporated. A detailed and quantitative method to estimate effective rainfall from total measured rainfall was outlined by Dastane (1974), based on monthly ET and precipitation values. It indicates that effective rainfall as defined for irrigation purposes is also related to the irrigation frequency.

Evapotranspiration can be calculated from measured meteorological data using one of the empirical equations adjusted for canopy coverage. All the difficulties of determining ET by those equations will thus prevail. A most serious one is having a soil profile mainly in the upper layers partly dry during extended periods of the growing season.

Estimates of rooting depth, which is also needed, have to be periodically obtained as they are changing throughout the season. This has to be obtained from field measurements or be available from earlier information. Allowable depletion of available soil water can be set at 50%, unless other information is known for a particular crop.

The depletion of water from the soil can also be determined using pan evaporation rates and an appropriate crop factor for a given crop. This has some limitation, as was pointed out earlier. In humid regions it is not only an evaporation pan, but also a rainwater collecting one. Such data could thus serve as a direct guide for irrigation timing and estimating the quantities of water to be added and not just as an information source for calculating soil water budget. The pan must thus be located within the irrigated area (and will be adequate for moving or fixed sprinkler systems only). It has to be leveled and filled to overflow the day after a rain or irrigation that filled the soil profile. The irrigation timing will be set by the allowable depletion of stored water which can be calculated as:

$$D_a = Q_s L_a K_c ,$$ (3.7)

and the amount of water to be added can also be calculated:

$$Q_i = [(Q_s L_a) - Q_r] K_c / E ,$$ (3.8)

where D_a is the depletion allowed in mm of stored soil water; Q_s, Q_i and Q_r are quantities of water in mm depth of total stored in the respective root zone volume, added by irrigation, and left free for a coming rain; L_a is the relative loss of allowable available water (usually 0.5); K_c is the crop factor which is approximately 0.4 at early growth stage, increasing up to 0.75–1.0 at full coverage (depending on crop), and E is the irrigation efficiency. The D_a values can be set on the measuring device of the pan for the particular crop and climatic region, so that only Q_i has to be calculated. If rainfall occurs, the pan will collect it and the excess will be lost by overflow, identical to losses by drainage.

Another approach is to determine irrigation timing based on measuring soil or plant water potentials or other plant parameters. These are supposed to respond to irrigation and effective rain water in a similar manner and can thus be used in humid as in dry areas. The amount of water to be added will, however, have to be determined by one of the methods outlined. In the case of the tensiometers, it could be determined on the basis of soil water retention curves determined with the tensiometer on that particular soil.

Small and ineffective rainfall may have some effect on decreasing ET due to cloudy and humid conditions. On the other hand, frequent rainfall as well as frequent irrigations may increase total ET as compared with infrequent rainfall, because ET rates tend to increase for some time after rainfall or irrigation.

References

Acevedo E, Hsiao TC, Henderson DW (1971) Immediate and subsequent growth responses of maize leaves to changes in water status. Plant Physiol 48:631–636

Boast CW, Robertson (1982) A "micro-lysimeter" method for determining evaporation from bare soil: description and laboratory evaluation. Soil Sci Soc Am J 46:689–696

Boyer JS (1970) Leaf enlargement and metabolic rates in corn soybean and sunflower at various leaf water potentials. Plant Physiol 46:233–235

Brutsaert WH (1982) Evaporation into the atmosphere. Theory, history and applications. Reidel, Boston

Cassel DK, Klute A (1980) Water potential: tensiometry. In: Klute A (ed) Methods of soil analysis, pt 1, 2nd edn. Agronomy 9:563–596

Cassel DK, Martin CK, Lambert JR (1985) Corn irrigation scheduling in humid regions on sandy soils with tillage pans. Agron J 77:851–855

Cohen Y, Fuchs M, Falkenflug V, Moreshet S (1987) Calibrated heat pulse method for determining water uptake in cotton. Agron J 80:398–402

Dastane NG (1974) Effective rainfall. FAO Irrigation and drainage 1, Pap 17:1–61

Doorenbos J, Kassam AH (1979) Yield response to water. FAO Irrigation and drainage, Pap 33:1–193

Doorenbos J, Pruitt WO (1977) Guidelines for predicting crop water requirements. FAO Irrigation and drainage 1, Pap 24:1–1411

Dubetz S, Krogman KK (1973) Comparison of methods of scheduling irrigations of potatoes. Am Potato J 50:408–414

Guinn G, Mauney JR, Fry EE (1981) Irrigation scheduling and plant population effects on growth bloom rate, boll abscission and yield of cotton. Agron J 73:529–534

Hansen VH, Israelsen OW, Stringham GE (1980) Irrigation principles and practices. John Wiley & Sons, New York, pp 145–170

Hatfield JL (1990) Methods of estimating evapotranspiration. In: Irrigation of agricultural crops. Agron Monogr 30. ASA-CSSA-SSSA, Madison WI, pp 435–474

Hiler EA, van Bavel CHM, Hossain MM, Jordan WR (1972) Sensitivity of southern peas to plant water deficit at three growth stages. Agron J 64:60–64

Hsiao TC (1974) Plant responses to water deficit, water-use efficiency and drought resistance. Agric Meteorol 14:59–84

Hsiao TC, Bradford KJ (1983) Physiological consequences of cellular water deficits: osmotic adjustment. In: Taylor HM, Jordan WR, Sinclair TR (eds) Limitations to efficient water use in crop production. ASA-CSSA-SSSA, Madison WI, pp 227–265

Iruthayaraj RM, Morachan YB (1978) Relationship between evaporation from different evaporimeters and meteorological parameters. Agric Meteorol 19:93–100

Jackson RD (1982) Canopy temperature and crop water stress. In: Hillel DE (ed) Advances in irrigation, vol 1. Academic Press, New York, pp 43–85

Jensen ME (ed) (1973) Consumptive use of water and irrigation water requirements. Am Soc Civ Eng, New York

Kaufman MR, Elfving DC (1972) Evaluation of tensiometers for estimating plant water stress in citrus. Hort Sci 7:513–514

Kittock DL (1979) Pima and upland cotton response to irrigation management. Argon J 71:617–619

Meyer WS, Green GC (1981) Plant indicators of wheat and soybean crop water stress. Irrig Sci 2:167–176

Monteith JL (1985) Evaporation from land surfaces: progresses in analysis and prediction since 1948. In: Advances in evapotranspiration. ASAE, St Joseph, MI, pp 4–12

Morgan TH, Biere AW, Kanamasu ET (1980) A dynamic model of corn yield response to water. Water Resour Res 16:59–64

Musick JT, Dusek DA (1971) Grain sorghum response to number, timing and size of irrigation in the Southern High Plains. Trans ASAE 14:401–410

Nobel PS (1991) Physiochemical and environmental plant physiology. Academic Press, San Diego

Plaut Z (1983) Application and practice of cotton fertigation in Israel. In: Proc 3rd Conf Irrigation and fertigation. Agritech Tel-Aviv, Israel

Plaut Z, Carmi A, Grava A (1988) Cotton growth and production under drip irrigation restricted soil wetting. Irrig Sci. 9:143–149

Plaut Z, Ben-Hur M, Meiri A (1992) Yield and vegetative growth as related to plant water potential of cotton irrigated wtih a moving sprinkler system at different frequencies and wetting depths. Irrig Sci 13:39–441

Ritchie JT, Johnson BS (1990) Soil and plant factors affecting evaporation. In: Irrigation of agricultural crops. Agron Monogr 30. ASA-CSSA-SSSA, Madison, WI, pp 363–390

Sakuratani T (1984) Improvement of the probe for measuring water flow rate in intact plants with the stem heat balance method. J Agric Meteorol 40:273–277

Salter RJ, Goode JE (1967) Crop response to water and different stages of growth. Common Agric Bur Farnham Royal, Bucks, UK

Shalhevet J, Mantell A, Bielorai H, Shimshi D (1981) Irrigation of field and orchard crops under semi-arid conditions, 2nd edn. IIIC. Bet-Dagan, Israel

Sojka RE, Stolzy LH, Fischer RA (1981) Seasonal drought response of selected wheat cultivars. Agron J 73:838–845

Stegman EC (1982) Corn grain yield as influenced by timing of evapotranspiration deficits. Irrig Sci 3:75–87

Stegman EC (1983) Irrigation scheduling: applied timing criteria. In: Hillel D (ed) Advances in irrigation, vol 2. Academic Press, New York, pp 1–30

Stewart JI, Misra RD, Pruitt WO, Hagan RM (1975) Irrigation, corn and grain sorghum with a deficient water supply. Trans ASAE 18:270–280

4 Irrigation Techniques and Evaluations

A.J. CLEMMENS and A.R. DEDRICK

4.1 Introduction

The primary motivation for the development of farm irrigation systems is almost always economic. Water is considered an input to the farm enterprise just as are seed, fertilizer, labor, equipment, etc. In more humid regions, irrigation water serves to maintain production in years of drought. In slightly more arid regions, irrigation produces higher yields in a majority of years. In extremely arid regions, crop production relies entirely on water from irrigation. In the second case, the decision to supplement rainfall with irrigation water is an economic decision. If the expected benefit of increased yields more than offsets the cost of the irrigation system plus the cost of water, then development of the irrigation system is justified. In many cases, this can be considered a short-term investment. In the third case, development of a farm irrigation system is dependent on the economic feasibility of the entire farming enterprise. This is nearly always a long-term investment decision.

Farming is filled with uncertainty associated with weather, insects, disease, yields, crop prices, etc. Farmers use water as a management tool, which in part helps them deal with some of this uncertainty. Water is used for frost control, germination, leaching, etc. These are considered management uses of water, which differ from consumptive use by the crop. Considerations of water conservation are not always paramount in the farmer's mind. Rather, management focuses on the factors which have the most impact on economic production. Where water is inexpensive and plentiful, excess water is often applied, partly as insurance against low crop production and partly to focus management efforts on more critical issues. Applying water both efficiently and in an adequate amount requires management effort. While there may be economic benefits to applying water efficiently in terms of both water cost and higher yields (e.g., excess water applied can often reduce yields), these benefits must outweigh the management cost (and often capital cost) of applying irrigation water efficiently.

Adv. Series in Agricultural Sciences, Vol. 22
K.K. Tanji/B. Yaron (Eds.)
© Springer-Verlag Berlin Heidelberg 1994

4.1.1 Farm Irrigation Methods

There are three basic types of irrigation methods: surface irrigation, sprinkler irrigation, and micro-irrigation. A number of irrigation methods are hard to categorize since they cross over the traditional boundaries of these three general types.

Surface irrigation methods are the earliest known methods employed for large-scale agricultural production. They have been practiced longer than recorded history. More than three-fourths of the irrigated agricultural land in the world is surface irrigated (FAO 1981), with the remaining land predominantly sprinkler irrigated since micro-irrigation represents less than one tenth of 1% (Nakayama and Bucks 1986). In the USA, surface irrigation comprises roughly 58% of the irrigated land, sprinkler 40% and micro-irrigation about 2% (USDA 1990; Irrigation Association 1991). Sprinkler irrigation has spread rapidly since World War II. Micro-irrigation has experienced several surges of growth – in the late 1960s and early 1970s and again in the mid 1980s. Under certain conditions, each method would be preferred over the others.

Surface irrigation methods are so named because water is distributed across the field by flowing over the field surface. Thus soil infiltration and soil and crop flow resistance have a major influence on the distribution of water. By contrast, under sprinkler and trickle irrigation a majority of the water distribution is accomplished in closed pipelines. Their distribution is affected by changes in pressure and by variations in outlet properties (e.g., sprinkler head or emitter manufacturing variations). Once the water leaves the pipeline, it is distributed over the area served by the outlet in a variety of ways depending on the system and the crop.

4.1.2 Definitions of Efficiency and Uniformity

Evaluation of different irrigation system types is relevant only once definitions are made for various types of efficiency and uniformity. Unfortunately, because the objectives of irrigating are often different under the conditions most applicable to each system type, numerous definitions of efficiency and uniformity have been developed over the years for these various types. Attempts at unifying these definitions have not been entirely successful. Some of the difficulties stem from conflicting objectives, while others arise from differences between academic and practitioner needs. Some basic definitions of efficiency and uniformity are offered below, followed by discussion of the implications for management.

ASAE (1987, 1991) suggests the following definitions for application efficiency, E_a, and distribution uniformity of the low quarter, DU_{lq}, which apply to a single irrigation event:

$$E_a = \frac{\text{average depth of water stored in the root zone}}{\text{average depth applied}}$$

$$DU_{lq} = \frac{\text{average low quarter depth of water infiltrated}}{\text{average depth applied}}.$$

The first equation needs clarification. The numerator refers to water which is available for consumptive use by the plant, and is eventually used for that purpose. This includes water added to root zone soil moisture and water which evaporates in lieu of transpiration.

When small amounts of water are applied, the irrigation and application efficiencies can be high, while not providing sufficient water for crop production. Thus another measure of irrigation performance is needed to assess how much water is usefully stored in the crop root zone. The storage efficiency is defined as:

$$E_s = \frac{\text{average depth of water stored in the root zone}}{\text{available depth of root zone storage}}.$$

Irrigation management suggests that a target depth of application be determined prior to an irrigation. In arid areas, this is often the soil moisture deficit, or the amount of water consumptively used since the last irrigation. In more humid areas, a target depth to apply is often less than the soil moisture deficit, which allows for effective use of rainfall in replacing soil moisture between irrigations. In evaluating the performance of a single irrigation event, this "target depth" replaces the "available root zone storage" for evaluating storage efficiency and the "average (over the field) of the depths infiltrated up to the target amount" and replaces the "average depth of water stored in the root zone" in the application and storage efficiency.

Figure 4.1 is useful for understanding the various efficiency and uniformity terms for a single irrigation event. If a small volume of water is applied, little or no excess water or deep percolation results, the application efficiency is high, but the storage efficiency is low. As more water is applied, the application efficiency decreases, while the storage efficiency increases. An example of these relations is shown in Fig. 4.2. If too much water is applied, some will be lost for use due to surface runoff, deep percolation below the root zone, or the inability to effectively use rainfall (e.g., even if amount above target depth is stored in the soil, it may not be of benefit). If too little is applied, the next irrigation will need to occur sooner. The impact of applying too little water thus depends on how well irrigations are scheduled. If plant stress or soil moisture indicators are used to schedule irrigations, applying too little water during a particular irrigation event may have little effect on production. An exception to this is irrigation systems which become more non-uniform as less water is applied (e.g., some surface irrigation systems).

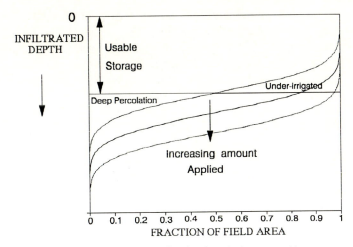

Fig. 4.1. Uniformity of water application in relation to usable storage

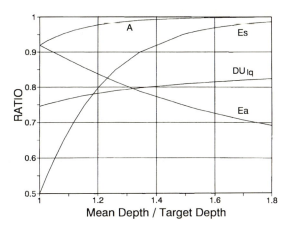

Fig. 4.2 Relationship between application efficiency (E_a) storage efficiency (E_s) low quarter distribution uniformity (DU_{lq}) and fraction of field adequately irrigated (A) in relation to amount of water applied (for constant standard deviation of depth = 0.2 times target depth, e.g., surface irrigation)

Then either more crop stress results, or upon sooner irrigation, excess water is lost through deep percolation since part of the field would still have sufficient water.

If we define the target depth as Z_n and the average depth applied as Z_g, then the application efficiency becomes:

$$E_a = E_s \frac{Z_n}{Z_g} .$$

Blair and Smerdon (1988) have proposed a single measure of irrigation performance where the volume usefully infiltrated and stored is divided by the sum of total volume applied and the amount of deficit. A weighting can be applied to the volumes which represent runoff, deep percolation, and deficit, to reflect the differing values placed on these. The researcher's rationale for this approach is the development of search procedures for optimizing design and management recommendations for surface irrigation. This work points out the inherent problem of describing irrigation efficiency and uniformity, particularly when comparing irrigation systems of different types. Performance cannot be described with a single parameter, and the tradeoffs among the various parameters (or weights on these volumes) depend on the site-specific conditions.

A clear distinction should be made between measures of seasonal performance and single event performance. Single event performance evaluations (i.e., the above equations) can be very misleading and may not represent a true picture of seasonal performance. The irrigation efficiency, E_I, is often used to determine seasonal efficiency, namely:

$$E_I = \frac{\text{average depth of water beneficially used}}{\text{average depth applied}}.$$

For a particular irrigation event, it is difficult to know if the water infiltrated and stored in the root zone is beneficially used. Beneficial use has an extended meaning over consumptive use or evapotranspiration. It includes such factors as leaching, germination, temperature control and frost control. See Burt et al. (1988) for a good analysis of seasonal performance. Neither E_a nor E_I provides useful measures for evaluating a seasonal volume balance for water supply and drainage considerations. Such a measure would have to include considerations of changes in soil moisture during fallow periods and of multiple crops, which can be very site specific.

All irrigation systems are inherently non-uniform; thus for each irrigation event the farm manager is faced with a tradeoff between the total amount of water to apply and the amount of deficit to allow. Most farmers do not make such decisions based on analysis, but rather develop rules of thumb or use judgement that suit their particular needs. Such analysis or judgement must evaluate what effect each irrigation has on total yield. While some attempt is made to express this tradeoff in terms of seasonal water applied and yield (i.e., a simple economic model), the problem is not that simple. First, there is a great deal of uncertainty about the relationship between water applied and yield. Some of this uncertainty is related to irrigation uniformity, some to yield potential for a given year, and some to the effects of irrigation timing. In the face of uncertainty, farmers tend to apply more water than would be justified by simple economic analysis based on certain known factors.

Part of the difficulty with evaluating irrigation uniformity is the number of variables which contribute to non-uniformity of water application. Such factors include:

Surface Irrigation: opportunity time differences caused by advance and recession, spatial variability of soil infiltration properties, variations and undulations in soil surface topography (including side fall), and variations in flow rate at the supply end, variations in application time for different parts of a field;

Sprinkler Irrigation: variations in pressure caused by pipe friction and topography, sprinkler pattern effects, differences in wind drift, runoff due to excessive application rates, and variation in timing; and

Micro-irrigation: variations in pressure caused by pipe friction and topography, variations in hydraulic properties of emitters or emission points (from manufacturing or clogging), variations in soil wetting from emission points (e.g., line source along crop row), and variations in timing.

Many of these factors are difficult to quantify, even in a research setting, which adds to uncertainty. If these factors can be quantified, techniques are available for determining the influence of each on overall irrigation uniformity (e.g., Clemmens 1991).

4.1.3 Project or River Basin Considerations

Over 200 million ha of land are irrigated worldwide, representing between 15 and 20% of the total cultivated land (Framji 1984). However, irrigated land provides about one-third of the world's food production, and thus irrigation represents significant economic benefit.

Worldwide, irrigated agriculture uses about 80% of water withdrawals (Framji 1984). However, this is a somewhat misleading view of water balance since more than half of the water diverted is returned for reuse. In the upper end of a river basin, excess use of water does not reduce the supply, and improvements in irrigation practices cannot be justified in terms of the conservation of water volume (but they may be justified in terms of crop production or water quality degradation). Excess irrigation water may not be recoverable lower in a river basin, where any improvements in irrigation practices may indeed provide water for other uses.

Excess irrigation can result in drainage problems either locally, e.g., single farm or basin-wide, or at or near the natural drainage outlet of a large basin. Better irrigation practices are necessary for maintaining the viability of all arable land. Unfortunately, those farms causing the drainage problem may not be those whose lands are affected.

Since crop transpiration uses essentially pure water, drainage is needed to remove the dissolved salts which are left behind in the soil. Thus the use of water in agriculture tends to concentrate these salts where drainage flows return to the natural watercourses. Once the soil in the root zone has achieved a natural chemical balance, excess water applications will not alter the salt loading of rivers and streams unless salts are leached from underlying soil and rock. In fact, higher irrigation efficiencies will result in increases in drainage water salinity.

The relationship between improvements in farm irrigation efficiency and water conserved for other uses depends highly on the specific location of the particular farm within the hydrologic basin. Efforts at conserving water through improvements in irrigation practices need to focus on areas where water is truly conserved, i.e., either more water becomes available, or water quality is not degraded.

4.2 Surface Irrigation

4.2.1 Classification of Types

Surface irrigation methods can be grouped in the following categories: continuous flood (paddy), basin, border-strip, and furrow. Even within these categories, there are a number of further distinctions that affect how such systems are designed and operated.

Continuous flood or paddy irrigation is commonly associated with rice production. Here, small basins are flooded essentially during the entire growing season. Often, water is supplied to one end of a series of basins, and water flows through each downstream basin in turn. In some cases, water is supplied continuously to maintain a near constant ponding depth, while in others the water is supplied intermittently so that the water level fluctuates. This is a unique method applicable to paddy crops; however, it represents a large portion of irrigated land over the world (Fig. 4.3).

Basin irrigation is a method of intermittently supplying water to crops. Water supplied to the basin advances across and then ponds over the entire surface. It differs from continuous flood irrigation in that water does not remain ponded for long periods (usually less than one day). The basin may be level in all directions, have a small downhill slope in the direction of water flow, or be only roughly leveled. Basins with a small downhill slope are often considered low-gradient border strips; usually they are individually leveled. Basins range in size from less than 1 ha to more than 15 ha, the latter associated with precision level basins (with laser-controlled leveling) and large flow rates (up to 0.7 m^3/s) as practiced in the United States. Crops can be raised on beds or furrow ridges within a basin, or can be planted on-the-flat (e.g., as in border strips).

Under border-strip irrigation, water is applied to one end of a rectangular strip of land which is sloping. The water advances down the slope and either runs off the end or is ponded behind a dike. The inflow is generally cut off before the water reaches the end to reduce runoff or depth of ponding. Once flow is cut off, the water level at the head is reduced to zero as the water moves downhill and recession moves down slope as the water flows off. Side

Fig. 4.3. Contour paddy irrigation of rice in Southeast Asia

slope is often used with border strips, where the strip is not level perpendicular to the direction of flow. This allows border-strip irrigation to be used on sloping land without leveling within individual border strips.

Furrow irrigation provides a means for controlling and guiding water on steep land. Soil from the furrow bottom is used to make ridges of beds on which the crop is generally planted. Furrow irrigation is practiced on steep, undulating land and on very level land. Furrows can be oriented down slope, on the contour, or somewhere in between. When land grading equipment is used to smooth out the slope for furrow irrigation, it is often referred to as graded-furrow irrigation; however, it implies only sloping furrows. Level furrow irrigation is another practice, which is distinguished from furrows in level basins. In the former, water is distributed and guided to individual furrows rather than collectively (i.e., all furrows are open and connected at the ends). Low gradient furrows can also be distinguished from sloping furrows and level furrows where some slope exists, but water is ponded at the lower end of the furrows (Fig. 4.4). Sloping furrows usually produce some runoff.

Standard sloping furrows can be modified to limit runoff. Water can be ponded at the lower ends of the furrows as the slope diminishes, and the lower ends of rain-fed fields may receive runoff from the upper ends, which

Fig. 4.4. Low-gradient furrow irrigation of cotton with siphon tubes in Arizona, USA

are fully irrigated. This system uses water very efficiently where supplies are limited.

4.2.2 Conditions of Applicability

Surface irrigation systems are particularly applicable where investment in infrastructure and total crop values are low. Surface-irrigated systems are usually the lowest in capital investment. Surface systems are also applicable where surface water is available from rivers, streams and reservoirs. Irrigation projects which utilize primarily open canals are best suited to surface irrigation. Some effort is required to convert from open to closed systems, regardless of system type, which cause problems in keeping debris out of pipelines and matching supply and demand. Although surface irrigation systems function better with a well-controlled water supply, in general they do not require it (i.e., they can be operated with whatever supply of water is available).

Surface irrigation systems are most applicable in arid areas where evaporative demand is high, little rainfall occurs, and evaporation and wind drift from sprinklers are high. Surface irrigation is best applied where soils are uniform, since non-uniform soils will reduce efficiency and uniformity of crop growth.

While surface irrigation systems can be applied on steep or undulating terrain, they are best suited to land which can be effectively graded to a plane surface (at least in small blocks). Benching and terracing is frequently done, particularly for paddy irrigation, even where slopes are great. Depth of topsoil becomes an issue where large cuts and fills are needed.

Of the non-paddy surface methods, sloping furrows, followed by border strips and basins, seem to be the easiest to adapt to difficult topographic conditions, non-uniform soils, and water supply limitations. Even level furrows are more easily adapted than level basins (i.e., non-perfect conditions have less impact). While under ideal conditions (uniform soils, good control over water supply, etc.), level basins are potentially easier to manage and have greater potential efficiency.

The choice of surface system type is often dictated by the crop grown, rather than personal preference. Crops which are broadcast, such as wheat and alfalfa, are generally grown in basins and border strips, while crops which are grown in rows, such as corn and cotton, are generally irrigated with furrows. In rare instances, alfalfa has been grown on beds to control surface drainage. Wheat is often grown on corrugation; however, such furrows do not serve as water control furrows. Many row crops are grown in furrows and beds within level basins (e.g., cotton, lettuce, melons), but some have been grown effectively in rows with no furrows.

4.2.3 Practices

Most surface irrigation systems are somewhat labor-intensive, using generally unskilled or semi-skilled labor. Even where equipment is used for other farming operations (e.g., cultivation, planting, harvesting), irrigation requires manual labor. Furrow irrigation is the most labor-intensive, since water must be guided into individual furrows. Methods for guiding water into furrows include siphons, spiles, canal side weirs for open canals, and gated pipe. Gated pipe is most applicable on mild to steep slopes where flow rates are relatively small. The other methods usually require considerable shovel work to keep water guided to individual furrows. In some areas, extensive shovel work is used to move water from furrow to furrow within the field.

For border-strip and basin irrigation, borders or dikes are constructed. Where benches or changes in elevation exist, the dikes are somewhat permanent (i.e., not reconstructed each year). Intermediate borders are constructed by either borrowing soil on either side or by dragging soil from the middle of each strip and depositing it at the edge (called bucking-up). For sloping and low-gradient border strips, bucking-up is preferred, since it does not create guide furrows on the side of the border. Flow-down in these guide furrows needs to be stopped to allow uniform water advance across the width of the strip. Small dikes, perpendicular to the borders, are used to force the water back toward the center of the strip. These dikes are also used where side-fall causes water to channel down one side of the strip. With level basins, the guide furrows down the sides help to spread water over the basin more quickly. These are often constructed larger than needed to facilitate water advance.

Field sizes are greatly varied. Lengths of run vary with the soil type, field slope, and degree of leveling precision, with low intake soils, steeper slopes, and greater leveling precision typical of longer run lengths. For sloping furrows and border strips, run lengths range from 200 to 800 m, with 400 m being common in the USA. For level basins, lengths vary from 10 m for poorly leveled fields in developing countries to 200 to 400 m for precision leveled basins in the USA.

Advance rates may vary because of equipment wheels which compact the soil in some furrows. Weights dragged behind the equipment, called torpedoes, are used to compact and firm non-wheel rows so that advance is more uniform. This is also done simply to speed up the advance rate, for example on high infiltration rate soils or on level basins. Other methods of infiltration rate modification are not in common practice. Paddy irrigation generally results in a plow pan which restricts infiltration. Compaction layers also result in non-paddy fields where heavy equipment is used. Deep tillage or ripping are used to reduce this compaction.

Most non-paddy methods apply water at a nearly constant rate for a fixed time duration. Basins and border strips have set times that are dictated by the irrigation system design, and cannot be fully controlled by labor schedules. For furrows, the duration is often fixed by labor schedules (e.g., 12- or 24-h sets) which can make efficiencies poor in many cases. For border strips, uniformity is obtained by trying to match advance and recession. For basins, uniformity is obtained by making advance as rapid as possible. For sloping furrows, uniformity is obtained by relatively rapid advance and either significant runoff or a cutback stream. Reducing the flow rate (cutback) after advance is complete is relatively labor-intensive and not widely practiced.

Pump-back, tail-water reuse, or return flow systems have been in use in some areas for decades. These take the excess water from furrow irrigation runoff, collect it in a sump, and pump it back into the supply channel. Two relatively new advances, surge flow and cablegation, are attempts to improve the uniformity of furrow irrigation while reducing runoff. Surge flow uses valves to cycle the water on and off. By wetting and dewatering the soil, infiltration rates are often reduced and advance is more rapid. After advance is complete, quick cycling of the valves provides a cutback stream. This system is particularly applicable to the first irrigation of the season where infiltration is high and advance is slow. Cablegation uses a plug inside a gated pipe. Flow begins as the plug moves by an opening. The further the plug moves by (downhill), the lower the pressure on the outlet and the lower the flow. Thus cablegation provides a gradual cutback over the irrigation event. The total flow is dictated by the plug speed and the pipe slope. Both have found modest successes. Several other methods of achieving cutback have been developed, but few are in common use. A new method, called the level-basin drain-back system was developed for lighter, more uniform applications for level basins. Here, the supply channel is below the field grade. Blocking the field channel results in a rise in water level and irrigation of the field, while opening the channel results in drainage of excess water applied to the field and subsequent irrigation of lower lying basins. The method has only been tried on only a limited basis, but offers some opportunity for rotation of paddy and non-paddy crops.

4.2.4 Constraints to Adoption

Terrain places the most severe limitations on the adoption of surface irrigation systems. Undulating terrain with shallow topsoil, with either broadcast crops or orchard crops, is the least suitable combination for surface irrigation. Extremely non-uniform soil is another conditions under which surface irrigation has limited potential. Where water is scarce, some crops such as vegetables are difficult to irrigate efficiently with surface methods. Germination of certain crops can also cause efficiency problems with surface systems.

4.2.4.1 Constraint to Efficient Use of Water

Experience indicates that if farmers need to be efficient, they can and will find ways to do so. Generally, the pressure to conserve stems from expensive water, limited amounts of water, or restrictions on the amount of drainage water allowed. Attempts to convince framers to be efficient with water for the sake of conservation have not been highly successful. As discussed above, farmers use water for convenience. If using extra water makes other jobs more convenient, they will use more water. The pressure to conserve water will not prompt farmers to conserve overnight. A period of learning is required.

The primary constraints on the efficient use of water for irrigation systems are: labor, water supply, irrigator decisions, field design and layout, water measurement and accounting, irrigation scheduling and infiltration variability. These are all factors in controlling water use. Where labor schedules dictate irrigation set times, irrigation efficiency will likely suffer. Where the water supply is not flexible enough to provide needed water at the proper rate, frequency, and duration, water conservation will be difficult. Fluctuation in flow rates to farms from open channel distribution systems is a common problem.

Many of the new, more modern surface irrigation systems tend to take the key decisions away from the unskilled irrigation laborer and place these decisions with the manager or foreman. Irrigators tend to err on the side of excess water application. In many instances, field designs are not appropriate for efficient surface irrigation. Large fields with long run lengths are desired for machinery efficiency, often at the expense of lower water application efficiency.

Water measurement under surface irrigation is often inaccurate or missing entirely. Even where measurement exists, farmers do not often keep track of water use and may not be aware of the amounts used. Careful scrutiny of water use per irrigation is a good first step towards improving irrigation system operations. Deciding when to irrigate and how much to apply is a key component of good water management for irrigation. This is more of an education problem than one of adoption of modern technology. With only a general knowledge of consumptive use and water requirements, and with simple methods (visual, soil moisture feel, etc.), farmers can learn both the proper amount of water to apply and the proper scheduling of irrigations. There is a real question whether the more sophisticated technologies can significantly improve irrigation scheduling, i.e., whether improvement will justify the cost of the technology.

Where soils are variable, little can be done to overcome the effect on the application efficiency. Dividing the fields to separate dissimilar soil types can be done to a degree, but it does not solve the whole problem. Soil swapping has also been used to remove lenses of dissimilar soils. In most cases this is prohibitively expensive. Variations in soil water holding capacity have also been shown to have an impact on water use and yield. Such variations have an impact on irrigation scheduling and on how scheduling practices affect total yield.

4.2.5 Evaluation Methods

Most methods for evaluating surface irrigation systems focus on the advance and recession of the irrigation stream. These methods are useful for evaluating the irrigation system, its operation, and its potential for improvement. These methods are only partly effective in determining the real distribution of water. Generally, such evaluations do not consider the soil infiltration variability or variations in land grading. Moreover, they tend to look at a single irrigation set, rather than a field or a farm. Finally, they are often not considered in the context of seasonal water management.

To evaluate the effectiveness of an irrigation system, one should first look at seasonal water applied versus estimated water requirements (e.g., consumptive use), and compare this ratio to yields obtained versus potential or expected yield, keeping in mind whether water or other factors have limited yields. Next, water applied versus water required can be examined within a shorter time frame, e.g., monthly or irrigation to irrigation. This comparison often indicates areas of weakness for the irrigation system; e.g., surface systems are often inefficient early in the irrigation season when infiltration and surface roughness are high and the needed depth of application is low. Unfortunately, such information is often not available, and the evaluator has only a single irrigation event which can be evaluated.

Single event evaluations usually include measurement of inflow rate, application time, advance, recession, and some estimate of infiltration. From this information the distribution of infiltrated water can be determined (Merriam and Keller 1978). Techniques for determining distribution uniformity in basins can be found in Clemmens and Dedrick (1981). Furrow irrigation evaluation methods are available from ASAE (1987). Additional methods are given in Walker and Skogerboe (1987) and Jensen (1980). These techniques all use volume balance concepts, with data collected to determine the variation in infiltrated water over the field. An alternative is to do extensive soil moisture sampling, where soil moisture contents are measured before and after the irrigation. A large number of samples is needed to accurately define the distribution of infiltrated water because of soil variability. Any water that percolates below the root zone would likely not be measured. Thus this method does not give an accurate picture of distribution uniformity, but may give an accurate picture of application efficiency, provided that the applied water is measured.

Application efficiency can be determined only by knowing the soil moisture deficit (usually the target amount to apply). Such information can come from several sources, including: soil moisture sampling, a combination of climatic data, crop coefficients (Jensen et al. 1990), and time since the last irrigation, or published consumptive use data (e.g., Erie et al. 1982) plus the time since the last irrigation. Even where good data on advance and recession are obtained, it is useful to verify that sufficient water has penetrated in that area of the field receiving the shortest infiltration opportunity time. If this area has received insufficient moisture, then rough calculations of application efficiency from soil moisture deficit divided by depth applied will estimate too high an application efficiency (i.e., according to storage efficiency).

Field practitioners often need rough, approximate methods to guide farmers and irrigators toward more effective use of water. Theoretical methods often require too much data. Burt et al. (1988) provide such approximate methods for routine field use. Although available, such procedures usually have to be adapted to local conditions.

4.3 Sprinkler Irrigation

Sprinkler irrigation dates back to the early 1900s. However, prior to about 1920 sprinkler irrigation was limited to orchards, nurseries, etc. (Keller and Bliesner 1990). Continual improvements in sprinkler nozzles, fittings, pumps and the availability of power have caused sprinkler irrigated acres to expand rapidly within this century. In particular, they have allowed lands to be irrigated which could not be irrigated practically by surface methods. In addition, sprinkle irrigation often allows more efficient supplemental irrigation where only a small amount of water is needed to meet crop needs.

4.3.1 Classification of Types

The main distinction among sprinkler irrigation methods is their use of fixed or moving nozzles during the irrigation. Under each category of fixed or moving, there are a number of subcategories.

4.3.1.1 Fixed Sprinklers

Some sprinkler systems are fixed in place for their useful life. These are generally referred to as permanent solid-set sprinklers. These can be fixed on risers coming up from buried lines or on lines suspended above the crop, such as over-tree sprinklers. Risers can bring the spray head to the surface with pop-up heads typical of landscape applications. They can also bring the heads fully above the crop (e.g., for wheat) or anywhere in between. These are common for perennial crops such as orchards, pastures, etc.

Another category of fixed sprinklers is the hand-move sprinklers, which are disassembled, moved, and reassembled between irrigations. These can be in place for a season, for part of a season (e.g., vegetable germination), or for a single irrigation event (Fig. 4.5). When in place for part or all of a season, they are usually referred to as portable solid set systems.

There are several categories of fixed sprinklers that can be moved periodically without being disassembled. One is a lateral line which is dragged from one end, called an end-tow sprinkler system. While the end-tow is similar to the hand-move system, the side-roll and side-move systems are quite different. For these systems, the lateral lines move sideways on wheels. The side-roll lateral lines are on a large wheel, with the lateral line at the hub of the wheel. The set widths thus must be even multiples of the wheel circumference (i.e., to put nozzle to top of pipe), unless the nozzles are automatically leveled. For the side-move systems, the wheels are mounted on a separate framework from the lateral line. Training lines can be connected to the lateral line to increase the area covered.

4.3.1.2 Continuous-Move Sprinklers

For these systems, the sprinkler moves continuously during the irrigation. This tends to greatly increase the uniformity of application, since water

Fig. 4.5. Hand-move sprinkler irrigation of carrots in Portugal

application in the direction of movement is not significantly affected by the sprinkler pattern.

The traveling-gun and rotating-boom systems are considered moving-sprinkler irrigation systems. Under these systems, a high pressure rotating sprinkler nozzle or sprinkler nozzles on a rotating boom are dragged along with water supplied from a flexible hose. These systems can be left stationary, but more typically move during the irrigation.

Center-pivot irrigation is the most common moving irrigation system. Here water is supplied at a central point and the lateral line rotates-around this center. This is the most efficient water supply connection for continuous-move sprinklers. Also, the system returns to its starting position after each rotation. Center pivots have a series of towers (on wheels) which support the structural spans which in turn support the lateral pipe. Because each length of pipe covers a larger and larger area as it moves away from the center, the flow rate per unit length must be adjusted to obtain even coverage over the field area. A guidance system is needed to keep the various towers in line and to shut the system down if one tower gets stuck or a drive mechanism fails.

The linear-move system is similar to the center-pivot in size and style. Here, however, the lateral moves in a straight line, with a supply along one side of the field. Water supply pickup can be from an open channel (Fig. 4.6), a flexible hose, or a buried pipeline with automatic moving riser connectors. Guidance is somewhat more difficult than for a center pivot in that neither end is fixed. An advantage is that nozzling is uniform; however, the machine

Fig. 4.6. Linear-move irrigation system with water supplied from a concrete-lined canal on cotton in Arizona

must be moved back across the field in order to return it to its starting position.

4.3.1.3 Types of Sprinkler Outlets and Nozzles

A number of sprinkler outlet types are used with the different methods described above. The most common types for agricultural purposes are the spray nozzle and the rotating head sprinklers where the opening rotates in a circular path. The spray nozzle has a fixed opening which can represent a variety of patterns (circular, rectangular, etc.), while the rotating head can cover sections of a circle, but over a much larger area with the same flow and pressure.

The energy crisis of the mid-1970s caused a shift from high-pressure sprinkler systems (400 kPa and above) to lower-pressure sprinkler systems (e.g., below 300 kPa). Low-pressure spray nozzles (70–200 kPa) are commonly used on continuous-move systems. For these systems, water exiting an opening impacts a small plate which deflects the water outward to create the spray pattern. Droplet sizes are generally smaller for the low-pressure spray nozzles. Spitters and misters are another type of low application, small-droplet-size sprinkler. These are typically used for orchard type crops.

In perforated pipe systems, water sprays from a series of closely spaced holes which extend the full length of the pipe. Because of variations in droplet and spray distribution with pressure down the length of the tube, these are not in common use for agricultural purposes.

4.3.2 Conditions of Applicability

Sprinkler irrigation systems are most applicable where irrigation water is supplemental to rainfall in meeting crop water needs. While sprinkler irrigation is used successfully in extremely arid conditions, it is generally not the most preferred method in terms of either water use or economics. There are a number of exceptions. Where the topography is steep or highly undulating, particularly when topsoil is shallow, sprinklers are generally preferred. Also, under extremely arid conditions and poor water quality, many crops' leaves are sensitive to salt damage when sprinklers are used.

Non-movable, fixed sprinklers are applicable to many orchard crops where spray patterns are often a minor consideration. While fixed sprinklers are common for landscaping applications (turf) and are sometimes used in permanent pastures, they are generally not the most efficient or economical sprinkler system for field crops such as small grains (wheat, barley, etc). The more arid the conditions, the less efficient fixed sprinklers are for these types of crops. The hand-move systems are most economical for short-duration needs, such as germination of vegetables, which might be inefficient with surface methods. In humid areas, they are used where irrigation is not normally practiced to minimize the impact of drought. The high labor requirements make them unattractive for most large-scale field crops, unless labor costs are low relative to equipment costs.

The movable, fixed sprinklers (e.g., side roll) have been successfully used for field crops such as wheat and low growing forage. Side-roll and side-move systems are used only on low-growing crops so that the lateral line will pass above the crop. In sandy soils these systems offer some advantage over surface irrigation methods. Generally, such moving systems need rectangular field shapes and can handle only mildly rolling topography with no obstructions. Wind drift and wind damage to these systems have been a problem, and wind braces are recommended (Keller and Bliesner 1990). Automatic levels for sprinkler heads are also recommended. Labor needs to be available to move these systems periodically (once or twice per day). While end-tow systems require less labor, they generally have restricted applicability (e.g., even topography, lower growing crops) and are more subject to damage by operators while being moved.

The continuous-move sprinkler irrigation systems developed within the last half of this century have been the cause of the large increases in irrigated acreage in the United States over the last 30 years. The center-pivot and linear-move systems are applicable to large areas and require little, but more highly skilled, labor. Center-pivots can be run on extremely undulating topography, although such terrain may reduce overall uniformity of application and aggravates soil traction problems. They work best when there is a full, unobstructed circle, although partial circles are also used. Linear-move systems are more restricted in their areas of application. They are practical only on relatively even terrain because of the increased difficulty in guiding the towers. They have been adapted more to arid conditions and specialty crops such as vegetables, for example, where surface irrigation has traditionally been applied.

4.3.3 Practices

Sprinkler irrigation systems vary widely in the type and amount of labor required. The fixed, movable systems require the most labor, largely unskilled. The continuous-move systems require the least labor, but generally demand a higher skill level for repairs.

In more humid areas, sprinkler systems are turned on intermittently to cover periods of water stress caused by insufficient rainfall. In more arid areas, center pivots (for example) are often run nearly continuously during peak crop water use periods. Linear-move and fixed, movable systems need to be designed with enough capacity so that they can be stopped and moved back to their starting position, which for some soils may require a period for the soil to dry out.

Hand-move and end-tow laterals are generally laid out parallel to the crop rows, whereas side-roll and side-move systems are often run in either direction. Linear-move systems usually travel perpendicular to the row direction, while center pivots travel parallel. Either straight or circular rows are commonly used for center pivots with an access road to the center.

Where sprinkler application rates are high, topography is steep, and soils are erosive, furrow diking is often practiced to keep water from running down the furrows. Furrow dikes are constructed to hold water in small furrow lengths which vary depending on topographic conditions (typically 3 to 5 m).

The fixed, movable systems must be drained before moving in order to prevent structural damage, since these systems are not designed to be moved with the extra weight of the water in the lines. Quick flush drains should be provided with these systems to assure that they are properly drained.

The continuous-move system must have good guidance systems to keep the lateral towers in line. Also, semi-permanent tracks are made for the tower wheels to follow. These are made naturally by the weight of the towers on moist soil. However, these should not be allowed to sink so deeply that tower movement is impaired. Some maintenance of these tracks may be required. In other situations, bridges have been built to allow the towers to pass over small washes or depressions in the natural topography.

Lateral lines for most sprinkler systems run up to about 400 m. Shorter lengths are more common for the hand-move, side-roll, and other systems. The travel distance for these linear systems can be quite long. For practical purposes, however, travel distances are typically less than 1.7 km.

Center-pivot laterals, typically 400 m long, irrigate a 50-ha circle out of a 65-ha square. With an end gun at the end of the lateral, an additional hectare can be irrigated in the corners. Special corner units which fold out to irrigate more of the corner are also available. These significantly complicate the hardware which controls the pivot and have been troublesome in the past, but are continually being improved.

4.3.4 Constraints to Adoption

The biggest constraint to the application of sprinkler irrigation is the cost of equipment – that is, whether the capital expenditure can be repaid in a reasonable length of time. With current farm economic difficulties, lenders are looking at shorter payback periods for capital improvements, making such large equipment expenditures difficult to justify. Leasing of sprinkler equipment is a common alternative.

The second largest constraint is the energy required to pressurize the system. Availability of a reliable energy source is critical. The cost of energy also plays an important part in overall irrigation cost and must be considered when comparing alternative systems.

Maintenance of equipment is an important consideration in adopting sprinkler irrigation. Maintenance requires semi-skilled labor to repair minor malfunctions as well as sources for spare parts and hardware items, which may not be available in some parts of the world. In more arid areas, such maintenance is critical since equipment breakdowns for even short periods can result in significant crop stress.

Sprinkler irrigation systems are typically designed for low applications of water (typically less than 25 mm for continuously moving systems). In arid areas where transpiration demands are high, this requires either more frequent applications or greater application amounts.

Soil erosion from sprinkler irrigated land is becoming an increasing problem on continuous-move systems, particularly center pivots. Since application rates are higher at the outer edge of the pivots to cover a greater area in the same amount of time, the resulting runoff may cause significant erosion in highly undulating topography.

Sprinkler irrigation requires good control over the water supply. Where supplies are unreliable, reservoirs are needed to store sufficient water over the irrigation period. Where water is supplied from open channel delivery systems, supply and demand must be balanced. Water should be taken from a canal with flow in excess of the capacity of the sprinkler pumps to keep them from shutting off automatically. Control should also be in place so that if a pump shuts off or a continuously moving system breaks down, the water flow in the canal can be stopped or diverted to other uses.

Sprinkler irrigation requires power for pumping unless water is supplied from a pipeline pressurized by pumping or gravity flow. Ensuring a dependable power supply is important to the success of a sprinkler irrigation system, where pumping is required. Continuous-move systems also need power to drive the irrigation system. For center-pivots, electrical power is primarily used to power the towers. For linear-move systems, the electrical power for the individual towers is usually provided by a diesel engine powered generator.

4.3.4.1 Constraints to Efficient Use of Water

Sprinkler systems apply water non-uniformly because of the spray pattern of the sprinkler heads. This is less of a concern for both orchard crops and continuously moving systems. For solid-set systems it is a major concern, e.g., balancing cost of system with amount of pattern overlap. Any non-uniformity affects application efficiency, crop production, and/or water requirements. Wind drift has been one of the largest contributors to the inefficiency of sprinkler irrigation systems. Evaporation and drift losses increase with higher wind, more arid conditions, and smaller droplet sizes. While lower pressure for the same nozzle generally means larger droplets, the designs for low-pressure heads cause smaller droplets. Sprinkler heads can be lowered closer to the canopy to reduce drift and evaporation losses. Other methods are discussed under micro-irrigation (i.e., LEPA-Low Energy Precision Application).

Sprinkler irrigation in some cases can lead to poor water application efficiencies. It is simply so convenient to irrigate that not enough thought goes into the timing or amount of water applied. In some cases, center pivots are simply turned on and left running. This can lead to very poor use of water and the ineffective use of rainfall, especially early and late in the season when irrigation demand is low.

4.3.5 Evaluation Methods

There are two aspects of sprinkler irrigation systems that need to be evaluated: the pattern of water application caused by the overlapping sprinkler patterns, and the change in patterns and total sprinkler flow caused by pressure variations within the system. The effect of patterns of sprinkler overlap can be estimated with catch cans which intercept water that enters the crop canopy. Any type of small can may be used, but the same type should be used for a given evaluation. The number and layout of cans depend somewhat on the type of sprinkler irrigation system. The cans should be placed at the top of the crop canopy to eliminate interference with the crop itself. A small amount of light-weight oil should be placed in the catch cans to eliminate evaporation losses.

For solid-set irrigation systems, a two-dimensional array of cans is needed to clearly define the pattern. The array should fully cover the distance between two lateral lines and include several sprinkler heads. The cans should uniformly cover an even area of spray overlap so that they truly represent the overall sprinkler pattern. The same layout can be used for periodically moved systems. Evaluation should also be made of the variations in flow caused by pressure variations along lateral lines. It may be necessary to test the sprinkler pattern at several places along the lateral to determine any changes in flow rate and pattern uniformity (if pressure variation is sufficient to affect sprinkler pattern).

For continuously moving systems, it is necessary to have only a single line of catch cans perpendicular to the direction of travel. Because of the higher degree of overlap for continuous-move systems, overlap patterns are less of a concern than for fixed systems. For center-pivots there are two main concerns: pressure differences along the lateral line caused by either pipe friction or elevation changes, and matching nozzle flow changes with area changes over the length of the lateral. For linear-move systems, only the former is of concern. If pressure changes are too large, pressure regulators for individual nozzles should be installed. Installation of pressure regulators is more common on low-pressure systems, since line losses and elevation differences represent a higher percentage of the pressure. Alternatively, flow control nozzles can be used that have relatively uniform discharge over a fairly broad range of pressures. Catch cans can be evenly spaced over the length of the lateral line to determine the distribution along the lateral line. A grouping of catch cans at a smaller spacing to determine pattern effects is also recommended.

For center-pivots with large elevation changes as the pivot moves around the circle, a general idea of the effects of these changes of uniformity can be determined by simply measuring the flow rate over time. Significant changes in flow rate would indicate a corresponding effect on overall uniformity.

In general, the volume caught by catch cans per unit area will be less than the average water applied by the sprinkler system per unit area. The differences result from direct evaporation of the spray and wind drift. Whether these amounts of water are considered losses is the subject of much debate. Some of these losses undoubtedly offset transpiration during irrigation, and wind drift moisture may be beneficially used elsewhere.

In evaluating the potential for runoff from sprinkler systems, particularly center pivots, it is important to determine the time to ponding at which runoff can start, the amount of surface storage available, and the infiltration relation of the soil. Simple rain simulators can be used to test soil properties and surface storage amounts. This will help in determining whether the application rate of the sprinkler system will cause runoff. Furrow dikers (implements used to create small dikes in the furrows) can be used to increase the amount of surface storage.

Problems with pressure, flow and the distribution of water are usually caused by improper hardware or installation, improper water filtration or wear (insufficient maintenance). Burt et al. (1988) provide good guidelines on identifying such problems and make recommendations for correction.

4.4 Micro-irrigation

Micro-irrigation is a general category to handle a variety of types of low emission rate devices. These have been called various names including drip irrigation, trickle irrigation, subsurface irrigation, bubbler irrigation, LEPA, etc. The idea is not to wet the entire soil surface or volume, but only that portion which needs to be wetted.

4.4.1 Classification of Types

There are several basic types of micro-irrigation which relate to the way in which water is distributed over the field. As with sprinkler irrigation, there are fixed systems, which are applied to orchards and row crops, and moving systems, which are typically applied to row crops. Some of the earliest developments of micro-irrigation were for the watering of orchard crops, such as citrus. Individual emission devices were developed to deliver water to individual trees at low rates of flow, thus presenting surface ponding of water (Fig. 4.7). These emitters can be constructed in-line in the tubing or can be inserted into the soft plastic tubing with a barb connection. Line source micro-irrigation is another common system, where flexible plastic tubing is constructed with small perforations (e.g., 1 mm holes) at a fixed interval (e.g., 0.3 m). A dual chamber system is used, where one set of holes releases water from the main tube into the outer tube and another set of holes in the outer tube releases water to the soil. These lines are either laid on the ground surface or buried from 0.2 to 0.5 m below the soil surface, the latter referred to as subsurface irrigation.

Bubbler irrigation is a form of micro-irrigation which employs higher rates of flow. Small basins are constructed around each tree to hold the water that ponds there. Bubbler emission devices are somewhat larger than trickle irrigation emitters, but serve the same purpose – to regulate the rate of flow. Another bubbler scheme was developed to eliminate the emission device entirely. Here the tube is left open but is set at a fixed distance from the hydraulic grade line to regulate flow. This system requires significantly

Fig. 4.7. Typical micro-irrigation lateral with emitters on grapes in Arizona

lower pressures than other micro-irrigation methods, and usually requires small basins to prevent runoff.

Micro-irrigation also encompasses low flow rate sprinklers, misters, line-source mister tubing, and low rate under-tree sprinklers, or micro-spray. The distinction between micro-irrigation sprayers and sprinklers is the rate of flow (typically less than 10 l/h) and the area covered (typically a single tree or a part of the tree canopy). Line-source mister tubing is also available.

The Low Energy Precision Application (LEPA) system is another example of micro-irrigation. For this application, a continuous-move sprinkler system (e.g., center pivot or linear move) is adapted so that low rate emission devices are lowered to the soil surface. Emission devices can be small sprayers, emitters, open tubes, or trailing perforated tubes. Higher rate open tubes usually require pressure regulators and furrow diking to keep the water in the furrow locally.

4.4.2 Conditions of Applicability

Micro-irrigation primarily by individual drip emitters, bubblers, or micro-spray systems is particularly applicable to orchards and vineyards or other perennial crops for which the capital investment can be justified. Micro-irrigation is common in greenhouses and for ornamental crops where uniform crop quality is essential and is also well adapted to high value crops which are sensitive to moisture and salinity stress, such as sugar cane, strawberries, melons, tomatoes, and other vegetables. For these, line source emission is preferred, frequently underground. Subsurface micro-irrigation has been used on row crops such as cotton and corn, but with less success than LEPA systems. Fixed microsystems require a relatively high value crop to make an adequate return on an investment which is usually higher for micro-irrigation than for sprinkler irrigation (Fig. 4.8).

Micro-irrigation generally requires a good quality water supply. Most natural waters need treatment to remove physical particles, to keep dissolved salts in solution, and to keep microbial activity from plugging emitters or emission points. Emitter clogging is one of the main limitations of micro-irrigation and has caused the failure of many installations. In areas where water is saline, subsurface irrigation has been applied successfully to keep salts away from plant roots. This requires careful management.

Micro-irrigation is suitable for soils and topographies which are difficult to adapt to other methods. Micro-irrigation can be adapted to extremely sandy soils or to rocky hillsides. Micro-irrigation is frequently used in settings where other methods might interfere with the landscaping.

Micro-irrigation is also applied where water is scarce or expensive. Micro-irrigation is usually expected to be the most efficient method, although this is not always the case in practice. Where labor is expensive or scarce, micro-irrigation offers a good alternative. While micro-irrigation typically utilizes less labor than many other systems, it requires a more highly skilled labor force.

Fig. 4.8. Micro-irrigation system being installed for orchards in Arizona

4.4.3 Practices

Under micro-irrigation, water is applied more frequently than under surface and sprinkler irrigation. Irrigation frequencies range from daily to weekly, depending on the conditions. For some crops this improves the environment for plant growth and reduces the influence of high soil salinity on plant water uptake. Chemicals can be injected directly through the irrigation water, thus spoon feeding nutrients to the crop.

Micro-irrigation systems commonly employ automatic timers which allow the grower to easily and conveniently apply only the necessary amount of water. The automatic timer settings need to be changed at least monthly to track transpiration needs.

Standard cultural practices, such as spraying, weeding, thinning or harvesting, can often be conducted without interrupting the irrigation cycle. Micro-irrigation often minimizes weed growth since the soil surface is not fully wetted, and for some subirrigation systems, not wetted at all.

Micro-irrigation is applied to highly undulating and rough land. This reduces potential erosion from other irrigation methods and thus uses water more effectively.

Water treatment facilities are necessary for most micro-irrigation systems, except for some of the higher rate systems (e.g., bubblers, microsprays, etc). This may add considerable capital and maintenance costs to the system. Even with proper water treatment, emitter clogging is a common problem. Line-source systems are typically replaced annually or biannually. Lateral lines should be flushed at least annually to remove any sediments or debris in the line. (Nakayama and Bucks 1986, recommend every 6 months for tree crops and three times per season for row crops.) Filters also need periodic cleaning or backflushing when filter clogging causes an increase of 20 to 30 kPa in the pressure across the filter. Typical operating pressures for most micro-irrigation systems are 100 to 250 kPa, with the higher pressures associated with microspray. Low-head bubbler systems have pressures of 20 to 30 kPa.

Micro-irrigation system lateral lines usually do not exceed 200 m in length. Mainlines and submains are usually buried, while lateral lines are buried only if they interfere with equipment operation. For subirrigation, the lines are buried to keep the soil surface drier.

For orchard crops, it is a common practice to use several emitters per tree and to spread these emitters over the tree canopy. This reduces the variability of flow from tree to tree and helps to spread the roots over a larger soil volume. Spreading the flow over a larger soil area also reduces the potential of deep percolation on high infiltration rate soils.

Various manufacturers have developed pressure-compensating emitters to remove variation in discharge from emitters at one end of a lateral line to the other. These have been only partially successful, and should not be relied upon to completely remove pressure effects (Nakayama and Bucks 1986). Some pressure-compensating emitters have a greater tendency to clog.

With line-source emission, where a series of holes are spaced a distance apart, the assumption is made that capillary forces will move water into the soil between the holes, thus bridging the gap between holes. The extent to which this occurs depends on the soil type. The importance of this lateral movement depends on the extent of crop roots. For relatively large mature row crops such as cotton, this poses little difficulty. However, for germination of crops or for crops with small root systems (e.g., some vegetables), hole spacing can become important.

4.4.4 Constraints to Adoption

The main drawbacks of micro-irrigation are the high initial cost and the high operation and maintenance costs. For perennial crops and long service life micro-irrigation systems, maintenance is a constant struggle. Some growers have an employee walk the irrigation system daily to look for leaks and clogged emitters. While this might be excessive in most cases, Nakayama and Bucks (1986) recommend this be done at least monthly, and in some cases weekly.

The cost of installing subsurface irrigation tubing is sufficiently high that growers often try to obtain several crops from one installation, even though tubing performance and uniformity and efficiency of application may suffer.

Micro-irrigation requires a different style of management, which may be more intense than with surface or even sprinkler irrigation. While such intense management often provides payoffs in terms of better crop production, such efforts do not come without time and cost. Growers using other methods need time to fully adapt their attitudes and management style to micro-irrigation, as well as to learn the potential advantages.

Low head bubblers have significantly lower equipment costs than other micro-irrigation systems; however, the labor cost of installation is quite high. Where labor is expensive or unavailable, this method will likely not be adopted.

LEPA systems provide an alternative to the high operating costs of high-pressure sprinkler irrigation and offer some advantages in uniformity of application for row crops. The benefits of LEPA need to outweigh the added cost of hardware for these systems plus any added maintenance efforts. Most LEPA systems have been retrofitted to existing sprinkler systems. Commercial availability of LEPA systems should aid adoption.

4.4.4.1 Constraints to Efficient Use of Water

On finer textured soils, irrigations which are too frequent can cause root aeration or disease problems which can severely limit plant growth and yield. Under these conditions, less frequent irrigations may mean longer durations.

Frequent micro-irrigation of a limited soil volume can restrict root development. If too small a volume is wetted, plant growth can be restricted for some crops. Wetting a limited soil volume can also cause soil salinity to build up close to the plant roots. Rain events can then wash these salts into the root zone causing plant stress. The usual procedure in areas where soil salinity is a problem is to continue irrigating during rain events to keep the salts out. Current design procedures attempt to spread water over larger surface areas to ensure proper root development and to reduce potential salinity problems. Note that such procedures somewhat reduce the advantages of micro-irrigation by not using rainfall effectively and by allowing the wetted soil volume to expand (resulting in greater soil surface wetting).

As discussed earlier for sprinkler irrigation, when irrigation becomes automatic and simple there is a tendency to ignore it in favor of other pressing problems. Periodically altering the amount of water takes some effort, and the tendency is to maintain higher rates of application when the crop no longer requires them. Higher rates are often applied early to build up soil moisture, but not raised again until soil moisture has dropped. Without proper soil moisture monitoring, the tendency is to apply too much water during low water use periods and not enough water during high use periods (i.e., the same tendency as with other methods).

Some micro-irrigation systems are best used in combination with other methods. For example, deep subirrigation sometimes requires surface or sprinkler irrigation for germination since the subirrigation water does not wet the soil surface adequately without excessive applications of water. Surface line-source emission can cause salts to accumulate in the furrow ridges, which may need to be leached by other methods between crops.

Emitter clogging can cause potentially significant increases in water application amounts. The tendency when clogging occurs is to add extra water to the entire system to make up for lower flows by the clogged emitters. Taken to an extreme, micro-irrigation can have very poor uniformities and efficiencies when significant clogging occurs. This brought on the failure of many micro-irrigation systems early in the development of the technology. Even though the hardware has improved, the tendency to cut corners on water treatment and maintenance can lead to poor water use efficiencies.

4.4.5 Evaluation Methods

For systems with individual emitters, evaluation is relatively straight forward. A sample of emitters is chosen and the flow rate of each emitter is measured. Rate of flow can be determined by noting the time required to fill a small container of known volume or by measuring the volume collected over 1-min interval in a graduated container (Merriam and Keller 1978). Statistical considerations suggest than at least 30 measurements be taken (Merriam and Keller 1978, recommend 32). Emitters should be chosen randomly over the area of interest; however, Bralts (Nakayama and Bucks 1986) recommends choosing two adjacent emitters at each of four locations on four different lateral lines. With the assumption that rate of flow corresponds to volume applied, uniformity and efficiency values can be determined. It is also recommended that the pressure be measured at various points within the system to determine whether flow variations are caused by pressure differences or by emitter variation. While Merriam and Keller (1978) recommend eight measurements, it would be useful to measure pressure at each emitter for which discharge was measured.

Similar procedures can be used for microspray, bubblers, LEPA, etc. Although the physical mechanics of collecting these flows in a container is more difficult, this is the only practical way to conduct an evaluation. For line-source emission, Bralts suggests flow over a 1-m collection trough. The length of this trough should vary depending on the spacing of holes and the scale of plant roots, such that the length represents an integral number of holes and an effective emission point.

In very high infiltration rate soils, it is important to confirm that water distributed by the micro-irrigation system largely stays within the plant root zone. This can be done by comparing applied water with estimates of consumptive use. The applied water can, however, exceed evapotranspiration (ET) while the plants are under stress if significant leaching is taking place. Measurements of plant stress and soil moisture can aid in this determination, which is necessary to quantify application efficiency and determine actual crop water use (i.e., not just theoretical ET).

A good indication that emitter clogging is occurring is that flow within the system decreases while pressure increases. Continuous records of flow rate and pressure will help to forestall major clogging between detailed evaluations.

4.5 Considerations in Irrigation Method Selection

The factors affecting selection are so numerous and varied that it is extremely difficult to make concrete recommendations on which method is preferred for a particular setting. Rather, it is usually appropriate to compare a number of alternatives before making a final selection. Sometimes it is necessary to prepare at least a partial design of several systems before making a reasonable comparison between them. The final selection is usually up to the discretion of the owner or financier of the irrigation system, rather than the designer or consultant. Also, new techniques or continual refinements to old techniques potentially expand the range of conditions for these methods. Thus any guidelines given here are intended only for preliminary consideration and are not meant to restrict practices.

Prior to examining selection conditions, it is important to define the objectives and expectations of the irrigation system. Economic return is usually one of the primary considerations, but other factors may influence the acceptance of a system that does not produce maximum expected profit. All farming enterprises have an element of risk, and the perceived risks of the decision maker or system selector may alter his selection. Future energy prices are always uncertain, and selection may depend on the expectation of these prices. Environmental goals, both from a personal and societal point of view, may prompt the choice of one particular design over another. Environmental assessments are becoming the rule rather than the exception in new irrigation development. The security of water supplies is always an issue, particularly with increasing pressure from municipal users and fish and wildlife needs, which may influence investment decisions.

The selection process can be broken down into three main areas of consideration: physical site conditions, social or institutional considerations, and economic considerations. The general site conditions to be considered are listed in Table 4.1 and discussed more thoroughly by Blair et al. (1990).

Table 4.1. Physical site conditions to consider in irrigation system selection. (After Blair et al. 1990; Keller and Bliesner 1990)

Crops	Land
Crops grown	Field shape
Crop rotation	Obstructions
Crop height/root volume	Topography
Cultural practices	Soil – texture
Disease potential	– uniformity
Pests	– depth
Water requirements	– intake rate
Climate modification	– water capacity
	– erodibility
Water supply	– salinity
Source	– drainability
Quantity	– bearing strength
Quality – salinity	Flood hazard
– sediments	Water table
– organics	
Reliability	*Climate*
Delivery – schedule	Precipitation
– frequency	Temperature
– rate	Frost conditions
– duration	Humidity
	Wind
	Energy
	Availability
	Reliability

Table 4.2. Suggested guide for selection of irrigation methods. (0) indicates no influence on selection, (+) indicates possible reasons for preference, and (−) indicates possible reasons for choosing alternate method

	Surface					Micro				Sprinkler						
	Furrow	Modern furrow[a]	Border strip	Basin	Basin paddy	Bubbler low-head	Micro spray	Micro point	Micro line	LEPA	Linear move	Center pivot	Traveling gun	Side roll	Hand move	Solid set
Physical conditions																
Crop type																
corn	0	0	0	0	−	−	−	−	0	0	0	0	0	−	−	−
cotton–humid	0	0	0	0	−	−	−	−	0	0	0	0	−	−	−	0
cotton-arid	0	0	0	0	−	−	−	−	+	0	−	−	−	−	−	0
alfalfa	−	−	+	+	−	−	−	−	0	−	0	0	0	0	0	0
small grain	0	0	0	0	−	−	−	−	−	−	0	0	0	0	0	0
potatoes	−	−	−	−	+	−	−	−	0	0	+	+	−	0	0	+
rice	−	−	−	−	+	−	−	−	−	−	−	−	−	−	−	−
vegetables	0	0	0	0	−	−	−	−	+	0	+	0	−	0	0	0
other row	0	0	0	0	−	−	−	−	0	0	0	0	−	0	0	0
orchards/vineyards	0	0	0	0	−	+	+	+	0	−	−	−	0	0	0	+
rotation	0	0	−	+	−	−	−	−	0	+	0	0	−	0	0	0
climate modification	−	−	−	0	−	0	+	0	0	0	+	+	0	0	0	+
cultural/machinery operations (equip. in the way, field length, etc.)	0	0	0	−	−	0	0	0	0	−	−	0	−	−	−	0
Land																
odd-shaped fields	0	0	0	+	+	+	+	+	+	−	−	−	+	−	0	0
obstructions in field	0	0	0	0	0	0	0	0	0	−	−	−	+	−	0	0

Criterion	T1	T2	T3	T4	T5	T6	T7	T8	T9	T10	T11	T12	T13	T14	T15	T16
high water table	o	o	o	o	o	o	o	+	+	+	+	o	o	−	−	−
undulating slope with shallow soils	+	+	o	o	+	+	+	o	o	o	o	−	−	o	o	o
steep slopes	+	+	o	o	o	o	o	+	+	+	−	−	−	o	o	o
steep, rocky slopes soils	+	o	−	−	−	−	−	+	+	+	−	−	−	−	−	−
sandy, high infiltration soils	+	+	+	+	+	+	+	+	+	+	+	−	−	o	o	o
loamy moderate infiltration soils	o	o	o	o	o	o	o	o	o	o	o	−	o	o	o	o
clay, low infiltration soils	o	o	o	−	−	−	−	o	o	−	o	−	o	o	o	o
highly non-uniform soils	+	+	+	+	+	+	+	+	+	+	+	+	−	−	−	−
low water-holding capacity soils	+	+	+	+	+	+	+	+	+	+	+	o	o	o	o	o
saline soil	−	−	−	o	o	o	+	+	+	+	o	o	o	−	−	−
poorly drained soils	o	o	o	o	o	o	+	+	+	+	+	o	−	−	−	−
highly erodible soils	o	o	o	−	o	o	o	+	+	+	+	o	o	o	−	−
low-bearing capacity soils	o	o	o	−	−	−	−	o	o	o	o	o	o	o	o	o
Water supply																
groundwater well	+	+	+	+	+	+	+	+	+	+	+	o	o	o	o	o
delivered surface water – flexible	−	−	−	−	−	−	o	−	−	−	o	o	o	o	o	o
delivered surface water – rigid	−	−	−	−	−	−	−	−	−	−	o	o	−	o	o	o
delivered surface water- continuous flow rate	+	+	+	+	+	+	+	+	+	+	+	−	−	−	−	−
unreliable rate and timing	−	−	−	−	−	−	−	−	−	−	o	o	o	o	o	o
high sediment load	−	−	−	−	−	−	−	−	−	−	−	o	o	o	o	o
high organic matter	o	o	o	o	o	o	−	−	−	o	o	o	o	o	o	o
high salinity	−	−	−	−	−	−	o	−	−	o	o	o	o	o	o	o
wastewater	o	o	o	o	o	o	o	−	−	o	o	o	o	o	o	o
large rate of flow	−	−	−	−	−	−	−	−	−	−	−	o	+	o	o	o
small rate of flow	+	+	+	+	+	+	+	+	+	+	+	o	−	o	o	o
Climate																
high rainfall	+	+	+	+	+	+	o	o	o	o	o	o	−	o	o	o
low rainfall	−	−	−	−	−	−	o	o	o	o	o	o	o	o	o	o
high temp. – humid	+	+	+	+	+	+	o	o	o	o	o	o	o	o	o	o
high temp. – arid	−	−	−	−	−	−	o	o	o	o	o	o	o	o	o	o

Table 4.2. *Contd.*

	Surface					Micro					Sprinkler					
	Furrow	Modern furrow[a]	Border strip	Basin	Basin paddy	Bubbler low-head	Micro spray	Micro point	Micro line	LEPA	Linear move	Center pivot	Traveling gun	Side roll	Hand move	Solid set
windy	0	0	0	0	0	0	0	0	0	0	−	−	−	−	−	−
frost conditions	−	−	−	0	−	0	+	0	0	0	0	0	0	0	0	+
Energy																
scarce or unreliable	0	0	0	0	0	0	−	−	−	−	−	−	−	−	−	−
Social/institutional conditions																
low labor skills	0	−	0	0	0	0	0	−	−	−	−	−	−	−	0	0
low parts availability	0	−	0	0	0	−	−	−	−	−	−	−	−	−	−	−
little tech. assistance availability	0	−	0	0	0	0	0	0	0	−	−	−	0	−	0	0
low labor availability	−	−	−	+	+	+	0	0	0	+	+	+	0	−	−	+
automation potential	+	0	0	0	0	0	0	0	0	+	+	+	0	0	0	+
vandalism potential	0	−	+	+	+	0	+	+	+	+	+	+	0	0	0	+
low management skills	0	0	0	−	+	0	0	0	0	0	0	0	0	0	0	0
environmental concerns	−	−	−	−	−	0	0	0	0	−	−	−	−	−	−	0
land transformation	−	−	−	−	0	+	+	+	+	+	0	0	0	0	0	0
chemical use	0	0	0	0	0	+	+	+	+	+	0	0	0	0	0	0
sustainability	0	0	0	+	+	0	−	−	−	−	−	−	−	0	0	0

[a] Modern furrow irrigation refers to irrigation with a cutback stream, surge flow, cablegation, etc.

Table 4.3. Typical economic lives and maintenance costs for irrigation system components (adapted from Jensen 1980; Keller and Bliesner 1990)

System type and component	Economic life, yr[a]	Annual maintenance, % of cost
Sprinkler		
Lateral		
Hand move	15	2
End tow	10	3
Side roll	15	2
Side move	15	4
Hose-fed/pull	5/20	3
Traveling gun	10	6
Center pivot		
Standard	15	5
with/corner	15	6
Linear moving	15	6
Solid set		
Portable	15	2
Permanent	20	1
Trickle		
Point-source		
Drip	10/20	3
Spray	10/20	3
Bubbler	15	2
Hose-basin	7/20	2
Line-source		
Reusable	10/20	3
Disposable	1/20	3
Other components		
Buried PVC main line	20–40	1
Steel main line	10–20	1
Aluminum main line	10–20	2
Electric pumps	15	3
Diesel/gas pumps	10	6
Wells	25	1
Open farm ditches (perm.)	25	1.5
Concrete structures	30	0.8
Concrete pipe	40	0.5
Pipe, aluminum, gated, surface	10	2
Sprinkler heads	8	6
Trickle emitters	8	6
Trickle filters	12	7
Land grading	indefinite	2
Reservoirs	indefinite	1.5

[a] Where two lives are shown with a slash, the first number is for above-ground components and the second for below-ground components.

Table 4.4. Typical capital and operating costs (in US$/ha) for selected types of irrigation systems. Water costs were considered too variable and are not included here

	Surface irrigation					Micro-irrigation		
	Furrow	Furrow modern	Border strip	Basin	Basin paddy	Bubbler low-head	Micro spray	Micro-point
Capital costs								
Equipment/installation	50	150–250	0	0	0	2500–4000[a]	2000–3500	2000–3500
Water supply/convey[b]	200–800	200–800	350–800	500–1200	200–400	0–300	0–200	0–200
Land preparation[c]	100–200	100–200	100–250	250–1000	200–400	25–125	0–100	0–100
Total	350–1050	450–1200	450–1050	750–2200	400–800	2525–4425	2000–3800	2000–3800
Operating costs[d]								
Repair and maintenance	0	0–5	0	0	0	10–15	50–200	50–200
Land preparation[c]	30–50	30–50	40–50	40–50	10–30	20–30	0–5	0–5
Energy[e]	0	0	0	0	0	0–15	50–125	50–125
Labor[f]	40–60	20–40	40–60	10–20	5–10	15–25	20–40	20–40
Total	70–110	50–95	80–110	50–70	15–40	45–85	120–370	120–370

4 Irrigation Techniques and Evaluations

Table 4.4. *Contd.*

	Micro-irrigation		Sprinkler irrigation					
	Micro-line	LEPA	Linear move	Center-pivot	Traveling gun	Side roll	Hand move	Solid set
Capital costs								
Equipment/installation	1150–3000	1200–2200	1100–2050	800–1500	950–1200	750–1200	500–675	2500–3500
Water supply/convey[b]	0–200	0–200	0–200	0–200	0–200	0–200	0–200	0–200
Land preparation[c]	50–100	0–100	0–100	0–100	0–100	0–100	0–100	0–100
Total	1550–3300	1200–2500	1100–2350	800–1800	950–1500	750–1500	500–975	2500–3800
Operating costs[d]								
Repair and maintenance	400–600	70–150	65–125	35–70	55–95	15–35	10–25	50–80
Land preparation[c]	0–5	15–20	0–20	0–20	0–5	0–5	0–5	0–5
Energy[e]	50–125	25–62	50–125	50–125	175–250	50–100	50–100	50–100
Labor[f]	20–40	4–8	4–8	2–8	15–30	25–50	35–70	5–20
Total	470–770	114–240	119–278	87–223	245–380	90–190	95–200	105–205

[a] Predominantly labor for installation at US prices.
[b] Water supply and conveyance costs are extremely variable and depend on the source of supply. They include farm canals, pipelines and structures, booster pumps, reservoirs, etc. to bring the farm water supply to the field to be irrigated.
[c] Land preparation costs are typical costs in the US based on 1990 fuel prices.
[d] Based on 1 m of water use.
[e] Based on US$.05 kWh and water available at the high point of the field. Additional energy costs are incurred if pumping from groundwater. These additional costs increase as the application efficiency decreases (e.g., while shown here as zero energy cost, inefficient surface irrigation may result in more energy to pump ground water than efficient sprinkler and micro-irrigation systems).
[f] Based on US labor conditions: low end of range is based on h/ha to apply 1 m at $5/h.

Some detail has been provided in the foregoing discussion on the various methods. Table 4.2 has attempted to classify the suitability of the various methods under different physical and social conditions. This table, which is meant to give only general guidance and should not be taken too literally, is based on information from Thorne and Peterson (1949), Booher (1974), James (1988), Nakayama and Bucks (1986), Balir et al. (1990), Keller and Bliesner (1990). For a particular condition defined by a row in Table 4.2, a 0 for a given column means that the stated condition does not influence the selection of that particular irrigation method. A plus (+) means there are reasons that this method might be preferred to other methods. A minus (−) means that this condition is often not well suited to this method and other methods are probably preferred. This should not be interpreted to mean that this irrigation method is never suitable for the given condition, since the rationale used to suggest preference of other methods may not always apply.

Because the selection of an irrigation system must consider all of these factors simultaneously, it is not possible to examine all the interactions between factors that are relevant. One way to use Table 4.2 is as follows. Define the particular conditions of interest. Examine the scores (−, 0, +) for different methods. Those methods with the most minus signs should be moved to the bottom of the list of considerations, while those with a significant number of pluses should be moved higher on the list. This table is insufficient to provide a realistic selection since a number of social and economic issues are not considered.

An important issue in the selection of irrigation methods is the familiarity of the owner, management, and the labor force with the method chosen. Choosing appropriate technology is not strictly a matter of having equipment, parts, and labor available; it must encompass local knowledge. The legal issues related to water rights and laws must also be considered. What responsibilities does the manager have in measuring and recording water use, and how does this affect the type of system and layout? How certain is the legal right to the water supply, or how senior is the right under prior appropriation laws? These questions need to be addressed when examining the desired longevity of the system chosen. If an irrigation system is currently in place, the cost of adopting a new system will be influenced by existing facilities.

Economic considerations also play an important role in irrigation system selection. The economic life of various irrigation system components plus the annual maintenance costs as a percentage of initial cost are shown in Table 4.3. Capital cost and annual operating costs for various irrigation methods are given in Table 4.4. These data were taken from a variety of sources, predominately from the USA (Jensen 1980; Hansen et al. 1980; Nakayama and Bucks 1986; Stringham 1988; Blair et al. 1990; Keller and Bliesner 1990). In many cases, costs were shown in different forms for the different categories of methods. Some assumptions were needed to translate these data into common units. One of the most difficult factors to estimate is the water supply and conveyance component of the capital cost. This refers to farm reservoirs, canals, pipelines, pumps, etc., that are needed to bring the water from the farm supply point to the field to be irrigated and to pressurize it if necessary. For surface systems, this included farm ditches, gates, outlets, etc. For the pressurized systems, much of this component is already included in equipment costs. Labor is another cost which is difficult to compare for different systems. Some labor is seasonal and unskilled, while other groups represent full time employees. Some laborers irrigate only, while others do a variety of chores. Table 4.4 shows a wide range of costs for all methods, indicating that site-specific conditions can have an overriding influence on actual costs and appropriate selection of methods.

The availability of money for investment in irrigation improvements may limit the grower's ability to select a system which is economically optimum on paper. If lenders are not willing to invest in certain levels of technology, then other systems will likely be chosen.

There is always considerable debate on the efficiencies of the different methods. Poor management and poor design complicate the issue and widen the range of actual performance substantially for all methods. What is relevant here is the type of performance possible for these systems if they are reasonably well designed and managed. A range of values for a number of selected methods is shown in Table 4.5 (Adapted from Dedrick 1984; Blair et al. 1990; Keller and Bliesner 1990; USDA 1991 and others). In general, such efficiencies are achieved only when management has an incentive to be efficient – for example high water cost or limited supplies. Field-observed

Table 4.5. Typical potential application efficiencies for well-designed and managed irrigation systems

	AE
Furrow	50–70
Furrow: modern	60–80
Border-strip	55–80
Basin (w or w/o furrows)	65–90
Basin: paddy	40–60
Bubbler: low-head	80–90
Microspray	85–90
Micro: point-source	85–90
Micro: line-source	85–90
LEPA	80–90
Linear move	75–90
Center pivot	75–90
Traveling gun	60–75
Side roll	65–85
Hand move	65–85
Solid set	70–85

application efficiencies are usually within these ranges for these various systems (i.e., not based on theoretical considerations alone).

Finally, the selector must consider the length of the investment period and the long-term sustainability of the irrigation enterprise. For example, will the irrigation system selected tend to degrade the downstream water supply? Will drainage of this water be restricted in the future for environmental reasons? Will the system selected help to maintain soil tilth and crop productivity? These questions can be answered only through a thorough investigation of site-specific conditions, which is beyond the scope of this chapter.

References

ASAE (ed) (1987) Evaluation of furrow irrigation systems. ASAE EP419, Standards 1991. Am Soc Agric Eng, St Joseph, MI, pp 644–649

ASAE (ed) (1991) Soil and water engineering terminology. ASAE X526. Am Soc Agric Eng, St. Joseph, MI, Draft Copy

Blair AW, Smerdon ET (1988) Unimodal surface irrigation efficiency. J Irrig Drain Eng 114(1):156–168

Blair AW, Bliesner RD, Merriam JL (eds) (1990) Selection of irrigation methods for irrigated agriculture. ASCE on-farm committee report. Am Soc Civil Eng, New York, Draft Rep, 85 pp

Booher LJ (1974) Surface irrigation. FAO Agric Dev Pap 95. FAO, Rome, 160 pp

Burt CM, Walker RE, Styles SW (1988) Irrigation system evaluation manual. Dep Agric Eng Cal Poly, San Luis Obispo, CA

Clemmens AJ (1991) Irrigation uniformity relationships for irrigation system management. J Irrig Drain Eng 117(5):682–699

Clemmens AJ, Dedrick AR (1981) Estimating distribution uniformity in level basins. Trans ASAE 24(5):177–1180 and 1187

Dedrick AR (1984) Cotton yields and water use on improved furrow irrigation systems. Water today an tomorrow. In: Spec Conf Proc Am Soc Civil Eng, Flagstaff AZ, July 24–26, 1984, pp 175–182

Erie LJ, French OF, Bucks DA, Harris K (1982) Consumptive use of water by major crops in the Southwestern United States. USDA, Agric Res Serv, Conserv Res Rep 29, Washington, DC 42 pp

FAO (ed) (1981) Agriculture: Toward 2000. FAO, Rome

Framji KK (1984) Second ND. Gulhati memorial lecture for international cooperation in irrigation. ICID Bull 33(2):1–58

Hansen VE, Israelsen OW, Stringham GE (1980) Irrigation principles and practices, 4th edn. John Wiley & Sons, New York, 417 pp

Irrigation Association (ed) (1991) 1990 Summary – US irrigated acreage. Irrig J Jan/Feb 40(1):24–34

James LG (1988) Principles of farm irrigation system design. John Wiley & Sons, New York, 543 pp

Jensen ME (ed) (1980) Design and operation of farm irrigation systems. Am Soc Agric Eng, St Joseph, MI, 829 pp

Jensen ME, Burman RD, Allen RG (eds) (1990) Evapotranspiration and irrigation water requirements. ASCE manuals and reports on engineering practice 70. Am Soc Civil Eng, New York

Keller J, Bliesner RD (1990) Sprinkle and trickle irrigation. Van Nostrand Reinholt, New York, 652 pp

Merriam JL, Keller J (1978) Farm irrigation system evaluation: a guide for management. Agric Irrig Eng Dep, Utah State Univ, Logan, 271 pp

Nakayama FS, Bucks DA (1986) Trickle irrigation for crop production: design, operation and management. Elsevier, Amsterdam, 383 pp

Stringham GE (ed) (1988) Surge flow irrigation: final report of the Western Regional Research Project W-163. Res Bull 515, Utah Agric Exp Stn, Utah State Univ, Logan, 92 pp

Thorne DW, Peterson HB (1949) Irrigated soils: their fertility and management. Blakiston, Philadelphia, 288 pp

USDA (ed) (1990) Agricultural statistics 1990. US Gov Print Off, Washington, DC, 517 pp

USDA (ed) (1991) Farm irrigation rating index (FIRI): a method for planning, evaluating and improving irrigation management. USDA, Soil Conserv Serv, West Natl Tech Center, Portland OR, 56 pp

Walker WR, Skogerboe GV (1987) Surface irrigation: theory and practice. Prentice-Hall, Englewood Cliffs, 386 pp

Other Suggested Reading and Background

Cuenca RH (1989) Irrigation system design: an engineering approach. Prentice Hall, Englewood Cliffs, 552 pp

Hagan RM, Haise HR, Edminster TW (eds) (1967) Irrigation of Agricultural Lands. Agron Ser 11. Am Soc Agron, Madison, 1180 pp

Hart WE (1975) Irrigation systems design. Dep Agric Eng, Colorado State Univ, Fort Collins, 150 pp

Hoffman GJ, Howell TA, Solomon KH (eds) (1990) Management of farm irrigation systems. Am Soc Agric Eng, St Joseph, MI, 1040 pp

Holzapfel EA, Marino MA, Chavez-Morales J (1985) Procedure to select an optimal irrigation method. J Irrig Drain Eng 111(4): 319–329

Kay M (1986) Surface irrigation: systems and practice. Cranfield, Bedford, UK, 142 pp

Tecle A, Yitayew M (1990) Preference ranking of alternative irrigation technologies via a multicriterion decision-making procedure. Trans ASAE 33(5): 1509–1517

Walker WR (1989) Guidelines for designing and evaluating surface irrigation systems. FAO irrigation and drainage paper 45. FAO, Rome, 137 pp

5 Drainage and Subsurface Water Management

R.W. SKAGGS and C. MURUGABOOPATHI

5.1 Introduction

Drainage of water from the soil profile is an important hydrologic process in most agricultural soils. Natural drainage processes include groundwater flow to streams or other surface outlets, vertical seepage to underlying aquifers, and lateral flow (interflow) which may reappear at the surface at some other point in the landscape. In many soils the natural drainage processes are sufficient for the growth and production of agricultural crops. In other soils artificial drainage is needed for efficient agricultural production.

Soils may have poor natural drainage because they have low surface elevations, they are far removed from a drainage outlet, they receive seepage from upslope areas, or they are in depressional areas. Water drains slowly from soils with tight subsurface layers, regardless of where they are on the landscape; so soils may have poor natural drainage due to restricted permeability or hydraulic conductivity of the profile. Climate is another important factor affecting the need for artificial drainage. A natural drainage rate that is sufficient for agriculture at a location in Iowa where annual rainfall is 800 mm may be inadequate in southern Louisiana where the annual rainfall is 1500 mm. Nowhere is this factor more evident than in irrigated arid and semi-arid areas. Lands that have been farmed for centuries under dryland cultures often develop high water tables and become water-logged after irrigation is established. By contrast, natural drainage may be adequate in other arid region soils and the several-fold increase in the amount of water applied to the surface due to irrigation will not result in poorly drained conditions. Seepage from unlined irrigation canals or from man-made reservoirs may also result in poorly drained soils in areas where they did not previously exist.

In summary, most soils require drainage for efficient agricultural production; many need improved or artificial drainage. In most cases improved drainage practices can be used to satisfy agricultural requirements. Drainage requirements and methods used to design systems to satisfy those requirements are discussed in the following sections. Drainage, either natural or artificial, may be excessive, resulting in loss of water that could be used by

Adv. Series in Agricultural Sciences, Vol. 22
K.K. Tanji/B. Yaron (Eds.)
© Springer-Verlag Berlin Heidelberg 1994

the crop and increased movement of pollutants to receiving streams. Thus, drainage should be considered as one component of an agricultural water management system. Controlled drainage and sub-irrigation are being used in some areas to provide both irrigation and drainage needs for crop production. Methods for analyzing and designing drainage and subirrigation systems are discussed in this chapter. Examples are presented to demonstrate the effects of drainage and subirrigation on yields and profits, and to illustrate the use of simulation models to optimize the design of agricultural water management systems. This chapter concludes with a discussion of the off-site impacts of agricultural drainage and the need to consider those impacts in design and management of drainage and related water table control practices.

5.2 Drainage Requirements

There are basically three reasons for the installation of agricultural drainage systems: (1) for trafficability so that seedbed preparation, planting, harvesting and other field operations can be conducted in a timely manner; (2) for protection of the crop from excessive soil water conditions; and (3) for salinity control.

5.2.1 Trafficability

The effect of good drainage on timeliness of farming operations is discussed in detail by Reeve and Fausey (1974). Soils with inadequate drainage may experience frequent yield losses because essential farming operations cannot be conducted in a timely fashion. The result may range from complete crop failure if planting is delayed too long to reduced yields if tillage, spraying, harvesting or other operations are not performed on time.

5.2.2 Protection from Excessive Soil Water Conditions

It is well recognized that one of the major effects of excessive soil water on crop production is the reduction in exchange of air between the atmosphere and the soil root zone. Wet soil conditions may result in a deficiency of O_2 required for root respiration, an increase in CO_2, and the formation of toxic compounds in the soil and plants. Under field conditions, both soil water and plant conditions vary continuously. Evaluating the effects of water content and aeration status on plant growth requires integration of these conditions over time during the entire growing season. One of the parameters that gives a certain integration of these factors in soils requiring artificial drainage is the water table depth (Wesseling 1974). Although the depth of the water table has no direct influence on crop growth, it is an indicator of the prevailing soil water status, water supply, aeration and thermal conditions of the soil.

Numerous laboratory and field experiments have been conducted to determine the effect of water table depth on crop yields. Probably the main reason that yield has been related to water table depth is that water table depth is easier to measure than other variables such as the distribution of oxygen in the profile. Most experiments have been directed at relating yields to constant water table depths (Williamson and Kriz 1970). However, such steady-state conditions rarely occur in nature. The effects of fluctuating water tables and intermittent flooding on crop yields depend on the frequency and duration of high water tables as well as the crop susceptibility. Simulation models that predict the response of the water table and soil water status to both weather conditions and drainage design include methods to predict crop yields (see reviews by Skaggs 1987; Feddes et al. 1988). Models are discussed in more detail and a sample application is given in subsequent sections of this chapter.

5.2.3 Salinity Control

It seems somehow unjust that the dry lands of arid and semi-arid regions, when irrigated, often require artificial drainage. Luthin (1957) documents drainage problems that have beset irrigators since the earliest recorded times in history. Accumulation of salts in the surface soil layers caused the once fertile Tigris and Euphrates River valleys of ancient Mesopotamia to return to desert. A present-day example is the annual loss of thousands of acres in the San Joaquin Valley of California due to salt accumulation and water logging.

Practically all irrigation water contains some salt. Evaporation and consumptive use of water by plants concentrates the salts in the residual soil water resulting in a solution that is usually more saline than the irrigation water. Repeated irrigations continually increase the salinity of the soil water, even if the irrigation water is of relatively good quality. In order to prevent the buildup of soil water salinity to the point where it harms plant roots and reduces productivity, irrigation water in excess of the amount needed for evapotranspiration is applied to leach the concentrated soil solution from the root zone. If drainage is adequate, the excess irrigation water will carry the concentrated salt solution out of the root zone. If it is inadequate, the water table will rise in response to the excess irrigation water, the salinity will continue to increase in the root zone, and crop yields will be reduced because of both high salinity and high water tables.

Artificial drainage systems are needed on many soils to remove excess irrigation water from the soil profile. Because the salinity below the water table is usually several times that of the irrigation water, drainage systems are normally designed to hold the water table well below root depth.

It is important to recognize that drainage needs in irrigated lands are very dependent on the irrigation component. Drainage requirements may be dramatically reduced by improving the efficiency and management of irrigation. Thus drainage for irrigated lands must be treated as a component of the water management system and its design should depend on the design and management of the other components.

5.2.4 Limiting Factors and Constraints

The trafficability objective is more important in humid regions where the frequency and amount of rainfall is unknown, compared to irrigated arid lands where field operations and irrigation events can be coordinated. Conversely, control of soil salinity is not usually a problem in humid regions but may constitute the primary drainage need in arid irrigated lands. Normally, drainage systems would be designed to satisfy the above objectives subject to cost constraints. Since there is interaction between drainage and irrigation systems, additional constraints include minimizing over-drainage and the amount of irrigation water required.

Another factor of critical importance in some cases is the impact of agricultural water management on quality and quantity of water leaving the field and the effects downstream. An example where such impacts have caused major problems is the contamination of the Kesterson National Wildlife Refuge via selenium in irrigation-induced drainage water in California's San Joaquin Valley (Committee on Irrigation-Induced Water Quality Problems 1989). In humid regions, drainage waters have been found to cause high nutrient loading to surface waters and are partly responsible for degrading water quality in some streams and lakes (Baker and Johnson 1977; Logan et al. 1980; Gilliam 1987). The environmental impacts of drainage and irrigation system design are often neglected, especially in developing countries where large irrigation projects are being constructed or planned. The impact on drainage water quality, specifically salinity, is well recognized, however, as it affects reuse of the water for irrigation downstream. Whether attaining a given drainage water quality is viewed as an objective of the drainage system design or as a constraint on the total water management system depends on the situation. In any case, it is an important factor that is affected by both irrigation and drainage system design and management. Consideration of this factor will be essential in future decisions regarding design and operation of drainage and irrigation systems.

5.3 Subirrigation

The same drainage system that removes excess water during wet periods can also be used in some soils for irrigation during periods of deficient soil water. Subirrigation involves raising the water table and maintaining it at a position that will supply water to the growing crop. Subirrigation has been practiced in scattered locations for many years (Clinton 1948; Renfro 1955) and has advantages over other alternatives for certain conditions. However, until recently this method has not been widely used because of the lack of established design criteria and information characterizing the operation of systems in the field. Research in recent years has provided answers to many of these questions, and the use of subirrigation is increasing rapidly in some areas.

In order for subirrigation systems to be practical, certain natural conditions must exist: an impermeable layer or a permanent water table at a rather shallow depth (within 6 m of the surface) to prevent excessive seepage losses; relatively flat land; a moderate to high soil hydraulic conductivity so that a reasonable spacing of ditches or drain tubes will provide subirrigation and drainage; and a readily available source of water. These topographical and soil conditions exist in several million acres of land in the humid regions of the USA. Where suitable conditions exist, combined subirrigation – drainage systems offer a number of advantages and can play a significant role in water management strategies. Probably the most outstanding advantage is the cost; both drainage and subirrigation can be provided in one system, often with a considerable cost reduction in comparison to separate systems. In addition, energy requirements for pumping irrigation water may be considerably lower than for conventional irrigation systems (Massey et al. 1983). While salt buildup at the soil surface poses no problem in humid regions, subirrigation may not be feasible in arid and semi-arid areas because of this problem.

The most critical aspect of design and management of subirrigation drainage systems is the interaction between the irrigation and drainage functions. It is nearly impossible to determine, a priori, what the most critical sequence of weather events might be for a given management strategy. The most effective way of analyzing the performance of such systems is to use simulation models, as will be discussed in subsequent sections.

5.4 Design and Analysis of Drainage Systems

5.4.1 Drainage Methods

Drainage materials and installation methods have improved tremendously in the last 25 years. The development of corrugated plastic tubing, the drain tube plow, laser grade control for both surface and subsurface drainage machines, and synthetic envelope materials represents some of the advances that have occurred. Both subsurface and surface drainage systems can be installed quickly and efficiently using modern technology. Numerous devices for controlling water level elevations in the outlets and for introducing subirrigation water have been developed. The challenge, from the engineering and scientific prospective, is to tailor the design and operation of drainage and associated water management systems to soil, crop, and climatological parameters and specific site conditions.

Both surface and subsurface drainage systems are used to meet drainage requirements of poorly drained sites. A schematic of a drainage system is shown in Fig. 5.1. Subsurface drainage is provided by drain tubes or parallel ditches spaced a distance L apart. Most poorly drained soils have a restrictive layer at some depth, shown here as a distance, d, below the drain tubes. When rainfall occurs, water infiltrates at the surface, raising the water content of the soil profile. Depending on the initial soil water content and the amount of infiltration, some of the water may percolate through the profile, raising the water table and increasing the subsurface drainage rate. If the rainfall rate is greater than the infiltration rate, water begins to collect on the surface. If good surface drainage is provided so that the surface is smooth and on grade, most of the surface water will be available for runoff. However, if surface drainage is poor, a substantial amount of water (average depth, s, in Fig. 5.1) must be stored in depressions before runoff can begin. After rainfall ceases, infiltration continues until the water stored in surface depressions is infiltrated into the soil. Thus, poor surface drainage effectively lengthens the infiltration event for some storms, permitting more water to infiltrate and a larger rise in the water table than would occur if depressional storage did not exist. Once excess water enters the soil profile it may be removed by evapotranspiration, by natural drainage processes via deep and lateral seepage, and through man-made systems consisting of drainage tubes, ditches or wells.

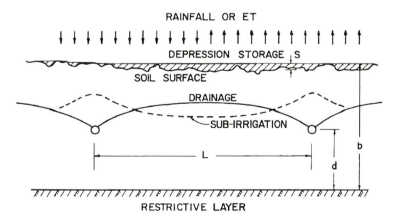

Fig. 5.1. Schematic of a drainage and subirrigation system showing the important hydrologic components for shallow water table soils

For conventional drainage the water level in the outlet is maintained at or below the drain tubes. Subirrigation is activated by using a control structure to raise the water level in the outlet to force water back through the drains and raise the water table as shown in Fig. 5.1. Note that the water table profile during subirrigation is its maximum elevation over the drain and at its lowest elevation midway between drains, which is the reverse of the profile during drainage. Properly designed and managed, a subirrigation system should maintain the water table at a depth that will supply evapotranspiration requirements of the crop.

5.4.2 Drainage Theory and Simulation Models

Numerous drainage theories, based on both steady-state and transient analysis, have been developed (van Schilfgaarde 1974) to relate drain spacing and depth to soil properties and site conditions. Generally, these theories allow determination of drain spacings and depths that will provide a specified steady drainage rate, or a specified rate of water table drawdown. While there are many steady-state and transient methods available to treat a wide range of boundary and site conditions, practically all are based on criteria that are only indirectly related to the actual objectives of a drainage system. Simulation models developed in the last 20 years provide a more direct link between design parameters and objectives of drainage and related water management systems. Simulation of the performance of a drainage system over a several years of climatological record can be used to determine if a given design will satisfy trafficability and crop protection requirements. By considering a range of alternatives the design of the system can be optimized for a given crop, soil, and location.

Several computer models have been proposed to simulate the performance of drainage and irrigation systems (see reviews by Skaggs 1987; Feddes et al. 1988). These range from numerical solutions to complex differential equations to approximate methods for conducting a water balance in the soil profile. The most theoretically rigorous method for modeling soil water movement and storage is the so-called exact approach. This method involves the solution of the Richards equation for combined saturated-unsaturated flow in two or three dimensions. It is based on the Darcy-Buckingham equation for soil water flux and the principle of conservation of mass. By solving the Richards equation subject to the appropriate boundary and initial conditions, the system can be described for most conditions of interest. The solution provides soil water contents and pressure heads as a function of time and space, the water table position, drainage or subirrigation rates, infiltration and evapotranspiration rates and excess water available for surface runoff. Flow paths can be determined and the time that water resides in various sections of the soil profile can be predicted. This capability is important in modeling effects of water management practice on water quality.

The biggest problem with this approach is that the equations are non-linear and numerical methods must be used to obtain solutions. Such solutions are both difficult and expensive. The requirement of unsaturated soil water properties for each profile layer may also limit the applicability of this approach. These functions are difficult to measure and are compiled and available for only a few soils.

Until now, application of the exact approach has been mostly confined to events of short duration. Exceptions are the works of Zaradny (1986), Harmsen et al. (1991), and Munster et al. (1991), who simulated conditions for growing seasons of 3- to 6-months durations. As faster computers reduce computational limitations, long-term simulation models based on the two-dimensional Richards equation will be used more frequently. However, the application of this approach is still difficult and will likely be confined to research scientists.

An approach less complex than the one discussed above, but which still predicts water contents in the unsaturated zone with good resolution, is based on numerical solutions to the one-dimensional Richards equation for vertical flow. Lateral water movement due to drainage is evaluated using approximate equations that are imposed as boundary conditions on the solutions to the Richards equation. The SWATRE model developed by Feddes and colleagues (Feddes et al. 1978; Belmans et al. 1983) is an example of this approach. The model can consider up to five soil layers having different physical properties. Climatological data including daily rainfall and potential evaporation and transpiration are used to specify the boundary condition at the top of the profile. Flux in the saturated zone at the bottom of the profile is calculated by a steady-state equation developed by Ernst. Six other bottom boundary conditions can be considered. SWATRE has been linked with other models to predict trafficability, crop emergence, growth and production (van Wijk and Feddes 1986). Karvonen (1988) introduced very efficient numerical techniques to this approach and added the capability of simulating the effects of soil temperature differences and frozen conditions. Bronswijk (1988) modified this approach to model the effects of soil swelling and shrinkage on the water balance. Workman and Skaggs (1990) used a modification of this approach to include the effects of preferential flow in cylindrical macropores.

The advantage of this approach is that is based on sound theory for vertical water movement in the unsaturated zone. Since most of the unsaturated water movement tends to be in the vertical direction, even in drained soils, this approach should provide reliable predictions of the soil water conditions above the water table. Another advantage is that it can also be applied for soils without water tables. A disadvantage of this approach is the requirement of the unsaturated soil water properties as discussed above for the two-dimensional model.

A third approach to modeling drainage systems is based on numerical solutions to the Boussinesq equation. This approach is normally applied for watershed scale systems or where the horizontal water table variation is critical. Examples are the models developed by de Laat et al. (1981) and Parsons (1987). Approximate methods are used to conduct a water balance in the unsaturated zone above the water table. These models may be used in large nonuniform areas where lateral differences in soils, surface evaluations and water table depths are important (Parsons et al. 1990).

Analyses in this chapter were conducted with the water management simulation model DRAINMOD (Skaggs 1978). This model is based on a water balance in the soil profile. Simulation models based on this approach were first derived as an extension of methods for predicting water table response to precipitation and drainage (Krayenhoff van de Leur 1958; van Schilfgaarde 1965). DRAINMOD was developed for design and evaluation of multicomponent water management systems on shallow water table soils. It includes methods to simulate subsurface drainage, surface drainage, subirrigation, controlled drainage and surface irrigation. Input data include soil properties, crop parameters, drainage system parameters, and climatological and irrigation data. The model may be used to simulate the performance of a water management system over a long period of climatological record. Approximate methods are used to simulate infiltration drainage, surface runoff, evapotranspiration (ET), and seepage processes on an hour-by-hour, day-by-day basis. Water table position and factors such as the ET deficit are calculated to quantify stresses due to excessive and deficient soil water conditions. Stress-day-index methods (Hiler 1969) are used to predict relative yields as affected by excessive soil water conditions, deficit or drought conditions, and planting date delay (Hardjoamidjojo and Skaggs 1982; Evans et al. 1990a, 1992). The validity of the model has been tested with good results for a wide range of soil, crop and climatological conditions (Skaggs et al. 1981; Skaggs 1982; Chang et al. 1983; Gayle et al. 1985; Rogers 1985; Fouss et al. 1987; Susanto et al. 1987; McMahon et al. 1988).

The major advantage of the water balance approach is that the approximate methods used for the hydrologic components allow simplification of the inputs so that the model is easier to apply than the other approaches. Computational requirements are also substantially less than with other approaches.

5.5 Simulation Analyses

Results for an eastern North Carolina soil, Rains sandy loam (fine-loamy, siliceous, thermic, typic paleaquult), will be presented as an example of the use of simulation models for the analysis of drainage and sub-irrigation systems. This soil has a nearly level surface with a hydraulic conductivity of 1.0 m/d to a depth of 1.1 m and 0.2 m/d for depths from 1.1 to the restrictive layer at 1.4 m below the surface. It requires artificial drainage for traffic-ability and protection from excessive soil water during wet periods. Details of the soil properties and other input data are given in Table 5.1. Simulations were conducted for continuous corn for a 37-year period (1950–1986) of climatological record from Wilson, NC.

Table 5.1. Summary of soil input data and drainage system and crop parameters for Rains sandy loam

A.   Soil properties
1. Hydraulic conductivity, K: $K = 1.0$ m/d for depth < 1.1 m
                                      $K = 0.24$ m/d for depth > 1.1 m
2. Depth to impermeable layer   1.4 m
3. Water content at saturation: 0.37 cm^3/cm^3 (root zone)
4. Water content at wilting point: 0.09 cm^3/cm^3
5. Relationship between water table depth, drainage volume and upward flux:

Water table Depth (cm)	Drainage vol. cm^3/cm^2	Upward flux cm/h
0	0	>1.0
20	0.8	0.08
40	2.3	0.011
60	3.6	0.003
80	5.2	0.001
100	6.9	0.0004
150	12.5	0.0
200	20.0	0.0
500	53.6	0.0

B.   Trafficability inputs
1. Drained volume required for fieldwork: 3.9 cm (cm^3/cm^2)
2. Minimum daily rainfall to postpone fieldwork: 1.2 cm
3. Time after rain before fieldwork can resume: 2 days

C.   Crop parameters for crop in North Carolina
1. Desired planting date: Not later than April 15
2. Working days required for seedbed preparation: 5 days
3. Length of growing season: 130 days
4. Maximum effective rooting depth: 30 cm

D.   Drainage system parameters
1. Drain depth: 1.0 m
2. Drain diameter: 100 nm (4 in.)
3. Effective drain radius: 5.1 mm
4. Surface depressional storage: 5 mm, 25 mm
5. Drainage coefficient: 25 mm/day
6. Drain spacing: 5 to 100 m
7. Weir setting for subirrigation: 0.45 m below ground surface beginning 35 days after planting through growing season

5.5.1 Drainage

The average predicted relative corn yield for the 37-year simulation period is plotted as a function of drain spacing in Fig. 5.2. The overall relative yield for each year is calculated as the product of three components: relative yield as affected by excessive soil water stresses; relative yield as affected by deficit soil water stresses; and relative yield as affected by delay in planting. Each component is determined in the model with stress-day-index methods; 37-year average values are plotted in Fig. 5.2. The maximum overall relative yield of 75% was predicted for an 18-m drain spacing. Higher average yields were not obtained because of deficit soil water stresses. Excessive soil water conditions decreased yields for drain spacings greater than 15 m; planting date delays caused by lack of trafficable conditions further decreased yields for drain spacings greater than 25 m. Yields increased with increased drainage intensity (decreasing drain spacing) until the maximum was obtained. Further decreases in drain spacing cause average yield values to drop slightly because the higher drainage intensities remove some water that could be used by the crops, thereby increasing deficient soil water stresses. Thus a drainage intensity greater than that needed to satisfy trafficability and crop protection requirements would not only cost more, but would slightly decrease average yields.

Results given in Fig. 5.2 were predicted for a surface depressional storage of 5 mm, which is characteristic of relative intense or "good" surface drainage. The effect of surface drainage on the relative yield-drain spacing relationship is shown in Fig. 5.3. These results show clearly that yield

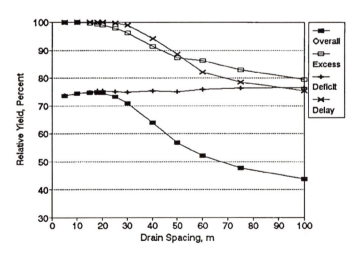

Fig. 5.2. Effect of drain spacing on predicted relative corn yield for the Rains soil with conventional drainage and good surface drainage (excess, deficit and delay represent relative yields due to excess soil water stress, deficit soil water stress and planting date delay)

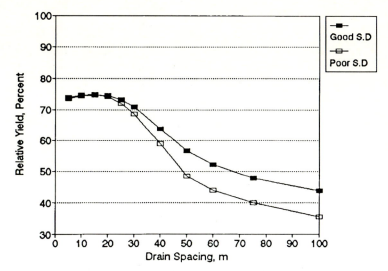

Fig. 5.3. Effect of drain spacing on predicted relative corn yield for the Rains soil with conventional drainage for good and poor surface drainage treatments (*S.D.* surface drainage)

response to surface drainage is dependent on subsurface drainage intensity. For intensive subsurface drainage (e.g., drain spacings less than about 25 m), surface drainage has little effect on predicted yields. However, surface drainage is very important when subsurface drainage is poor. For example, the predicted relative yields on the Rains soil for a 100-m drain spacing is 44% for good surface drainage versus 36% for poor surface drainage (depressional storage of 25 mm).

Annual relative yields predicted for years 1950 through 1986 are plotted in Fig. 5.4 for the Rains sandy loam with drain spacings of 25 and 100 m and good surface drainage (depressional storage of 5 mm). Average predicted relative yields were 0.74 and 0.47 for the 25- and 100-m spacings, respectively. However, differences in the predicted yields between the two spacings varied widely from year to year. In wet years, the closer drain spacing gave much higher yields. For example, in 1961, the relative yield was 90% for L = 25 m as opposed to 23% for L = 100 m. For this year the closer spacing allowed planting on time (by April 15) and the only decrease in yield was due to a short period when the water table was high early in the growing season. However, planting for the 100-m spacing was delayed until the last of May. High water table conditions after planting and dry conditions later on during the delayed growing season caused additional yield reductions. The cumulative effect resulted in a relative yield of 23% for the 100-m spacing.

During years when yields were limited by deficient soil water conditions, very little yield difference between the 25- and 100-m drain spacings occurred. Examples are 1952 and 1954. In a few years, e.g., 1964 and 1986,

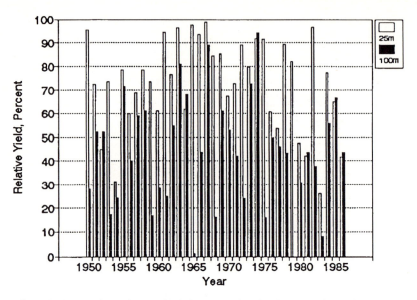

Fig. 5.4. Comparison of annual relative corn yield for drain spacings of 25 and 100 m on the Rains soil

predicted yields for the 100-m spacing were higher than for the 25-m spacing. This was caused by a sequence of weather events that allowed planting to be completed in time for the closer spacing but delayed planting for 20 to 30 days for the wider spacing. Subsequently, deficit soil water conditions occurred at a time when the early planted corn was most susceptible to drought. The later planted corn had a higher predicted yield because rainfall occurred before its period of maximum susceptibility.

Results given in Figs. 5.3 and 5.4 show that average yields are significantly increased by improved drainage. However, the benefits of drainage are widely variable from year to year. Improved drainage increased not only average yields but also the reliability of production for this case.

5.5.2 Subirrigation

The effect of subirrigation on predicted relative yields is shown in Fig. 5.5 for good surface drainage. Subirrigation was initiated 35 days after planting by raising the water level in the drainage outlet to within 0.45 m of the surface and holding it at that level during the growing season. Results in Fig. 5.5 show that the long-term average predicted yields were higher than predicted for conventional drainage for drain spacings less than 40 m. Subirrigation reduced stresses caused by deficit soil water conditions and increased yields. Closer drain spacings provided better water table control, and predicted

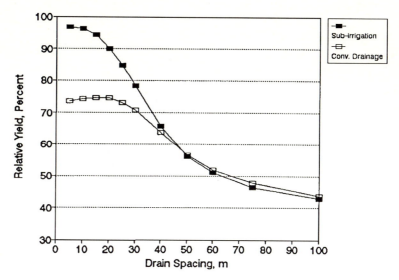

Fig. 5.5. Effect of drain spacing on predicted relative corn yield for the Rains soil with subirrigation and conventional drainage for good surface drainage treatment

yields increased with decrease in spacing to a maximum of 97% for drains placed 5 m apart.

5.6 Economic Analyses

Yield predictions such as those plotted in Fig. 5.5 can be used to design drainage or subirrigation systems to maximize yields. However, a more likely objective is to maximize profits. An example of an economic analysis to select drain spacings to maximize profits for both conventional drainage and subirrigation follows.

The economic analysis was conducted for the production of continuous corn. The potential corn yield for the Rains soil was taken as 11 000 kg/ha (175 bu/ac) and a price of $0.106/kg ($2.70/bu) was assumed. The average gross annual income was determined as the product of the relative yield, the potential yield and the price. For example, the average relative yield for conventional drainage with good surface drainage and a 100-m drainage spacing is 44% (Fig. 5.3), so the average gross income is:

$$0.44 \times 11\,000 \text{ kg/ha} \times \$0.106/\text{kg} = \$513/\text{ha} .$$

Annual costs were calculated as the sum of three components: (1) production costs (e.g., seed, fertilizer, pesticide, labor, machinery, etc.); (2) drainage system costs; and (3) equipment and operational costs for subirrigation. A

Table 5.2. Summary of production and drainage system costs for corn on Rains sandy loam soil

I.	Production costs	
	A. Variable costs: seed, fertilizer, pesticides, fuel, repair, labor, etc.	$491/ha ($199/ac)
	B. Fixed costs: machinery replacement, machinery interest, tax and insurance	$65/ha ($26/ac)
	Total production costs	$556/ha ($225/ac)
II.	Drainage system cost	
	A. Initial costs	
	Drain tubing (installed)	$2.62/m ($0.80/ft)
	Land forming for good surface drainage	$247/ha ($100/ac)
	B. Annual maintenance costs	
	Subsurface drainage	2% of annual amortized cost
	Surface drainage	$20/ha ($8/ac)

summary of corn production costs and drainage costs is given in Table 5.2. Drainage and subirrigation costs were based on a 100-ha system and were divided into initial costs and annual operating and maintenance costs. Initial costs for drainage were subsurface drain tubing (installed) and surface grading. Additional initial costs for subirrigation included water level structures and the well, pump and power unit.

Costs for the drain tubing and irrigation well were amortized over 40 years; the pump and control structure were amortized over 20 years, all at a 10% rate. The initial and operating costs of the well and pump were based on estimates given by Evans et al. (1987, 1988) for eastern North Carolina. Pumping costs were based on the volume of water subirrigation, which depended on drain spacing. The annual maintenance cost of the pumping system was taken as 1% of the initial cost. Examples of the costs for a range of drain spacings are given in Tables 5.3 and 5.4. Details of the costs are given by Murugaboopathi et al. (1991).

Average annual net profit was calculated by subtracting annual costs from annual income. The cost estimates did not include costs of land or management. Thus the profit calculated should be recognized as being profit to land and management. Average annual profits are plotted as a function of drain spacing in Fig. 5.6 for both subirrigation and conventional drainage. Good surface drainage was assumed in both cases. For drainage without subirrigation, the maximum profit of $139/ha was obtained at a 25-m drain spacing. The profit was nearly as high at $130/ha for a spacing of 30 m. Subirrigation can be used to substantially increase yields and profits. The maximum annual profit of $261/ha was predicted for a 20-m spacing. Thus, decreasing the drain spacing from 25 to 20 m providing subirrigation would increase average annual profits by 88% for this example. Subirrigation would be advantageous, even if the system had been designed for drainage with a 25-m spacing. The average annual profit for that case would be

Table 5.3. Summary of economic analysis of effects of drain spacing on average annual profit from corn on a Rains sandy loam soil with good surface drainage. Costs and income are given as $/ha

Drain spacing (m)	Drain per ha (m)	Drainage system[b] Initial cost	Annual cost	Annual maintenance cost[b]	Annual production costs	Total annual costs	Relative yield (%)	Annual gross income	Annual profit[a]
10	1000	2867.00	296.93	25.36	556.76	879.05	74.3	867.97	− 11.08
20	500	1557.00	162.97	22.68	556.76	742.41	74.5	870.31	127.90
30	333	1120.33	118.32	21.79	556.76	696.87	70.8	827.09	130.22
40	250	902.00	95.99	21.34	556.76	674.09	63.9	764.48	72.39
50	200	771.00	82.60	21.07	556.76	660.43	56.6	661.20	0.77
100	100	509.00	55.80	20.54	556.76	633.10	43.8	511.67	− 121.43

[a] Profit to land and management.
[b] Includes costs of both surface and subsurface drainage.

Table 5.4. Summary of economic analysis for subirrigation and drainage on a Rains sandy loam soil with good surface drainage. Costs and income are given as $/ha/yr

Drain spacing (m)	Drainage and Maintenance[b]	Irrigation volume (cm)	Pumping system cost	Pumping and maintenance	Production costs	Total costs	Relative yield (%)	Gross income	Profit[a]
10	338.11	20.69	22.96	26.04	556.76	943.91	96.4	1126.14	182.24
20	201.47	13.02	15.31	16.41	556.76	789.98	90.0	1051.38	261.40
30	155.92	8.09	15.31	10.35	556.76	738.38	78.3	914.70	176.33
40	133.15	5.05	7.65	6.41	556.76	704.01	65.7	767.51	63.50
50	119.48	3.22	7.65	4.16	556.76	688.10	56.1	655.36	− 32.74
100	92.16	0.7	7.65	1.06	556.76	657.67	42.9	501.16	− 156.51

[a] Profit to land and management.
[b] Includes amortized costs for surface and subsurface drainage plus maintenance costs.

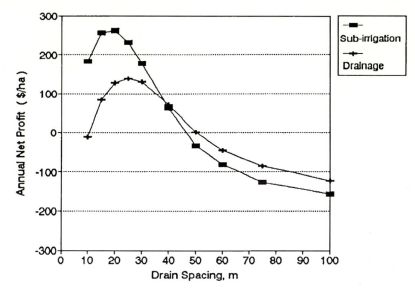

Fig. 5.6. Predicted average annual net profit for corn on the Rains soil with good surface drainage treatment

$230/ha, an increase of 65% over that for drainage alone. Results in Fig. 5.6 also show that subirrigation would negatively affect profits for drain spacings greater than 40 m for this soil.

Drainage is obviously essential for the profitable production of corn on this soil. Drains spaced farther apart than 50 m would result in a net loss, on the average. Details of the analyses including results for other surface drainage treatments, as well as for soybean and for a corn-soybean rotation, are given by Murugaboopathi et al. (1991).

5.7 Off-Site Impacts

The preceding sections have discussed drainage and water table management in terms of agricultural requirements. Traditionally, the efforts of engineers, technicians, farmers, and contractors have been aimed at one goal: to design and install systems that will satisfy agricultural drainage requirements at the least cost. However, recognition in recent years of agriculture's role in non-point source pollution of surface waters places additional constraints on the design and operation of drainage and related water management systems. This is particularly true where farms are located in environmentally sensitive areas, or where stringent water quality or quantity requirements are placed downstream. In most cases there are several design and operational alternatives that can be used to satisfy agricultural drainage needs. Some of these alternatives have different effects on the rate and quality of water leaving the fields than do others. The challenge is to identify those alternatives that will satisfy agricultural requirements while minimizing detrimental effects on the receiving waters.

Although research is by no means complete on this subject, considerable work has been done to determine water quality and hydrologic effects of drainage and associated water management practices. Studies on a wide range of soils, crops and site conditions have shown that increasing drainage intensity on agricultural lands may have both positive and negative impacts on hydrology and water quality. Improved subsurface drainage lowers water tables and increases the pore space available for infiltration of rainfall. This reduces the proportion of the total outflow occurring as surface runoff, which is rapid, and increases the proportion that is removed slowly by subsurface drainage over longer periods of time. Thus, improved subsurface drainage generally reduces peak outflow rates and sediment losses, while decreasing the losses of some pollutants and increasing the losses of others (Baden and Eggelsman 1968; Baker and Johnson 1977; Bottcher et al. 1981; Skaggs and Gilliam 1981; Bengtson et al. 1983, 1988; Istok and Kling 1983). For example, increasing the intensity of subsurface drainage generally reduces losses of phosphorus and organic nitrogen while increasing losses of nitrates and soluble salts (Baker and Johnson 1977; Bottcher et al. 1981; Gilliam and Skaggs 1986; Evans et al. 1990b). Conversely, improved surface drainage tends to increase phosphorus losses and reduce nitrate outflows. While exceptions have been reported in the literature for nearly all cases (e.g., Schwab et al. 1980; Bengtson et al. 1983; Culley et al. 1983), these general conclusions have been supported by the large majority of investigations. Thus, the water quality and hydrologic impacts of drainage cannot be simply and clearly stated. They depend on conditions prior to drainage improvements, drainage methods, soil properties, climatological variables, site conditions, and a host of other factors.

Research has shown that design and management strategies can be used to minimize pollutant loads from drained lands. These strategies range from water table management practices, controlled drainage and subirrigation, to cultural and structural measures. For example, controlled drainage has been found to reduce nitrogen and phosphorus losses by 45 and 35%, respectively, in some areas (Gilliam et al. 1979; Gilliam and Skaggs 1986; Evans et al. 1990). It is extremely important to select the appropriate management strategy for a given site. In some cases increasing subsurface drainage intensity to reduce surface runoff, sediment, and phosphorus losses would be a "best management practice" (BMP) for controlling non-point source pollution. In others, the BMP may be the use of controlled drainage to reduce nitrate outflows and to conserve water. While significant advances in our knowledge of environmental impacts have occurred and methods for managing these systems have improved in the last 20 years, there is much yet to be learned about the complex mechanisms governing losses of pollutants from drained soils. Research is continuing at a number of locations around the world to provide answers that will improve our ability to wisely manage poorly drained soils.

References

Baden W, Eggelsman R (1968) The hydrologic budget of the highbogs in the Atlantic region. In Proc 3rd Int Peat Cong, Quebec, pp 206–211

Baker JL, Johnson HP (1977) Impact of subsurface drainage on water quality. In: Proc 3rd Natl Drainage Symp, Chicago, ASAE Publ 77-1, 2950 Niles Rd, St Joseph, MI

Belmans CJ, Wesseling G, Feddes RA (1983) Simulation model of the water balance of a cropped soil: SWATRE. J Hydro 63:271–286

Bengtson RL, Carter CE, Morris HF, Kowalczuk JG (1983) Subsurface effectiveness on alluvial soil. Trans ASAE 26(2):23–425

Bengtson RL, Carter CE, Morris JF, Bartkiewicz SA (1988) The influence of subsurface drainage practices on nitrogen and phosphorus losses in a warm, humid climate. Trans ASAE 31(3):729–733

Bottcher AB, Monke EJ, Huggins LF (1981) Nutrient and sediment loadings from a subsurface drainage system. Trans ASAE 24(5):1221–1226

Bronswijk JJB (1988) Effect of swelling and shrinkage on the calculation of water balance and water transport in clay soils. Agric Water Manag 14:185–193

Chang AC, Skaggs RW, Hermsmeier LF, Johnson WR (1983) Evaluation of a water management model for irrigated agriculture. Trans ASAE 26(2):412–418

Clinton FM (1948) Invisible irrigation on Egin Bench. Reclam Era 34:182–184

Committee on Irrigation-Induced Water Quality Problems (1989) Irrigation induced water quality problems. National Academy Press, Washington, DC, 157 pp

Culley JLB, Bolton EF, Bernyk V (1983) Suspended solid and phosphorus from a clay soil. I Plot studies. J Environ Qual 12(4):493–498

de Laat PJM, Atwater RHCM, van Bakel PJT (1981) GELGAM – a model for regional water management. in: Proc Tech Meet 37 Versl Meded Comm Hydrol Onderz, TN027, The Hague, pp 25–53

Evans RO, Sneed RE, Skaggs RW (1987) Water supplies for sub-irrigation. NC Agric Extens Serv Tech Bulletin No AG-389, NC State Univ, Raleigh, 17 pp

Evans RO, Skaggs RW, Sneed RE (1988) Economics of controlled drainage and sub-irrigation systems. NC Agric Exten Serv Tech Bull AG-397, NC State Univ, Raleigh, 17 pp

Evans RO, Skaggs RW, Sneed RE (1990a) Normalized crop susceptibility factors for corn and soybean to excess water stress. Trans ASAE 33(4):1153–1161

Evans RO, Gilliam JW, Skaggs RW (1990b) Controlled drainage management guidelines to improve drainage water quality. NC Agric Extens Serv Bull AG-443, Raleigh, NC 27695, 16 pp

Evans RO, Skaggs RW, Sneed RE (1992) Stress day indices to relate corn and soybean yield to excess water stress. Trans ASAE (in press)

Feddes RA, Kawalik PJ, Zaradny H (1978) Simulation of water use and crop yield. Simulation Monogr, PUDOC, Wageningen, 188 pp

Feddes RA, Kabat P, van Bavel PJT, Bronswijk JJB, Halbertsma J (1988) Modelling soil water dynamics in the unsaturated zone – state of the art. J Hydrol 100:69–111

Fouss JL, Bengtson RL, Carter CE (1987) Simulating subsurface drainage in the lower Mississippi Valley with DRAINMOD. Trans ASAE 30(6):1679–1688

Gayle G, Skaggs RW, Carter CE (1985) Evaluation of a water management model for a Louisiana sugar cane field. J Am Soc Sugar Cane Technol 4:18–28

Gilliam JW (1987) Drainage water quality and the environment. In: Proc 5th Natl Drainage Symp, Chicago, ASAE, 2950 Niles Rd, St Joseph, MI 49085, pp 19–28

Gilliam JW, Skaggs RW (1986) Controlled agricultural drainage to maintain water quality. J Irrig Drain Div ASCE 112(3):254–263

Gilliam JW, Skaggs RW, Weed SB (1979) Drainage control to diminish nitrate loss from agricultural fields. J Environ Qual 8:137–142

Hardjoamidjojo S, Skaggs RW (1982) Predicting the effects of drainage systems on corn yields. Agric Water Manag 5:127–144

Harmsen EW, Gilliam JW, Skaggs RW (1991) Simulating two-dimensional nitrogen transport. ASAE Pap 912630, Am Soc Agric Eng, St Joseph, MI 49085-9659

Hiler EA (1969) Quantitative evaluation of crop-drainage requirements. Trans ASAE 12(4):499–505

Istok JD, Kling GF (1983) Effect of subsurface drainage on runoff and sediment yield from an agricultural watershed in western Oregon, USA. J Hydrol 65:279–291

Karvonen T (1988) A model for predicting the effect of drainage on soil moisture, soil temperature and crop yield. PhD Diss, Helsinki Univ Technol, Takentajanaukio, Finl

Krayenhoff van de Leur DA (1958) A study of nonsteady ground water flow with special reference to a reservoir-coefficient. Ingenieur 70B:87–94

Logan TJ, Randall GW, Timmons DR (1980) Nutrient content of tile drainage from cropland in the North Central Region. N Central Res Bull 268, Wooster, OH

Luthin JN (1957) Drainage of irrigated lands. In: Drainage of agricultural lands. Luthin JN (ed) Am Soc Agron, Madison, pp 344–371

Massey FC, Skaggs RW, Sneed RE (1983) Energy and water requirements for sub-irrigation versus sprinkler irrigation. Trans ASAE 26(1):126–133

McMahon PC, Mostaghimi S, Wright FS (1988) Simulation of corn yield by a water management model for a coastal plains soil. Trans ASAE 31(3):734–742

Munster CL, Skaggs RW, Parsons JE, Evans RO, Gilliam JW (1991) Modeling aldicarb transport under drainage, controlled drainage and sub-irrigation. ASAE Pap 912631, Am Soc Agric Eng, St Joseph, MI 49085–9659

Murugaboopathi C, Skaggs RW, Evans RO (1991) Sub-irrigation for corn and soybean production in North Carolina. In: Proc Int Conf Sub-irrigation and controlled drainage, Lansing MI

Parsons JE (1987) Development and application of a 3-dimensional water management model for drainage districts. PhD Thesis, NC State Univ, Raleigh, 476 pp

Parsons JE, Skaggs RW, Doty CW (1990) Simulation of controlled drainage in open ditch drainage systems. Agric Water Manag 18:301–316

Reeve RC, Fausey N (1974) Drainage and timeliness of farming operations. In: van Schilfgaarde J (ed) In: Drainage for agriculture, chap 4. Am Soc Agron, Madison, pp 55–66

Renfro G Jr (1955) Applying water under the surface of the ground. Yearb Agric USDA, pp 173–178

Rogers JS (1985) Water management model evaluation for shallow sandy soils. Trans ASAE 28(3):785–790

Schwab GO, Fausey NR, Kopcak DE (1980) Sediment and chemical content of agricultural drainage water. Trans ASAE 23(6):1446–1449

Skaggs RW (1978) A water management model for shallow water table soils. Tech Rep 134, Water Resour Res Inst Univ NC, NC State Univ, Raleigh 27695

Skaggs RW (1982) Field evaluation of a water management simulation model. Trans ASAE 25(3):666–674

Skaggs RW (1987) Model development, selection and use. In: Proc 3rd Int Works Land drainage, Ohio State Univ, Columbus, Dec 7–11, pp 29–3

Skaggs RW, Gilliam JW (1981) effect of drainage system design and operation on nitrate transport. Trans ASAE 24(4):929–934

Skaggs RW, Fausey NR, Nolte BH (1981) Water management evaluation for North Central Ohio. Trans ASAE 24(4):922–928

Susanto RH, Feyen J, Dierickx W, Wyseure G (1987) The use of simulation models to evaluate the performance of subsurface drainage systems. In: Proc 3rd Int Drainage Worksh, Ohio State Univ, pp A67–A76

van Schilfgaarde J (1965) Transient design of drainage systems. J Irrig Drain Div ASCE 91 (IR 3):9–22

van Schilfgaarde J (ed) (1974) Nonsteady flow to drains. In: Drainage for agriculture. Am Soc Agron, Madison, pp 245–270

van Wijk ALM, Feddes RA (1986) Simulating effects of soil type and drainage on arable crop yield. In: Proc Int Sem Land drainage, Helsinki Univ of Technol, pp 127–142

Wesseling J (1974) Crop growth and wet soils. In: van Schilfgaarde J (ed) Drainage for agriculture, chap. 2. Am Soc Agron, Madison, pp 1–37

Williamson RE, Kriz GJ (1970) Response of agricultural crops to flooding, depth of water table and soil composition. Trans ASAE 13(1):216–220

Workman S, Skaggs RW (1990) PREFLOW. A water management model capable of simulating preferential flow. Trans ASAE 33(60):1939–1948

Zaradny H (1986) A method for dimensioning drainage in heavy soils considering reduction in PET. Proc Int Sem Drainage, Helsinki Univ Technol, pp 258–265

6 Runoff Irrigation

J. Ben-Asher and P.R. Berliner

6.1 Introduction

The fraction of rainfall that flows over the landscape from higher to lower elevations is known as runoff. Runoff is usually associated with negative implications such as erosion, water loss, etc. It can, however, be used for the surface irrigation of agricultural crops during rainfall events. This is accomplished by channeling the runoff water into dike-surrounded plots where crops are grown. Concentrating runoff water allows agricultural activities in areas in which otherwise no such activities could take place. This technique has been called runoff agriculture or water harvesting; the latter is usually used to indicate water collection for domestic use. The amount of water that can be collected during a rainfall event depends on rainfall characteristics such as quantity, intensity and distribution, and on the generating area such as size, geomorphology and surface characteristics. The main difference between runoff irrigation and conventional irrigation is that the timing and the amount of the application cannot be determined a priori. The variability in available water can be minimized by adjusting the size of the plots receiving runoff. From this short description it is clear that in order to model the productivity of a crop in such a system it is necessary to model the diverse processes involved such as rainfall, infiltration, surface flow and consumptive water use.

6.2 Present Development and Expansion of Runoff Irrigation

A preliminary survey indicated that a conservation estimate of the area which is currently under runoff irrigation (RI) is about 500 000 ha. Table 6.1 shows the size of RI areas, the countries involved and the amount of rainfall associated with runoff agriculture.

Runoff irrigation is mainly practiced in arid or semi-arid areas with annual rainfall ranging from 100 to 600 mm. In these arid lands, in spite of the deficiency in rainfall, enormous amounts of water are lost through flash floods after heavy showers. In the Sahel, for instance, it is estimated that

Adv. Series in Agricultural Sciences, Vol. 22
K.K. Tanji/B. Yaron (Eds.)
© Springer-Verlag Berlin Heidelberg 1994

Table 6.1. Partial listing of countries with WH systems

Country	Total area (ha)	Rainfall (mm/yr)	Source of calculations
Africa			
Kenya	5 000	100–1000	Mutai (1986); Imbira (1986)
			Finkel and Gainey (1989)
Mali	70	400–600	Klemm and Prinz (1989)
Tunisia	200 000	150–300	Reij et al. (1988)
Niger	100 000	200–400	Reij et al. (1988)
Burkina	70	200–400	Prinz (1990)
Faso			(Pers. comm.)
Asia			
Israel	3 000	100–300	Ben-Asher (1988)
India	300	300–700	Grewald et al. (1989)
Yemen	90 000	100–300	Eger (1987)
North America			
USA	120 000	500–1000	Clayton et al. (1974)
Total:	518 440		

each year 200–500 million m^3 of water are lost through runoff, which if captured could be used to irrigate about 20 000–40 000 ha. In the Middle East, about 10 000–20 000 ha could be irrigated with runoff water (USSR National Committee for the International Hydrological Decade 1977). The present water situation does not allow us to neglect this resource, and methods must be developed to use runoff water in a more efficient manner in order to open marginal regions for agriculture production.

Water concentration techniques have been applied for thousands of years in Mediterranean agriculture (Evenari et al. 1982) and to a lesser extent in semi-arid tropic regions (M. Finkel 1986). Renewed interest in these techniques has resulted from the need to utilize more fully the available land and water resources. Using the same source of data in Table 6.1, Fig. 6.1 shows the trend of development of the water harvesting systems from 1982 to 1985.

The area under water harvesting is presently small compared to the 223 million ha of global irrigated land (Nakayama and Bucks 1986). Figure 6.1 indicates an annual increase of about 0.6%/yr; while expansion of irrigated lands from 1961 onwards was 2%/yr and is now only 1%/yr (Hoffman et al. 1990). However, both Table 6.1 and Fig. 6.1 represent the total area of water harvesting which includes runoff contributing and collecting areas. In order to calculate the cultivated area, one should assume a ratio of 1/10–1/30 for experimental plots and 1/100–1/1000 for commercial plots. Thus the productive or the run on area is only about 500–30 000 ha. A wide variety of design and construction concepts is available to establish the runoff collecting and storing system. The major techniques for RI have been compiled by

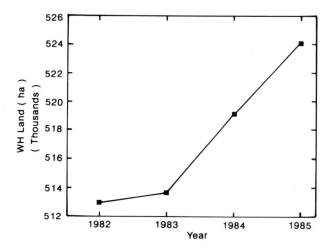

Fig. 6.1. Total WH land from 1982–1985 partial listing

Prinz (1990). These techniques are suitable for a variety of physical conditions (soil properties, slopes, etc.) and are used to store surface runoff in cultivated plots.

6.3 Generation of Runoff

6.3.1 General Principles

Runoff is generated on a local scale when the infiltrability of the soil is less than the rainfall intensity. The latter may decrease when the potential gradient decreases and/or a change in the physical properties of the upper soil layer reduces its hydraulic conductivity. The runoff thus generated flows over land towards the collection plot. The fraction of the generated runoff that reaches the collection plot is a function of the catchment size and is affected by hydrological parameters such as size, slope, roughness and vegetation.

A reduction in the hydraulic conductivity of the upper soil layer happens frequently in the bare soil or arid zones. In such regions rainfall limits plant productivity and the collection of runoff for irrigation becomes important. Even though this is not the only mechanism for runoff production, as mentioned earlier, we will emphasize this aspect throughout the critical review.

Two different approaches can be used to predict runoff volumes from a given catchment: (1) mechanistic: studying and modeling the microscopic level of runoff formation and progressively increasing the modeled scale and number of variables; and (2) empirical: studying the runoff producing patterns of catchments with different sizes and inferring from the input/ output relations the properties of the watersheds.

In the following sections we will describe the salient features of both approaches.

6.3.2 Rainfall Simulators

The study of the physical and physicochemical aspects which lead to runoff formation is easier under laboratory conditions than under field conditions. Runoff simulators are an efficient tool for this purpose because they facilitate controlled experiments in which the factors affecting runoff generation processes may be intensively studied (Agassi et al. 1987). These rainfall simulators have also been heavily used to study erosion and sediment production (Lane 1985). The three major rainfall variables (rain intensity, drop size distribution, and terminal velocity of the drops) in a simulator-produced rainfall should be similar to those of natural rain. Morin et al. (1967) recommended a "rotating disk" simulator. The three variables of simulated rain can be closely matched to natural rainfall over a very small area (1–2 m²) at a cost of about $5000. This simulator was described in the workshop on rain simulators held in Tucson, Arizona, in 1979 (USDA-SEA 1979). Another type of simulator is based on a rotating boom (Simanton and Renard 1982; Schiffler 1990). Both types of simulators perform well and have a relatively low price (less than $10 000). Their main disadvantage, however, is that they simulate rainfall over very small surface areas which cannot incorporate the effect of spatial variability of the soil surface, an extremely important factor under field conditions. To overcome this limitation Schram (1990) modified the rain simulator of Foster et al. (1982) and simulated rain over a relatively large area (22 × 4 m). The drop size distribution and intensity were obtained by using a system of oscillating nozzles. The area covered by this rain simulator can be increased by the addition of modular segments. Spatial soil variability can be incorporated with ease but at a very high cost (∼ $100 000). Most of the simulators are operated at a single intensity per simulation event. Studies requiring variable rainfall either relied on a manual controller, or specially developed control hardware to generate the desired intensity pattern. Recently, a microcomputer-controlled rainfall simulator has been fabricated. In this simulator (Hirschi et al. 1990) the problem of changing the intensity throughout the simulation run was solved by changing the number of oscillations of the nozzles per unit time. From the summary presented it appears that the means necessary to simulate actual rainfall are available. The simulators have been frequently used to study crust formation, as will be described in the next section. However, the information gained by the use of the aforementioned simulators has not been used to predict the actual hydrographs of even small catchments. This is the great challenge and should be the crux of future work in the field.

6.3.3 Runoff as a Result of Crust Formation

The formation of a soil crust on the surface is one of the main contributors to runoff generation in arid and semi-arid regions. The surface crust can be defined as a thin layer (0–3 mm) of higher density, finer pores and lower hydraulic conductivity than the underlying soil. Several hypotheses that explain crust formation have been put forward and can be summarized as follows (Shainberg 1991): the impact of falling drops causes the breakdown of the soil aggregates. The destruction of the aggregates reduces the average size of the pores of the surface layer. In addition, the impact of raindrops causes compaction of the uppermost layer of the soil. These factors produce the thin skin seal at the soil surface. Concurrently, the presence of cations adsorbed to the clay platelets can produce the dispersion of the soil clays which lead to the further disintegration of the aggregates. The small particles then move into the soil pores with the infiltrating water, eventually clogging the pores immediately beneath the surface. The first mechanism is mechanical in nature and the second physicochemical. The importance of both effects was dramatically demonstrated in a series of trials in which the composition of adsorbed cations, the electrolytic concentration of the rainfall and the energy of the rain drops were changed with a Morin-type simulator and with fog sprinklers. The results showed that absorbed cations and electrolyte concentrations in the rainfall water have a strong effect on crust formation in montmorillonitic soils. In Israel, particular attention has been paid to the effect of exchangeable sodium percentage (ESP) on crust formation in montmorillonitic soils (Shainberg 1991). The effect of clay and silt content on aggregate stability and crust formation has also been studied. Soils with 10–20% clay were found to be the most sensitive to crust formation and have the lowest infiltration rate (Ben-Hur et al. 1985). With increasing percentage of clay the aggregates are more stable and formation of crust was diminished. In soils with a lower clay content ($> 10\%$) the stability of aggregates diminished, but due to the limited amount of clay the clogging of the pores was not very pronounced. The resulting crust was less dense and had consequently higher infiltration rates. As mentioned earlier, crust formation depends on both rain (intensity and kinetic energy of drops) and soil properties (mineralogy, clay, and silt content, soil composition, and exchangeable ions). Zones of interest for the possible implementation of runoff agriculture on a large scale are the two subdivisions of the semi-arid zone: the Mediterranean and the sub-Saharan. In the Mediterranean area the rainy season coincides with the cool season of the year and in the sub-Saharan most or all of the annual rainfall occurs during the warm summer months. The soils in the two areas are also very different. Whereas the dominant clay in the Mediterranean soils is smectite, the predominant clay in the sub-Saharan zone is kaolinite. Colloidal properties of the smectites and the effect of exchangeable cations and electrolyte concentration on clay swelling and dispersion have been studied. Conversely, very little is known on the dispersion of kaolinitic soils and their crusting properties. In the Mediterranean area loess soils are the ones that crust with the greatest ease. These eolic soils have unstable aggregates, a clay content of 10–20%, and a varying proportion of silt and sand (usually a higher silt content).

6.3.4 Physical Models of Runoff Formation in Crusting Soils

Physical quantitative approaches were summarized by Ahuja and Swartzendruber (1991) and Mualem (1989). The latter claimed that, "... progress in quantitative modeling of soil sealing was tied by a short rope to a modified version of the Horton equation ..."

Many modeling approaches have been based on the approach of Morin et al. (1981). They assumed that when a rain drop hits the soil surface, the soil's hydraulic conductivity reduces to a value K_c which is much lower than the original conductivity Ks. The surface hydraulic conductivity is a weighted average of these two values, where the weight factors are the fraction of the area affected by rain drop impact, $K_c(g)$ and its complementary $(1 - g)$. If the rate of increase of the affected (or crusted) area is exponential, then the averaged surface hydraulic conductivity can be viewed as a Horton-type equation:

$$K(t) = Kf + (Ki - Kf)*exp(-gIt), \tag{6.1}$$

where $K(t)$ is the average surface hydraulic conductivity after time (t) from beginning of rainfall, Ki and Kf are the initial and final hydraulic conductivities, g is an exponent that originally relates the fraction of the impacted area to the rate that fraction increased, I is the rain intensity and t is the time.

Denoting the depth of the crusted layer as d and the depth of the underlying soil as D; and assuming a continuous variation of matrix potential at the interface, an expression for the flux in a crust-affected soil can be derived:

$$q = \frac{L}{D/K_c + (L - d)/Ks}, \tag{6.2}$$

where q is the infiltration rate and $L = d + D$. Mualem and Assoulin (1989) suggested that Eqs. (6.1) and (6.2) are related to each other through the functional form of D and K_c with time t.

Mualem and Assoulin (1989) presented a model which describes the crust as a disturbed soil layer in which the main factor that had changed and affected the rest of the changes is the bulk density. This change in bulk density occurs due to aggregate breakdown, compaction, and transport of solid particles along the soil pores until the pores are blocked by the moving solids.

Their model, even though it cannot be validated by direct measuring methods, fitted well to data collected by Morin and has the advantage of being a physically based model. The basic assumption of the model is that:

$$\frac{d(\Delta\rho)}{dz} = \alpha\Delta\rho \,, \tag{6.3}$$

where $\Delta\rho$ is the change in the bulk density, z is the depth variable and α is a soil parameter, which after integration leads to the change in soil bulk density with depth:

$$\Delta\rho_{(z)} = \Delta\rho_0 \exp(-\alpha z) \,, \tag{6.4}$$

where $\Delta\rho_0$ is the change in the density at the soil surface and is actually a constant for a given soil-rain system. The authors defined the crust depth as the depth at which the density changes are less than 0.1% of the original bulk density, such that the α becomes:

$$\alpha = \frac{\ln(0.001)}{d} \,. \tag{6.5}$$

The hydraulic properties with which the authors dealt are: saturated water content, residual water content, a pore size distribution parameter, and an air entry value originated by Brooks and Corey (1964).

The thickness of the crust was defined as:

$$d = nL^m \,, \tag{6.6}$$

where n and m are soil empirical constants.

Mualem et al. (1990) assumed that for a given soil-rain system the amount of solids that "enter" the crust (c) is constant, such that the change in the depth-dependent bulk density is given by:

$$C = \int_{z=0}^{z=d} \Delta p(z) dz \,, \tag{6.7}$$

$$C = 0.145 * \Delta p * d \,. \tag{6.8}$$

Using all the relations for the crust's depth and the averaged hydraulic conductivity in Eq. (6.3), the authors presented a five-parameter model that relates the final infiltration rate (q) to the hydraulic head under saturated conditions. The parameters which were estimated by an inverse method are:

ΔC, m, n, and α.

Using Morin's (1981) data, Mualem et al. (1990) obtained good results. However, in order to gain insight into Kc, the thickness of the crust has to be measured. This has been attempted by analyzing SEM photomicrographs. It is not easy to see a clear boundary to the crusted layer. Moreover, the variability of artificially formed crusts (Levi et al. 1988) casts some doubts on the validity of crust thicknesses derived from SEM microphotographs.

6.4 Modeling Watershed Runoff Production

Based on current practices, we will arbitrarily divide watersheds into two major systems: microcatchment water harvesting (MCWH) and small catchment watershed harvesting (SCWH). In the following we present currently available models for each system.

6.4.1 Microcatchment Water Harvesting

Three basic models were developed for the quantitative evaluation of microcatchment water harvesting (MCWH) systems.

The kinematic wave model (Zarmi et al. 1982) offers an analytic prediction of MCWH hydrographs on a short-term basis. The behavior of the solution can be applied to estimate hydrological parameters which are essential for design purposes. Results of hydrographs measured and predicted by the Kinematic wave equation are given in Fig. 6.2. Runoff is generated a short time after rain starts and it amounts to the net difference between rain intensity and final infiltration capacity a few minutes afterward. Thus, hydrographs of runoff generated throughout long storms may be approximated as a rectangular-shaped hydrograph in which the rising and the falling limbs are very short compared to the steady runoff flow which lasts throughout the entire storm and contributes the major part of the collected water volume. This part is linearly related to the storm duration and its associated rainfall depth as shown in Fig. 6.3, which resulted from measurements of Boers et al. (1986a). The linear model can be written as:

where: $\qquad\qquad\qquad R = \alpha(P - \delta)$ $\qquad\qquad$ (6.9)

R – runoff volume $\qquad\qquad$ (L^3)

α – runoff coefficient $\qquad\qquad$ (L^3/L^3)

P – storm volume $\qquad\qquad$ (L^3)

δ – threshold value to start flow $\quad$ (L^3).

In Fig. 6.3 the relationship between rainfall and runoff for a total of 28 storms is presented. It shows that the linear model fits the data quite well, with $r^2 = 0.81$. The model indicates that runoff production can be increased by two engineering measures. The threshold value d can be lowered by smoothing the runoff contributing surface to reduce the depression storage and the slope a can be increased by chemical treatment of the surface, which reduces the infiltration rate.

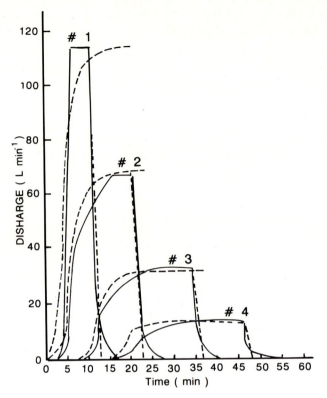

Fig. 6.2. Runoff discharge as a function of time. A comparison between measured and calculated hydrographs which occurred at four intensities of rainfall. The numbers *1* through *4* denote 59.4, 37.0, 18.1, and 10.7 mm/h, respectively

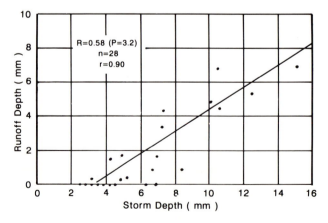

Fig. 6.3. Linear regression model for microcatchment

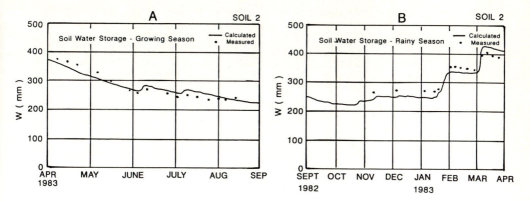

Fig. 6.4. Measured and simulated values of soil water storage for the hydrological year 1982/83 in Sede Boker

The numerical solution of Boers et al. (1986a) is capable of linking hydrological processes on the contributing area and the water uptake by a tree in the collecting area. It thus describes complete MCWH elements for short and long time scales. Results of measured and simulated soil water depth throughout the years 1982 and 1983 are presented in Fig. 6.4.

The model of Ben-Asher and Warrick (1987) takes into account spatial and temporal variability of soil properties and precipitation, and thus introduced stochastic considerations into the deterministic models that were previously mentioned.

6.4.2 Modeling Rainfall Runoff Relationships for an SCWH Element

The potential runoff production of a small catchment watershed harvesting system (SCWH) is the integral of the local runoff production over its area. The potential value will be attained when the rainfall is homogeneous over the watershed and surface detention, stoniness, and subsurface flow are negligible. These conditions are approximated under field conditions when the watershed is very small (a few hundred square meters). For arid zone watersheds, Karnieli et al. (1988) analyzed the runoff efficiency in relation to the catchment size (Fig. 6.5).

The results clearly indicate that catchment size affects runoff efficiency but it is not possible to determine the relative importance of the different factors involved. In modeling the rainfall runoff relationship for SCWH we found three major approaches: (1) deterministic one-dimensional modeling (Klemm 1990); (2) parametric modeling of the entire watershed (Dodi et al. 1987; Giraldez et al. 1987); and (3) modeling the watershed on the basis of the partial area contribution (PAC) concept (Karnieli et al. 1988; Humborg 1990; Ben-Asher and Humborg 1992).

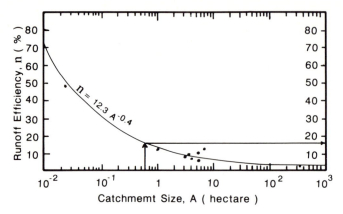

Fig. 6.5. Runoff efficiency as a function of catchment size

The parametric model assumes that runoff is contributed from the entire watershed. It therefore divides the area into cells which are connected to each other by lumped parameters. Runoff is calculated at each cell through a number of operations that transform input rainfall into a single output based on the unit hydrograph and is the cell's outflow hydrograph. This model was tested successfully in an arid watershed of 80 ha which contributed runoff to a cultivated area of 0.15 ha (a ratio of 1:550). The model requires a large number of parameters, is difficult to calibrate and it does not take into account the PAC phenomena and the processes on the cultivated area. However, dividing the area into independent compartments which differ from each other by their hydraulic properties was used in modeling the PAC concept. The empirical model of Karnieli et al. (1988) requires a large number of local data and therefore describes a specific situation. It provides a tool to design SCWH systems in the Israeli Negev and it includes the PAC concept. The PAC concept suggests that measurable runoff is contributed only from a part of the given watershed and not from all of it. This phenomenon is discussed in detail in Section 3.

Humborg (1990) studied runoff production of an SCWH system in Mali and concluded that the main factor affecting runoff yield is the PAC. The size of the contributing area was calculated from the ratio between results of rain simulation and actual runoff in the area. The location of the contributing area was determined from satellite imagery (see Fig. 6.6).

From Fig. 6.6 it can be seen that after 15 mm of rain, 119 pixels of 900 m² each generated runoff. An additional 45 runoff-generating pixels generated runoff after 20 mm of rain and so on, as detailed in the list in Fig. 6.6.

It is worth noting that approximately 50% of the potentially contributing pixels was active at 15 mm rain (this rain depth is very close to the runoff threshold value) and the rest was activated when rain depth increased to

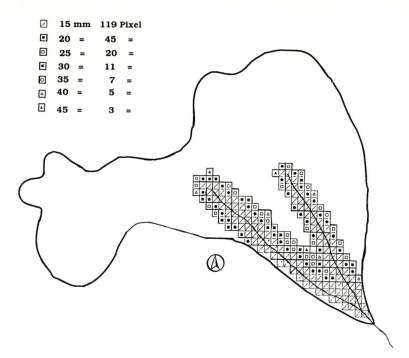

☑	15 mm	119 Pixel
▣	20 =	45 =
◙	25 =	20 =
▪	30 =	11 =
◎	35 =	7 =
◮	40 =	5 =
⊡	45 =	3 =

Fig. 6.6. The location and the shape of contributing area in Mali. (The *table* specifies the number of pixels that became activated after a given rain depth)

45 mm. The expected upper limit of runoff-contributing pixels (~ 250 pixels) is asymptotically approached as rain depth increases from 20 to 45 mm per storm.

The contributing area covers only about 20% of the total watershed. Rain on other parts of the watershed does not contribute to runoff, possibly because of large losses during overland flow and short storm durations which were too short to maintain continuous water flow from remote parts of the catchment to its outlet.

The catchment area was divided into runoff generating cells, each the size of a pixel (satellite picture element) and the total area runoff was simply calculated from the product of rain simulation and the total area of contributing pixels. The model addresses the complex problem of PAC by linking remote-sensing information with experimental results from rainfall simulators. In the near future this combination of remote sensing and rain simulation will have an important role in predicting runoff from SCWH systems. It is therefore imperative to include recent developments in the PAC concept and remote sensing in our review.

6.4.3 The Role of PAC in Water Harvesting

The partial area contribution (PAC) concept suggests that specific runoff yield is reduced as the size of the catchment is increased. This term was originated by Betson (1964), and the phenomenon is presented in Fig. 6.5. Van de Griend and Engman (1985) reviewed the various mechanisms that have been proposed for the PAC phenomenon with respect to remote sensing and commented that the concept has not been applied in practical hydrology. (Indeed, it was not mentioned in any one of the key books quoted in this review.) However, there is increasing evidence of the importance of this phenomenon (Yair et al. 1978; Boughton 1987). The major problem associated with it is the transferability of results from a small basin to a larger one (Pilgram et al. 1982). This is especially true when trying to apply results of rain simulation experiments (which are made on an area of only a few square meters) to small catchments. The error in extending results of rainfall simulation experiments to evaluate runoff yield from small catchments was demonstrated by Humborg (1990) (Fig. 6.7). He found that simulated runoff volume was much larger than the actual volume of runoff which was measured in the study area.

In Fig. 6.7 the total catchment area was estimated from both LANDSAT and SPOT. The solid lines represent the product of rainfall-runoff simulation on a microcatchment with the number of microcatchments in the watershed. The dots are measured volumes of runoff and the third solid line results from the best fit linear regression analysis of the measured runoff volumes. Although the observed data differ largely from their best fit line ($r^2 = 0.5$), they still support the major observation that actual runoff was not contributed by all elements in the area but only from part of them. Results are presented in Fig. 6.6.

Since many water harvesting studies have been carried out on MCWH systems, especially with rain simulators, the transfer of results to SCWH

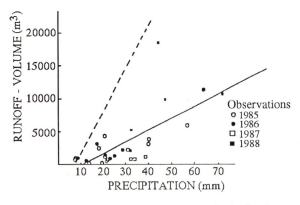

Fig. 6.7. Volume of runoff as a function of rain depth

systems is of obvious importance and its understanding may improve the scientific approach to SCWH.

Some explanations for the phenomenon are as follow: under humid conditions, the most effective areas that contribute to the peak hydrograph are the channels themselves, and belts that extend on both sides of the channels and near the watershed outlets (Beven and Kirkby 1983). The soils around these areas become saturated, or nearly saturated, during the rain event. Their infiltration capacities are reduced and thus create runoff.

Van de Griend and Engman (1985) suggested that the main reasons for the phenomenon are related to the temporal and spatial variability of the contributing areas and the storm.

The two explanations implicitly offer two mechanisms to generate runoff. One is soil saturation and the other is crust formation. The second one occurs in deserts where rainfall is characterized by low annual amounts but highly intensive single storms. (The crust formation mechanisms will be discussed later.)

In arid regions and as a result of PAC, the size of the contributing area which is required to satisfy crop water consumption use has to be amplified.

Figure 6.6 shows that PAC is increased to an asymptotic value with increasing amounts of rain. Under arid conditions, the amount of rain is small and often will not be sufficient even to bring the PAC to its potential maximum. Thus, it seems that in arid regions the size of the contributing area should compensate not only for low annual rainfall but also for smaller fractions of the contributing area. As a result, the ratio between rainfall in semi-arid and arid zones is smaller than the ratio between the contributing areas in arid zone and in semi-arid zone. For example, the rainfall in the experimental site in Mali is about five times more than the rainfall we reported for the experimental sites in Israel, while the required contributing area for SCWH in Israel is more than ten times larger than that in Mali.

Another result of the PAC phenomena is the need to develop and use remote-sensing information in order to identify the runoff contribution pixels in the basin and thus to improve our ability to quantify the phenomenon.

6.5 Runoff Irrigation and Remote Sensing

Remote-sensing information, which is usually associated with satellite data, has not been used extensively in runoff irrigation. On the other hand, in the last decade remote-sensing data have been used in hydrology. Van de Griend and Engman (1985) reviewed remote-sensing methods for measuring and monitoring soil water content.

Table 6.2 summarizes their findings and demonstrates the large number of possible applications of remote sensing to identify wet areas which are

Table 6.2. The regions of the E.M. spectrum and their potential for remote sensing of partial-area related phenomena and soil moisture differences

Domain of the E.M. spectrum	Wavelength	Physical background	Platform/ altitude	Resolution Spatial	Resolution Temporal	Selection of applications known from the literature
Microwaves (1) Active	1 mm–0.8 m	Determination of active microwave reflection. Reflection depends on surface relief, surface roughness, soil physical parameters and sensor characteristics	SEASAT SIR-B Space Shuttle	25 m 25 m	–	*Soil moisture:* Blanchard and Chang (1983) *Soil moisture:* Jackson and Schmugge (1981)
(2) Passive		Determination of emissivity. Emissivity depends on surface temperature, and dielectric properties. Dielectric properties depend on moisture content, etc.	Aircraft 1600 m and 300 m NIMBUS-5	400 m and 60 m 25 km	– –	*Soil moisture:* Engman et al. (1983); Burke and Schmugge (1982); Jackson (1982)
Thermal infrared	8–14 μm	Thermal infrared photography and thermal infrared scanning Determination of spectral emissivity coefficient and surface temperature	SMS 2 satellite HCMM satellite Aircraft	9 km 0.5 km	Twice daily Every 5 days 1 day and following night image	*Evapotranspiration and/or soil moisture:* Price (1980, 1982a, b); Nieuwenhuis (1981); Heilman and Moore (1982) *Soil moisture:* Schmugge et al. (1978); van de Griend and Engman (1985)

Visible + near infrared	0.4–0.8 μm 0.8–3 μm	Determination of spectral reflectance by infrared color photography, multi-spectral photography or multi-spectral scanning	LANDSAT (MSS) Aircraft	70 m —	18 days —	*Mapping land-cover differences:* Engman and Arnett (1977); Jackson et al. (1981, 1982); Jackson and Rawls (1981) *Spectral analysis to detect surface cover differences:* Engman and Arnett (1977)
Gamma radiation	μ6 ×10^{-6} μm	Measurement of natural terrestrial gamma radiation flux to infer areal soil	Aircraft 150 m	250 m	—	*Determination of soil water and/or snow water content:* Engman (1985); Jones and Carroll (1983); van de Griend and Engman (1985);

potentially runoff-contributing areas according to the mechanism suggested to explain the PAC phenomenon by the saturated zones near the watershed outlets. Two other aspects of application of remote sensing for runoff irrigation are the use of remote sensing for parameter estimation and data as input for hydrological models. Schultz (1986, 1987) demonstrated the use of radar rainfall measurements as input for a real-time runoff forecasting model and used it to estimate monthly values of runoff for design purposes.

In summary, the introduction of remote sensing into the discipline of hydrology may affect the development of runoff irrigation in several directions: (1) including a regional analysis of hydrological systems suitable for runoff irrigation, especially in remote areas; (2) reviewing of available rainfall-runoff models and reformulation of new models in which remote-sensing data will be used; and (3) using remote-sensing data for identification of hydrological parameters and as input for runoff predicting models.

6.6 Agricultural Use of Runoff Water

6.6.1 Engineering Description of Water Harvesting Systems

6.6.1.1 Microcatchment Water Harvesting

MCWH is a method of collecting surface runoff from a small runoff contributing area and storing it in the root zone of an adjacent infiltration basin to meet crop water requirements (Fig. 6.8 after Prinz 1990). In the infiltration basin there may be a single tree, bush or annual crop (Boers and Ben-Asher 1982). The aim of MCWH is to store runoff water in the soil profile below the basin during the rainy season. The size of the contributing area is designed to maintain annual water consumption of the crop in the basin.

MCWH is especially useful in arid and semi-arid regions, where irrigation water is either costly or unavailable. This method has been tested in a number of countries (National Academy of Science 1974), but specific research on the subject has been very limited.

Microcatchments are more efficient than larger-scale water harvesting schemes because conveyance losses are minimized. In light rains they provide runoff water, while others do not. It can be constructed on almost any slope, including almost level plane (1–2% slopes), enabling the farmer to use large flat areas. Reported sizes of a single MCWH element are 100–250 m^2 in Israel, 250–400 m^2 in India and 1000 m^2 in Mali.

During the past decade, MCWH has been studied extensively. Agronomic results in regions of rainfall larger than 200 mm/yr were good. In Israel, however, they were rather poor, because rainfall at the test sites (100–150 mm/yr) is not sufficient even when runoff water is harvested. On

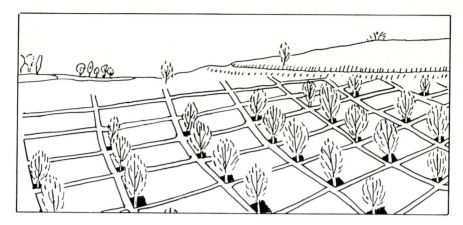

Triangular Microcatchment Construction

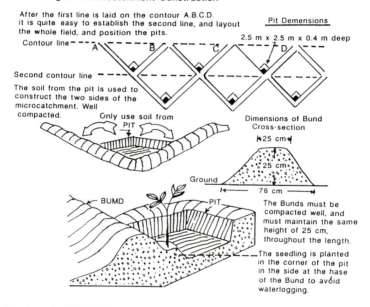

After the first line is laid on the contour A.B.C.D.
it is quite easy to establish the second line, and layout
the whole field, and position the pits.

Pit Demensions

2.5 m x 2.5 m x 0.4 m deep

Contour line

Second contour line

The soil from the pit is used to
construct the two sides of the
microcatchment. Well
compacted.

Only use soil from
PIT

Dimensions of Bund
Cross-section

25 cm

25 cm

Ground

76 cm

BUMD

PIT

The Bunds must be
compacted well, and
must maintain the same
height of 25 cm,
throughout the length.

The seedling is planted
in the corner of the pit
in the side at the base
of the Bund to avoid
waterlogging.

Fig. 6.8. A typical MCWH construction

the other hand, these studies have had a profound impact on the development of quantitative modern water harvesting science.

Major practical conclusions to be drawn from these studies are: (1) MCWH should be implemented in regions of larger annual rainfall, at least 250 mm/yr (Boers et al. 1986a); (2) an element of water harvesting should include a large contributing area (about 1000 m^2) and a cluster of several trees in the collecting area. Due to soil spatial variability, some MCWH elements contribute more water than consumption and others much less. A cluster of elements in which extra water from one would be transferred to the MCWH element with deficit water would improve yield (Ben-Asher and Warrick 1987); (3) when these approaches are adopted, MCWH can be economically feasible (Oron et al. 1983; Oron and Enthoven 1987); and (4) the main advantage of MCWH is the high specific runoff yield (15–90% of the rain) such that the ratio of collecting to contributing area is only 1:1 to 1:10. It can be practiced on moderate slopes (1–10%), where other water harvesting methods cannot be implemented.

6.6.1.2 Small Catchment Water Harvesting

Definition and Description. A typical SCWH system is shown in Fig. 6.9 after Evenari et al. (1982), and in Fig. 6.10. Catchment areas of the size of small watersheds were extensively studied by hydrologists (Haan et al. 1982), geographers (Kirkby 1978) and soil scientists dealing with soil erosion and conservation (El-Swaify et al. 1983). In this review we shall report only on studies which were oriented toward runoff irrigation. Reported sizes of SCWH elements are: in Avdat, Israel, an area of about 350 ha contributes runoff to an area of about 2 ha, such that the ratio of collecting to contributing area is about 1:175. About 200 existing SCWH systems are reported in the Negev Desert of Israel, with varying sizes of contributing areas from 1 ha to several hundred hectares (Ben-Asher 1988). Rain in the area is 100–250 mm/yr, and the agronomic results are acceptable. In Mali, Klemm (1990) has developed two SCWH elements. Their contributing area was 126.5 ha and the collecting area was 3.26 ha such that the ratio of collecting contributing areas was about 1:40, compared to Israel's 1:175. The reduced ratio was possible because rain in the region exceeds 500 mm/yr. As a result of water harvesting, the yield of sorghum was good (see Table 6.3).

Advantages of SCWH are: (1) the cultivated (collecting) area is a regular field with regard to planting space cultivation and agronomic maintenance. In contrast, MCWH spacing is large and cultivation of the field is relatively expensive; and (2) probability of floods is relatively large even in extremely arid areas of less than 100 mm/yr (Karnieli et al. 1988).

A quantitative approach suitable for SCWH engineering can be based to a certain extent on the models developed for small watersheds (Haan et al. 1982). In this monograph, Renard et al. (1982) counted 75 computer models that were ready for distribution. Most of them are suitable for predicting surface runoff but none of them are suitable for SCWH because they do not link runoff generation in the contributing area with evapotranspiration processes at the cultivated area. The only known model today in which these processes are linked is the model developed by Klemm (1990) which is an extension of the MCWH model of Boers et al. (1986a) to SCWH systems. It is a one-dimensional numerical solution of the water balance of cropped soil SWATER (Belmans et al. 1983) extrapolated to predict runoff on the basis of

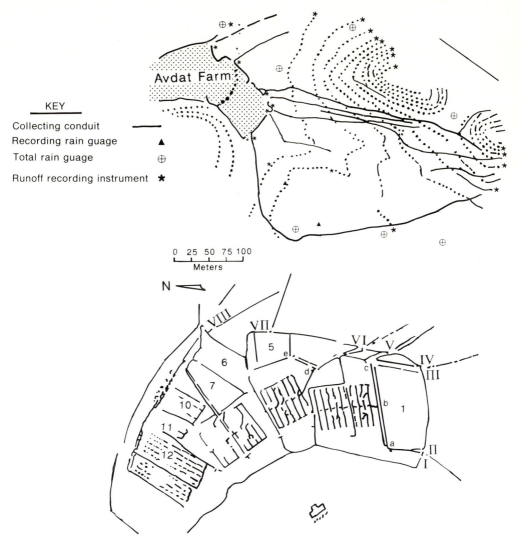

Fig. 6.9. Plan of Avdat farm showing runoff channels weirs water distribution system and terraced fields

daily rainfall data. The model was used to optimize seeding dates and for quantification of the required runoff water in order to determine the ratio between the cultivated and the runoff contributing areas.

6.6.1.3 Comparison of MCWH and SCWH Systems

With the progress in our understanding of the systems, several distinctions between MCWH and SCWH can be made. From a pedological or a

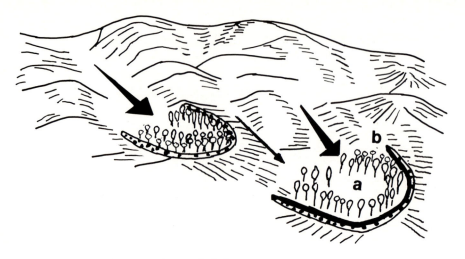

Fig. 6.10. A view of a liman close to Sede Boger

Table 6.3. Yields of several crops under water harvesting

Crop	Rain (mm)	Country	Yield with WH (kg/ha)	Yield without WH (kg/ha)	Source
Millet	300	Niger	600	N/A	Reij et al. (1988)
Cowpeas	300	Niger	600	N/A	Reij et al. (1988)
Almonds	140	Israel	400	N/A	Ben-Asher and Warrick (1987)
Sorghum	520	Mali	600	400	Klemm and Prinz (1989)
Sorghum	510	Botswana	1200	400	Carter and Miller (1991)
Pistachio	90	Israel	12300	1100	Boers et al. (1986a)
Soybean	–	India	2300	1500	Urkurkar et al. (1985)
Rice	–	India	2600	1500	Urkurkar et al. (1985)
Jojoba	560	India	1930	830	Sharma et al. (1982)

morphological point of view, MCWH occurs on a rather uniform surface, while in SCWH water flow from contributing to collecting area occurs on a very heterogeneous surface. It may commence on exposed bedrock, continue on stony watersheds and end up spreading on flat loessy soils.

From a hydrological point of view, SCWH differs from MCWH by the number of channels through which runoff is flowing before it reaches the cultivated area. In MCWH, a thin layer of water flows toward the basin on the surface, or in the first-order channels. In SCWH, flow may reach the cultivated area through the third, fourth or even higher order channels (see Figs. 6.9 and 6.10 for second and third order channels).

Another hydrological distinction is characterized by the effectivity of the partial area contribution (PAC) phenomenon. An MCWH element is not significantly affected by the PAC phenomenon, while the runoff efficiency of an SCWH element is affected. In MCWH, the flow path is too short to feel the effect on the cumulative runoff. The infiltration through the crust is also too small, compared to the infiltration rates through the boulders and the gravel in the channels of the SCWH systems. As a result, the changes in length of flow path combined with high infiltration rates are strongly pronounced in SCWH.

The two systems differ also in the nature of the rain distribution of the catchments. The cells of convective storms, which usually generate flash floods, are smaller than the size of the catchment (0.7–10 km in diameter), and rainfall is therefore not uniform over the area (Kutiel and Sharon 1980). Furthermore, the orientation of the slope (west or east) with respect to the direction of the storm also affects the increase in spatial variability of rain distribution over the catchment. On the other hand, an element of a MCWH system is wetted by a uniform rain, due to its small size.

The total amount of water that can be collected in the cultivated area of a single element is a few orders of magnitude larger than the expected water volumes on elements of MCWH systems. Thus, from an agronomic point of view, conventional agrotechniques can be practiced on an SCWH element but not on an MCWH element.

In both systems, however, water is stored in the root zone rather than in reservoirs behind dams; thus a hydrophysical definition of an SCWH element is based on all features described in the previous section. It can therefore be described as a runoff water harvesting system subject to large variability in soil, and hydraulic properties of the contributing area.

6.6.2 Agroforestry

The purpose of the collection of runoff water is to improve crop productivity. Surprisingly, this aspect has received comparatively little attention (Reij et al. 1988). There are very few reports on the consumptive use of water by crops in water harvesting schemes.

It is obvious that water harvesting is important in arid areas in which only meager crops would be obtained without supplementing rainfall with runoff. Arid and semi-arid lands (ASAL) occupy approximately one-third of the world's land surface. The sustained production of food, firewood and fodder is limited by the lack of an adequate water supply. Population increase has led to an increase in the demand for fuel for cooking and heating. The cutting of trees deprives the soil of its natural protection against the impact of rain drops, and leads to the formation of a surface crust. Water yield from the soil is therefore drastically reduced and the direct loss of water by evaporation from the soil surface increases due to the lack of cover. The result of these processes is that there is not enough water available in the soil profile to allow for tree regrowth. These processes are central to the process

Table 6.4. Total above-ground dry matter production after three years of growth for *Acacia salicina* (A) and *Eucalyptus occidentalis* (E) at dense (D-1250 trees ha^{-1}) and open 0–625 trees ha^{-1}) (Lovenstein et al. 1991)

Treatment	Yield (kg tree^{-1})	Yield (t ha^{-1})
AO	24.96 a	16.6
AD	15.36 b	19.2
EO	40.16 c	25.1
ED	22.72 a	28.4

Note: means followed by the same letter do not differ significantly at the 0.01 level.

of desertification. The feasibility of using runoff water for growing trees in arid areas was successfully tested by Zohar et al. (1988). Further studies (Lovenstein et al. 1991) indicated that relatively high yields could be obtained (Table 6.4) in an area in which the long-term mean annual rainfall is 115 mm.

The trees were harvested by cutting them slightly above the soil surface. The regrowth rate of the trees followed the same pattern as that of the original trees and the water use efficiency for the first year of regrowth was similar to the second year of growth. These results indicate that the harvesting of trees in a runoff system has no negative environmental effects as water is collected outside the plot irrespective of the activities within the plot. Moreover, crusting within the plot has no noticeable effect because water is ponded and only the time span for the infiltration process will be lengthened.

The water uptake patterns for the trees before harvest suggested that an intercrop could have been planted as water depletion in the upper soil layers was not complete. The suggested agroforestry system is depicted schematically in Fig. 6.11.

In such a system the effect of an intercrop and planting density of young saplings of *Acacia saligna* in an alley cropping design on their productivity was tested by Berliner and Rapp (1992). Their results (Fig. 6.12) indicated that the intercrop decreased the biomass production of the trees at all densities. The overall biomass production in the intercropped plots was much higher due to the higher productivity of the intercrop (*Sorghum sudanensi*).

Studies carried out presently focus on the quantification of the competition for water and solar radiation in mature agroforestry systems and their optimal management.

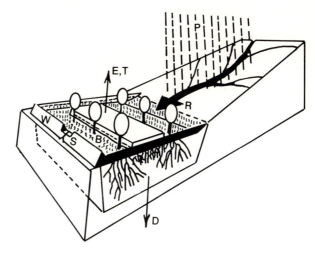

Fig. 6.11. Schematic presentation of an agroforestry system under runoff conditions. Precipitations (*P*) generate surface runoff (*R*) on hill slopes, which accumulates into natural tributaries feeding a leveled runoff catchment basin (*B*). The collected runoff water is trapped by a retaining wall (*W*) allowing the water to percolate into the soil profile (*D*). The stored water is transpired (*T*) by deep rooting trees and shallow rooting annuals, while losses by evaporation (*E*) and deep percolation are minimized. The build in spillway (*S*) controls surplus water

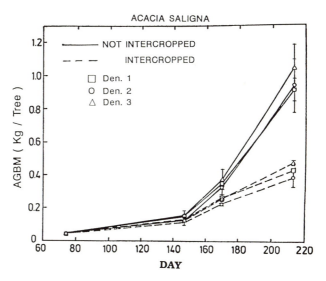

Fig. 6.12. Dry above ground biomass (AGBM) production of *Acacia saligna* at three densities (*den 1, 2,* and *3* distance in meters between trees in the row) intercropped and non-intercropped during the first growth year

6.7 Concluding Remarks

Rainfall is used more efficiently with runoff irrigation. This fact is especially important for arid and semi-arid countries where rain-fed agriculture is practiced. Runoff can be directly used on cultivated fields and water can be retained within the root zone. Reported data indicated that runoff irrigation (RI) can increase agricultural productivity two- to fourfold.

The total area under RI is not precisely known, but it probably ranges between 0.5 to 1.0 million ha; which is only a fraction of a percent of the world's total irrigated area. The rate of RI expansion, however, is about the same (0.5–1.0%/yr) as irrigated land.

The need to expand RI is obvious because it improves productivity in remote areas where it is most needed and funds least available. In these areas other alternatives such as surface or pressurized irrigation are not possible because of lack of infrastructure and means to develop it.

To improve and expand runoff irrigation projects, education and training of potential users are required, but most of all and more than ever the importance of research and development should be recognized.

Adaptive research for local needs, field trials and technology transfer deserve high priority especially in the following areas: (1) using remote-sensing techniques in (a) surveying suitable areas for runoff irrigation; (b) estimating the water balance of potential areas; (c) determining their hydraulic parameters; and (d) using available remote-sensing data input for real time runoff prediction; (2) developing better techniques for harvesting and storing the water; (3) developing new hydrological models which will include: available remote-sensing data, partial area contribution, and theories for crust formation; (4) formulating guidelines for runoff irrigation; and (5) investigating the improved productivity of agroforestry systems.

References

Agassi M, Beniamini Y, Morin T, Marish S, Henkin E (1987) The Israeli concept for runoff and erosion controlling semi-arid and arid zones in the Mediterranean basin. Misc Rep 30. Soil Eros Res Stn, Minist Agric, Israel

Ahuja LR, Swatzendruber D (1992) Flow through crusted soils: analytical and numerical approaches. In: Sumner ME, Stewart BA (eds) Soil crusting: chemical and physical processes. Advances in soil science. Lewis Publishers, London, pp 93–122

Aronson JA (1986) Runoff agriculture in arid and semi-arid lands. A review. Intern Pap, Ben-Gurion Univ Negev, Israel

Belmans C, Feddes RA, Wesseling JG (1983) Simulation model of water balance of a cropped soil: SWATER. J Hydrol 63:271–286

Ben-Asher J (1988) A review of water harvesting in Israel. World Bank Work Pap 2. World Bank Sub-Saharan water harvesting study

Ben-Asher J, Humborg G (1992) A remote sensing model for linking rainfall simulation with hydrographs of a small arid watershed. Water Resour Res (in press)

Ben-Asher J, Warrick AW (1987) Effect of variations in soil properties and precipitation on microcatchment water balance. Agric Water Manag 12:177–194

Ben-Asher J, Oron G, Button BJ (1985) Estimation of runoff volume for agriculture in arid lands. In: Singh A, Chawdhory GR (eds) Dryland resources and technology. Geoenviron Acad Jodphur 2:79–90

Ben-Hur M, Shainberg I, Bakker D, Keren R (1985) Effect of soil texture and $CaCO_3$ content on water infiltration in crusted soils as related to water salinity. Irrig Sci 6:281–294

Berliner P, Rapp I (1992) The effect of planting density and intercrop on the productivity of young *Acacia saligna* shrubs. Unpublished Report to PEF. The J Bleevstein Institute for desert research B.GU. of The Negev Sede Boger Campus, Israel

Betson RP (1964) What is watershed runoff? J Geophys Res 69:1541–1552

Beven KJ, Kirkby MJ (1983) Consideration in the development and validation of a simple physically-based variable contributing area model of catchment hydrology. J Hydrol 65:24–36

Blanchard BJ, Chang ATC (1983) Estimation of soil moisture from Seasat SAR data. Water Resour Bull 19(5):803–810

Boers ThM, Ben-Asher J (1982) A review of rainwater harvesting. Agric Water Manag 5:145–158

Boers ThM, de Graaf M, Feddes RA, Ben-Asher J (1986a) A linear regression model combined with a soil water balance model to design microcatchments for water harvesting in arid zones. Agric Water Manag 11:187206

Boers ThM, Zondervan K, Ben-Asher J (1986b) Microcatchment water harvesting (MCWH) for arid zone development. Agric Water Manag 12:21–36

Boughton WC (1987) Evaluating partial areas of watershed runoff. J Irrig Drain Engin ASCE 113:355–356

Brooks RH, Corey AT (1964) Hydraulic properties of porous media. Hydrol Pap 3, Colorado State Univ, Fort Collins

Burke HK, Schmugge TJ (1982) Effects of varying soil moisture contents and vegetation canopies on microwave emission. IEEE Trans Geosci Remote Sensing GE20(3):268–274

Burke HK, Schmugge TJ, Paris JF (1979) Comparison of 2.8 and 21 cm microwave radiometer observations over soils with emission model calculations. J Geogr Res 84(C1):287–294

Camillo PJ, Schmugge TJ (1981) A computer program for the simulation of heat and moisture flow in soils. NASA Tech Mem TM 82121

Carter DC, Miller S (1991) Three years experience with an on farm water catchment water harvesting system in Botswana. Agric Water Manag 19:191–203

Clayton LH, Neff EL, Woolhiser DA (1974) Hydrologic aspects of water harvesting in the Northern Great Plains. In: Proc Water harvesting Symp, Phoenix

Dodi A, Ben-Asher J, Adar E (1987) A parametric model for conversion of rainfall into runoff in small watershed in arid zones. A report. Ben-Gurion Univ Negev, 103 pp (Hebrew with English summary)

Eger H (1987) Runoff agriculture: a case study about the Yemeni Highlands. Reichert, Wiesbaden

El-Swaify SA, Moldenhauer WC, Andrew G (eds) (1983) Soil erosion and conservation. Soil Conserv Soc Am, 7515 Northeast Ankey Rd, Ankey, Iowa, pp 789

Engman ET, Arnett JR (1977) Remote sensing applications to a partial area model. NASA Rep, Contract NAS-5-233399, Rep NUS-3048, Goddard Space Flight Center, Greenbelt MD, 87 pp

Engman ET, Rogowski AS (1974) A partial area model for storm flow synthesis. Water Resour Res 10(3):464472

Engman ET, Jackson TJ, Schmugge TJ (1983) Implications of complete watershed soil moisture measurements to hydrologic modeling. IEEE Int Geo Sci Remote sensing Symp, San Francisco, 1(TA-1), pp 2.2–2.5

Evenari M, Shanan L, Tadmor N (1982) The Negev, the challenge of a desert. Harvard Univ Press, Cambridge, pp 427

Finkel HJ (1986) Semi arid soil and water conservation. CRC, Boca Raton

Finkel M, Gainey V (1989) A strategy for technical assistance to Turkana Kenya. In: Thomas DB (ed) Soil and water conservation in Kenya, pp 458–470

Foster GR, Neibling WH, Nattermann RA (1982) A programmable rainfall simulator. Pap 82-2570. ASCE, St Joseph MI, 49085

Giraldez JV, Ayoso JL, Garcia A, Lopez JG, Roldan J (1988) Water harvesting strategies in the semi-arid climate of SE Spain. In: van Hoorn JW (ed) Agrohydrology – recent developments. Elsevier, Amsterdam, pp 253–265

Grewald SS, Mittal SP, Agnihotri Y, Dubey LN (1989) Rainwater harvesting for management of agricultural droughts in the foothills of northern India. Agric Water Manag 10:309–322

Haan CT, Johanson HP, Brakensiek DL (eds) (1982) Hydrologic modeling of small watersheds. ASAE, 2950 Niles Rd, St Joseph MI, pp 523

Heilman JL, Moore DG (1982) Evaluating near-surface soil moisture using heat capacity mapping mission data. Remote Sensing Environ 12:117–121

Hirschi MC, Mitchell JK, Feezor DR, Lesikar BJ (1990) Microcomputer controlled rainfall simulator. Trans ASAE, pp 1950–1953

Hoffman GJ, Howell TA, Solomon KH (1990) Management of farm irrigation systems. Am Soc Agric Eng, p 1001

Humborg G (1990) Ermittlung und Darstellung des Wassererntepotentials zur Bemessung von Sturzwasserbewässerungsanlagen im Sahel unter Anwendung von Kenngrößen aus Fernerkundungssystemen. Diplomarbeit, Ins Wasserbau und Kulturtechnik, Univ Karlsruhe, Germany

Imbira J (1989) Runoff harvesting for crop production in semi-arid areas of Baringo. In: Thomas DB (ed) Soil and water conservation in Kenya. pp 407–431

Jackson RD (1982) Soil moisture inferences from thermal infrared measurements, vegetation temperature. IEEE Trans Geosci Remote Sensing GE-20(3):282–285

Jackson RD, Idso SB, Reginato RJ, Pinter PJ Jr (1981) Canopy temperatures as a crop water status indicator. Water Resour Res 17:1133–1138

Jackson RD, Cihlar J, Estes JE, Heilman JL, Kakle A, Kannemasu ET, Millard J, Price JC, Wiegand C (1978) soil moisture estimation using reflected solar and emitted radiation. In: Soil moisture Worksh, NASA Conf Publ 20703, Natl Oceanic and Atmospheric Administration, Greenbelt, MD

Jackson RJ, Rawls WJ (1981) SCS urban curve numbers from a Landsat data base. Water Resour Bull 17(5):857–862

Jackson TJ, Schmugge TJ (1981) Aircraft active microwave measurements for estimating soil moisture. Photogr Eng Remote Sensing 47(6):801–805

Jackson TJ, Schmugge TJ, O'Neill P (1983) Remote sensing of soil moisture from an aircraft platform In: IAHS Int Symp Hydrological application of remote sensing and remote data transmission, Hamburg

Jackson TJ, Schmugge TJ, O'Neill P (1984) Passive microwave remote sensing of soil moisture from an aircraft platform. Remote Sensing Environ 14:135–151

Janeau JL, Lamachere JM (1988) Caractérisation des principales surfaces élémentaires de la region d'Oursi, Programme d'évaluation préliminaire SPOT. PEPS 149, SPOT Oursi, Centre Natl d'Etudes Spatiales (CNES), SPOT IMAGE, Centre ORSTOM Ouagadougou

Jones WK, Carroll TR (1983) Error analysis of airborne Gamma radiation soil moisture measurements. Agric Meteorol 28:19–30

Karnieli A, Ben-Asher J, Dodi A, Issar A, Oron G (1988) Empirical approach for predicting runoff yield under desert conditions. Agric Water Manag 14:243–252

Kirkby MJ (ed) (1978) Hillslope hydrology. John Wiley & Sons, New York, 383 pp

Klemm W (1990) Bewässerung mit Niederchlagswasser ohne Zwischenspeicherung im Sahel. Inst Wasserbau und Kulturtechnik, Univ Karlsruhe, Germany

Klemm W, Prinz D (1989) Utilization of the factor "soil" through runoff farming – a case study from Mali. Natural Resources and Development, Institute for Scientific Co-operation, Germany, pp 76–88

Kutiel H, Sharon D (1980) Diurnal variations of rainfall in Israel. Arch Meteorol Geogr Bioklimator Ser A 29:327–395

Lane LJ (1985) Erosion on range lands: emerging technology and data base. In: Proc Rainfall simulation Worksh, Jan 14–15, 1985, Tucson

Levi GJ, Berliner PR, du Plessis HM, van der Wutt HvH (1988) Microtopographical characteristics of artificially formed crusts. Soil Sci Soc Am J 52:784–791

Lovenstein HM, Berliner PR, van Keulen H (1991) Runoff agroforestry in arid lands. For Ecol Manag 45:59–70

Morin J, Goldberg D, Seginer I (1967) A rainfall simulator with rotating disk. Trans Am Soc Agric eng 10:74–76

Morin J, Benyamini Y, Michaeli A (1981) The dynamics of soil crusting by rainfall impact and the water movement in the soil profile. J Hydrol 52:321–335

Mualem Y, Assoulin S (1989) Modeling soil seal as a non-uniform layer. Water Resour Res 25:2101–2108

Mualem Y, Assoulin S, Rohdenburg H (1990) Rainfall induced soil seal: C. A dynamic model with kinetic energy instead of cumulative rainfall as independent variable. Catena 17:289–303

Mutai SK (1986) Implementation of soil and water conservation in arid and semi-arid areas of southern Kiteri District. Eastern province. In: Thomas DB (ed) Soil and water conservation in Kenya. pp 243–253

Nakayama FS, Bucks DS (1986) Trickle irrigation for crop production. Elsevier, Amsterdam New York, pp 376

National Academy of Science (ed) (1974) More water for arid lands. Natl Acad Sci Washington, DC, 154 pp

Nieuwenhuis GJA (1981) Application of HCMM satellite and airplane reflection and heat maps in agrohydrology. Adv Space Res 1:71–86

Oron G, Enthoven G (1987) Stochastic considerations in optimal design of micro catchments layout of runoff water harvesting. Water Resour Res 23:1131–1138

Oron G, Ben-Asher J, Issar A, Boers ThM (1983) Economic evaluation of water harvesting in microcatchments. Water Resour Res 19:1099–1105

Pilgram HD, Cordery I, Baron BC (1982) Effects of catchment size on run-off relations. J Hydrol 58:205221

Price JC (1980) The potential of remotely sensed thermal infrared data to infer surface soil moisture and evapotranspiration. Water Resour Res 16(4):787–790

Price JC (1981) The contribution of thermal data in Landsat multispectral classification. Photogr Eng Remote Sensing 57(2):229–236

Price JC (1982a) Estimation of regional scale evapotranspiration through analysis of satellite-thermal-infrared data. IEEE Trans Geosci Remote Sensing GE-20(3):286–292

Price JC (1982b) On the use of satellite data to infer surface fluxes at meteorological scales. J Appl Meterol 21(8):1111–1122

Prinz D (1990) Soil and water conservation techniques for the Oved Mina watershed, Algeria. Intern Report Inst Wasserbau und Kulturtechnik, Univ Karlsruhe, Germany

Reij C, Maulder P, Begemann L (1988) Water harvesting for plant production. World Bank Tech Paper 91, Washington

Renard KG, Rawis WJ, Fogel MM (1982) Currently available models. In: Haan CT, Johanson HP, Brakensiek DL (eds) Hydrologic modeling of small watersheds. ASAE, St Joseph MI, pp 507–510

Rocheleau DWF, Field-Juma A (1988) Agroforestry in dry land Africa. Int Counc Res Agrofor (ICRAF), Nairobi

Schmugge TJ, Blanchard B, Anderson A, Wang J (1978) Soil moisture sensing with aircraft observations of the diurnal range of surface temperature. Water Resour Bull 14:169–178

Schultz GA (1986) Satellite data as input for long term and short term hydrological models. In: Hydrologic aspects of space technology. Proc Coca Beach Worksh. Fla. IAHS Publ 160, pp 297–306

Schultz GA (1987) Parameter determination and input estimation in rainfall-runoff modeling based on remote sensing techniques. Water for the future: hydrology in perspective. In: Proc Rome Symp, IAHS Publ 104, pp 435–438

Shainberg I (1992) Chemical and mineralogical components of crust. In: Sumner ME, Stewart BA (eds) Soil crusting: chemical and physical processes. Advances in soil science. Lewis Publishers, London, pp 39–53

Sharma KD (1986) Runoff behavior of water harvesting microcatchments. Agric Water Manag 11:137–144

Sharma KD, Pareek OP, Singh HP (1982) Effect of runoff concentration on growth and yield of Jojoba. Agric Water Manag 5:73–85

Urkurkar JS, Sastri ASRAS, Chandrawanski BR (1985) Rainfall harvesting for rain fed rice cultivation. Int Rice Newslett

USSR National Committee for International Hydrological Decade (ed) (1977) Atlas of world water balance. UNESCO, Paris

van de Griend AA, Engman ET (1985) Partial area hydrology and remote sensing. J Hydrol 81:211–251

Vicek J, King D (1983) Detection of subsurface soil moisture by thermal sensing: results of laboratory, close-range and aerial studies. Photogr Eng Remote Sensing 49(11):1593–1597

Yair A, Sharon D, Lavec H (1978) An instrumented watershed for the study of partial area contribution of runoff in the arid zone. Z Geomorphol (Berl) 29:71–82

Zarmi Y, Ben-Asher J, Greengard T (1982) Constant velocity kinematic analysis of an infiltrating microcatchment hydrograph. Water Resour Res 19:277–283

7 Efficient Use of Water in Rain-Fed Agriculture

E. RAWITZ and A. HADAS

7.1 Introduction

7.1.1 Definitions

7.1.1.1 Rain-Fed vs. Dryland Agriculture

In this chapter we shall consider "rain-fed agriculture" to be synonymous with "dryland agriculture", and deal with it in the context of crop production under the constraints typical of conditions in the semi-arid zone, the chief among them being inadequate and unpredictable precipitation (Hillel and Rawitz 1972; Brady 1988; Stewart 1989). While detailed definitions of semi-arid conditions may differ according to the classification system used, and thus also the estimates of the size of the area involved (Arnon 1972, 1980; Brady 1988), it appears that approximately half of the world's land area lies in the semi-arid zone, of which over 50% is in the developing world, most of whose inhabitants are very poor. The combination of projected population growth with prospects for further development of irrigated agriculture lead to the prediction that something like one half of the world's food production will have to come from dryland agriculture. The importance of optimizing the utilization of water resources in dryland farming is thus beyond dispute. Unfortunately, our ability to achieve substantial increase of crop production under these conditions has not been as great as it has been in irrigated agriculture.

Dryland farming can be highly productive even though the system is by definition extensive rather than intensive, and is naturally a more risky undertaking for the farmer. In order to maintain long-term high productivity under erratic precipitation patterns, a more complete, improved understanding of the interactive site-specific soil variability, cultural practices, and climatic resource inventory is a prerequisite for increased production, maintenance or even improvement of land use and fertility, while reducing risk elements in dryland farming (van Schilfgaarde and Rawlins 1983).

Adv. Series in Agricultural Sciences, Vol. 22
K.K. Tanji/B. Yaron (Eds.)
© Springer-Verlag Berlin Heidelberg 1994

7.1.1.2 Water Use Efficiency

The general concept of efficiency as a fraction, output per unit input, is subject to a number of interpretations in this context. Botanists have defined water use efficiency (WUE) as mass of dry matter production per unit mass of transpiration, which is equivalent to the inverse of the older concept of transpiration ratio. Agronomists have preferred the ratio of usable or marketable yield (which may be expressed as fresh weight) to unit mass of evapotranspiration (ET). Irrigationists have variously used the ratio of usable or marketable yield to total root zone water depletion, or to amount of irrigation water applied as indices of water use efficiency or of water application efficiency. Finally, economists may prefer to relate the value of the yield to the value of water input. Additional definitions may be possible, and the choice would depend to some extent on site-specific conditions and on the preference of the user.

7.1.2 Objectives of Dryland Farming

7.1.2.1 Subsistence Farmers

Subsistence farmers have no money to make investments, and their objective is to provide their family with the staple foods necessary for survival. They will strive to spread risk over several seasons by growing storable foods, and they will settle for stable, albeit low yields.

7.1.2.2 Farmers in a Market Economy

Provision of essential staples is the primary objective, but the farmer aims at having some marketable yield that can produce cash profit. This income can then be used for investment or for improving the standard of living.

7.1.2.3 National Economy (Institutional Interest)

This may be an increase in total food production and its effective distribution, not necessarily contingent on raising family income. Another national objective is, or should be, assurance of the sustainability of economically satisfactory farming practices.

7.1.3 The Relation Between Yield Production Objectives and Water Supply to Crops

Crop response to stored water and changes in its availability differs at various growth stages. The integral of crop responses is reflected in the final yield and its components (total vegetative and reproductive organ dry matter production). In order to describe and control crop development and yield production, it is appropriate to establish the relationship between the inputs and crop yield. This relationship is termed the "yield or crop response curve or production function". Establishing such a relationship facilitates analysis of the system efficiency, effects of tillage or crop rotation on the sustainability of a given crop, detect deterioration of crop performance under prolonged use of given dryland farming practices.

7.1.3.1 The Yield Response Curve

It has often been noted that dry matter production by many crop plants in the open field, where water is the limiting factor to growth, is a linear function of transpiration (de Wit 1958; Clempner and Solomon 1987, quoted by Helweg 1991). This has sometimes been applied to mean that yield production is likewise a linear function of transpiration, of evapotranspiration, or of water application. However, these situations can be quite different, and the above observation of linearity does not necessarily apply to them. Firstly, when the yield does not consist of the entire plant, but perhaps only of the reproductive organs, the basic linear relationship may not hold; furthermore, the plant's physiological activity may be manipulated so as to change the partitioning of metabolites in order to increase the economically important yield (Rawson and Turner 1982; Proebsting and Peretz 1986; Turner and Henson 1989), in which case again the linear relationship may not prevail. Even the existence of a linear relationship between yield production and transpiration is not necessarily proof of a functional relation between them. Only a tiny fraction of the transpired water is involved in the metabolic process, with most of the total transpiration merely passing through the plant at a rate dictated by atmospheric conditions. In fact, plants can thrive in a closed environment with a saturated or nearly saturated atmosphere with very low transpiration (Hillel 1971). Under open-field conditions, growth may well be a function of another factor, in this case carbon dioxide uptake from the atmosphere, which is affected by the same factors as transpiration. There is experimental evidence available that shows that even vegetative growth is not always a linear function of transpiration under water-limiting conditions (Rawitz 1969; Rawitz and Hillel 1969).

A crop yield response curve or production function is commonly determined by fitting empirically obtained data of resource input and resultant yields to a desired or selected function. The major problems involved are: (1) selection of a function, e.g., linear, polynomial, exponential or sigmoid function; and (2) estimation of its parameters which depend on the choice of independent variables (Hanks 1974; Stern and Bresler 1983; Helweg 1991). Choice of function is usually made on the basis of its intended use and quality of fit to the experimental data. It is commonly desired that the function be analytically differentiable at least to the second derivative if economic analysis is sought, whereas a linear form is acceptable if yield prediction is desired for a given seasonal mean rainfall. The general form of a "crop production function" (CPF) is as:

$$Y = f(P, D_n, F, I_j),$$

where P = seasonal precipitation, D_n = depth of wetting from each of n events, F = frequency of rainfall and I_j = amount of the jth nutrient or herbicide. Some of the CPF types used are that of Hanks (1974) based on de Wit (1958), which assumes a linear relation between water transpired or used by the crop and total dry matter or grain produced; an exponential "Mitscherlich" function used by Stern and Bresler (1983), a second degree polynomial (Zaslavski and Buras 1967), and composite functions (Shalhevet et al. 1981; Warrick and Yates 1987; English et al. 1990).

In the case where the input parameter is water used for yield production, then the derivative of the CPF with respect to water used yields the water use efficiency (WUE).

7.1.3.2 Water Use Efficiency vs. Water Use

Most importantly for agricultural application, only rarely is it possible to measure transpiration of a crop in the field, and the farmer is indeed more concerned with evapotranspiration or even with consumptive use. In such cases, the denominator of the WUE equation includes water losses, avoidable or otherwise, in addition to transpiration, and then the growth increment per unit water applied decreases as total water given increases and maximum yield is approached (English et al. 1990). When so much water is applied that growth is no longer limited by water, the slope of the production curve approaches zero, and eventually may become negative (Shalhevet et al. 1981; English et al. 1990; Helweg 1991).

7.1.3.3 Yield vs. Water Use Efficiency

Water use efficiency has been observed to increase with yield up to maximum yield (Hillel and Guron 1970; Hillel 1972; Shalhevet et al. 1981). This is inconsistent with the view that higher WUE may be obtained by imposing a suboptimal water supply on the crop (e.g., deficit irrigation). The apparent contradiction may originate in an imprecise use of the term WUE. In irrigation studies it is often of interest to evaluate the efficiency of irrigation water use. In a situation where some yield can be obtained from rain-fed farming provided a minimum threshold supply has been exceeded, any additional water supplied (e.g., by supplementary irrigation) results in a steep increase in yield, and this is associated with a high irrigation water use efficiency. However, if the total amount of water withdrawn from the soil by the crop is taken into account, the WUE may still be much lower than that associated with maximum or near-maximum yield.

7.1.3.4 The Effects of Increasing or Decreasing Consumptive Use (CU)

The above does not imply that water use efficiency may not be improved by eliminating avoidable losses, i.e., when water application exceeds ET. If it is possible to exercise any control over the water supply of a crop, then the objective is to decrease consumptive use (CU) if this is larger than the beneficial water use, and to increase CU if it is lower than beneficial use. In the context of open-field cropping as it is usually practiced, "beneficial use" would be taken as the sum of evapotranspiration, any essential leaching fraction, and possibly water applied to modify the microclimate, such as cooling the crop canopy or preventing frost damage. Unfortunately, in dryland farming the possibilities of manipulating the field water supply are very limited.

7.1.4 Constraints to Efficient Water Use

Dryland farming is essentially a rain-fed crop production system which is characterized by water deficiency which at times is coupled with low soil fertility. The major effort is naturally focussed on increasing WUE and avoiding or alleviating various constraints. In the following discussion various constraints to WUE will be considered and the need to (1) maximize the proportion of potentially available water used for transpiration by the crop; and (2) minimize water losses to runoff, evaporation, deep percolation and transpiration by weeds. This involves maintenance of proper soil-water characteristics, water retention in the soil, matching crop phenology and rooting habits to changes in water stored in the soil and precipitation pattern. In order to be able to meet these requirements there is a need to know these constraints, and manage properly the recommended sustainable cropping practices.

Various constraints to efficient water use have already been referred to above, and are listed below: (1) climatic factors: precipitation, temperature, humidity and evaporative demand; (2) soil infiltrability, evaporativity, layering and depth; (3) soil salinity: soil chemical constituents, salt amounts and distribution; (4) soil fertility: nutrient availability, micro and macro; (5) weed competition; (6) pests and diseases; (7) genetically determined crop yield potential; and (8) institutional and socio-economic factors.

The following discussion will include only points 1 to 5, and to some extent, point 6, whereas points 7 and 8 are beyond the scope of this chapter.

7.1.4.1 Climatic Constraints

Dry farming is not specific to any particular climatic region or precipitation regime. Dregne (1982), has listed the following precipitation distributions in particular: winter, summer, continental, and bimodal patterns associated with dryland regions.

In winter rainfall regimes, i.e., Mediterranean climate rainfall is concentrated in the cool period of autumn and winter, whereas no rainfall occurs during the rest of the year. Mean annual rainfall ranges between 200 and 600 mm. This pattern is typified by low evaporative demand during the rainy season, and thus soil water storage is relatively effective.

Summer rainfall patterns are generally associated with monsoon system activity during the hot summer months, with no rain falling during the rest of the year. Large rainfall amounts per event and high intensities occur commonly (Dregne 1982; Hatfield 1990).

Continental patterns are characterized by dispersed, local, intense rainstorms produced by thermal convection cells. Rain occurs mostly during the warm season, but may also fall in the other seasons. This pattern is subject to very large variations in monthly and annual rainfall amounts.

Bimodal patterns are characterized by two rainy seasons separated by a dry period. The duration and intensity of the rainy seasons vary from year to year.

Under such conditions of uncertainty, dryland farming must be aimed to cope with the negative impact of highly variable rainfall timing and amounts on crop production.

Temperature in the arid and semi-arid regions is not as variable as precipitation. Temperature variations affect evaporation rate from soil and crop, plant development and maturation, and when high temperature is coupled with low air humidity, water losses increase and may cause stress and reduce yields.

The water balance of an agricultural area is the net balance between precipitation, runoff, deep percolation, soil storage and evapotranspiration. According to UNESCO (1977), the ratio between precipitation (P) and evapotranspiration (ET) roughly delineates the severity of water deficit in various dryland farming zones. Various aridity indices are defined as follows:

Hyper-arid zone $P/ET \leq 0.03$

Arid zone $0.03 \leq P/ET \leq 0.20$

Semi-arid zone $0.20 \leq P/ET \leq 0.50$

Sum-humid zone $0.5 \leq P/ET$

These definitions are valuable in general terms for classification, but due to high variability in annual rainfall amounts typical of dry-farming regions, these indices based on mean annual rainfall are unreliable. Shorter-term seasonal rainfall patterns must be sought in order to identify the duration of "time windows" during which a crop can be grown with a given probability of successful yield. Such an approach was suggested by Reddy (1983a, b), Kanemasu et al. (1990) and Stewart (1989).

7.1.4.2 Soil Physical Properties

In its efforts to improve the P/ET ratio, dryland farming has directed special efforts to increase crop-available soil water by enhancing water infiltration into the profile, increase its storage, and to reduce evaporation. Cultural practices to improve infiltration include tillage operations that: (1) leave a coarse tilth on the soil surface; (2) create layers that cause the soil to self-mulch; and (3) leave a mat of crop residues on the surface to protect the surface tilth and to reduce direct evaporation (Greb 1979). Conservation of water stored in the soil from one rainy season to the next under fallow requires minimizing surface evaporation and deep percolation, and eradication of weeds (Greb 1979; Smika et al. 1986; Stewart and Steiner 1990; Whitman and Meyer 1990).

Since infiltration is to be kept as high as possible, any restriction due to crusting, existence of shallow plow pans or clay layers requires correction, possibly by plowing, chiseling or sweep plows. These measures may improve soil water storage if their effect is persistent either naturally or through the application of chemical stabilizers (Shainberg and Letey 1984). After infiltration and evaporation have been optimized, maximum exploitation of the stored water may require that soil resistance be modified mechanically to encourage proliferation of the root system, whose extent and density determine the final P/ET ratio and the efficiency of yield production. The efficacy of tillage operations for conserving water and improving soil water storage has been much debated in the literature. According to data presented in recent reviews there does emerge a consensus that tillage and mulching with water-stable aggregates or plant residues, applied prior to precipitation, can conserve water in the soil profile (Eck and Unger 1985; El Swaify et al. 1985; Steiner et al. 1988; Stewart and Steiner 1990; Unger 1990).

7.1.4.3 Soil Salinity

Many arid or semi-arid soils contain concentrations of soluble salts that have a negative impact on the efficiency of water use. In addition to the direct osmotic effect and possible toxicity of specific ions, soil salinity may have a deleterious effect on physical properties such as infiltrability, water rentention and aeration, especially if the soil is rich in exchangeable sodium. The surface soil is more easily dispersible by rain drops, and crusting and runoff will follow. Eluviation of sodium-affected clay particles may occur within the profile, and dense, impervious, sodic subsoil layers will form, restricting air and water movement and root exploration.

Reclamation of salt-affected soils under these water-deficient dryland conditions is very difficult, since removal of the excess salts by leaching is an essential part of the reclamation process. Gypsum, which is often applied in order to displace adsorbed sodium from the soil exchange complex, is itself a contributor of soluble salts. Unless it can be leached from the soil in the last stages of reclamation, it may thus further depress yields. Even if temporary relief may be obtained, this may also not be economical due to the cost of the amendments and their application (Bowen 1981; Bresler et al. 1982). Mechanical practices may in the long run be more beneficial.

7.1.4.4 Fertility and Nutrient Availability

Water use and uptake by plants depend on thorough exploration of the soil by the roots. Viets (1962) concluded that under limited water supply, management practices will determine yield, and thus soil fertility must be adjusted to the available water amount. Adequate soil fertility stimulates crop development and root proliferation. Well-developed plants can endure higher water deficits (Stewart and Steiner 1990) and reduce runoff due to earlier canopy closure and soil surface protection. However, at times all these advantages may come to naught, since such well developed plants may transpire more. Their root systems will explore the soil earlier in the season and exhaust the stored water by the time the crop arrives at a critical yield-forming stage such as grain filling (Stewart and Steiner 1990).

7.1.4.5 Weeds

Weeds compete with crop plants for light, water and nutrients, and especially under conditions of water deficit this leads to yield losses. The best way to avoid this would be to control or completely eradicate weeds prior to planting the crop. Any delay in weed control increases the probability of yield reduction. However, even if a field is weed-free at planting time, some will develop later, even from seeds that may have been dormant in the soil for several seasons (Radosevich and Holt 1984).

Weeds can be controlled by tillage, herbicides, and by crop rotation, singly or in combination. The best practice would be to destroy weeds by one of the operations needed to initiate or maintain the crop, but in many cases herbicides or additional tillage operations are required. The need to conserve as much water as possible for crop use makes weed control mandatory, even though its cost represents a loss of revenue.

Tillage operations help control weeds by deep burial of seeds, exposure of stolons to dessication by plowing, killing emerging weed seedlings or delaying their emergence. However, too many tillage operations may lead to increasing the risk of soil crusting, runoff and erosion (Unger 1990).

Herbicide application requires proper choice so as to kill weeds without damaging the crop. Some herbicides require incorporation by tillage to become effective, while the effectiveness of others may be impaired by plant residues. Operations using herbicides for weed control will thus often involve expenses additional to the actual cost of the chemicals (Radosevich and Holt 1984; Steiner et al. 1988; Unger 1990). These costs can be avoided or reduced if proper tillage and crop rotations are adopted.

7.1.4.6 Pests and Diseases

Some pests and diseases are carried over the dry, uncropped seasons and once a new, susceptible crop is established, become active again. Foci of infestation such as pathogen spores and pest eggs, etc., are often borne on crop residues, and if these are properly buried by tillage, their future incidence may be reduced. Crop rotation may improve control by denying hosts to these foci. Thus tillage systems designed to conserve water and eradicate weeds must also contribute to pest and disease control in order to reduce inputs and increase profitability.

7.2 Options for Improving Water Use Efficiency

7.2.1 Increase the Nominator of the Water Use Efficiency

Any management measure which will produce higher yield will have the effect of increasing the water use efficiency (WUE). Sinclair et al. (1984) have listed five options for improving the WUE: (1) biochemical modificiations; (2) control stomatal physiology; (3) improve the harvest index; (4) change the crop microenvironment; and (5) increase the fraction of transpired water out of the total amount of water lost during the growing season. Among these options, the first two are in the province of plant breeding and genetics and the third partly so, and will thus not be discussed here. The harvest index (HI), which is the ratio of usable reproductive organ yield to total dry matter production, is affected by sowing density and spatial distribution of the crop stand. The higher the density for a given amount of available water, or the lower the amount of available water for a given crop stand density, the lower will be the HI for a given variety. Changing tillage practices and stand density control the last two options mentioned above.

An increase in stand density will have two effects: (1) increased inter-plant competition for water, nutrients and light (the micro-environment); and (2) force an increase in rooting density of the crop stand and thus an improved WUE with respect to total dry matter production. Higher stand density reduces runoff, erosion, and direct soil evaporation and by denser stand rooting increases the fraction of the water balance that is transpired. The reduction of HI and increase in transpiration should be analyzed simultaneously so as to obtain the maximal value of their combined effect (Gardner and Gardner 1983; Stewart and Steiner 1990).

7.2.2 Decrease Denominator of Water Use Efficiency – Water Conservation

In addition to raising yield by management measures (control of nominator or numerator), WUE can be increased by minimizing root zone water losses (the denominator). This can be done by managing precipitation interception over the rainy season and the crop growing season, i.e., minimizing runoff, evaporation, and deep drainage while maximizing infiltration.

7.2.2.1 Eliminate Runoff Losses

Runoff will occur whenever the surface detention is full and rainfall rate exceeds soil infiltrability. Such conditions are exacerbated by: (1) poor soil tilth and structure management; (2) small surface retention (smooth surface, fine tilth); and (3) unstable surface soil structure and a tendency to form crust (no plant or residue cover, low organic matter content, high exchangeable sodium percentage).

Depending on the magnitude and uniformity of the land slope, runoff generated under these circumstances may take place either within field boundaries, and thus increase the soil water storage variability which at low rainfall amounts may lead to an increased yield, and for high rainfall to reduced yield (Zaslavski and Buras 1967); or, if runoff leaves the field and soil water deficit is not satisfied, a proportional yield loss is to be expected. These observations suggest that adequate surface residues or increased surface roughness (e.g., micro-basins, tied ridges) and aggregation to increase surface retention and infiltration, stabilization of surface soil structure to rainfall impact, may reduce runoff losses out of a field and also help in retaining a uniform soil water distribution.

7.2.2.2 Decreasing Evaporation Losses

Direct evaporation from the soil is a net loss to crop production. Soil tillage prior to rainfall and tilth stabilized against sudden wetting and the energy of raindrop impact have been found to reduce evaporation. However, many experimental results show that if tillage can be carried out only after the rainfall, then by the time it can be performed (after several days of drying), it is too late to substantially affect evaporation (Marshal 1959; Hadas 1974). Stewart and Steiner (1990) quote data claiming that only 45-47% of total precipitation was used for crop production, while 8–25% was lost by runoff. These observations and reviews also suggest the use of plant residues as water saving measures by reducing evaporation losses. In summary, any reduction of evaporation, runoff or deep drainage losses increases soil water storage and thus improves the potential for higher yields.

7.2.2.3 Limiting Deep Drainage

Crop development can not completely match precipitation interception patterns during the season. Under some conditions the soil profile might be fully wetted to its water-holding capacity to the whole potential root zone depth, and yet the roots may not explore the entire root zone volume. Even at maturity the stored water in the deeper soil layers is only partially exploited. Thus some of the water originally within the root zone may gradually percolate below its lower boundary. This water could be utilized only if rapid and dense root growth penetrates the soil deeply and utilizes the water before it is depleted by drainage. Forcing downward root proliferation by high-density sowing may be self-defeating in that stored water may be exhausted early in the growing season, causing stress damage towards the grain or fruit filling stages. Alternatively, drainage loss might be prevented by placement of an artificial impervious membrane at the lower root zone boundary (Erickson 1972). However, the cost of such a membrane would be prohibitive in the context of dryland farming. In some cases a similar effect may be achieved by choosing sites where natural soil layering impedes downward drainage. In sandy soils this problem of deep drainage is especially acute due to their limited water storage capacity. Thus incorporation of clay or organic colloids or residues into the profile may increase its storage capacity and thus improve WUE and yield. Altogether, however, not very much can be done to improve the field water regime by controlling deep drainage under dryland farming conditions.

7.2.2.4 Use of Anti-Transpirants

Water taken up by crop roots from soil storage is transpired from the crop canopy. Any measure, such as application of anti-transpirants may contribute to crop yield. However, transpiration also acts to cool the canopy. If transpiration is inhibited either by anti-transpirants or by water stress, the leaves will heat up and thermal damage may impair net assimilation of photosynthates and indirectly affect yield. Furthermore, anti-transpirants acting by effectively restricting water vapor flux through stomata will have a similar, though not necessarily proportional, effect on CO_2 uptake, thus restricting photosynthesis. Finally, the cost of applying anti-transpirants would at present disqualify their use in dryland farming.

7.3 Integrated Strategies for Efficient Water Use – Farming Systems

Many factors and management measures have been recognized as effective in improving WUE. In the course of daily and seasonal activities, the farmer has to combine many if not all of these factors into a systematic strategy or farming system. Some approaches to this task have been reviewed by Unger (190); El-Swaify et al. (1985), Lal (1991), Eck and Unger (1985), Steiner et al. (1988).

7.3.1 Conservation Tillage

The dryland farmer strives to minimize expenditures and to develop a system most suitable to his needs. "Conservation tillage" as a practice is one of the best means to conserve water under adverse supply conditions by using tillage practices as the principal management tool at his disposal. The term conservation tillage means any tillage sequence that minimizes soil and water losses, and consequently every region, crop and climate will require a somewhat different set of practices. Broadly speaking, the objectives are to achieve a soil surface with a high infiltrability and adequate retention storage that will retain these properties over extended time periods, and will provide a favorable seedbed and rooting medium for agricultural crops. In addition to specific tillage operations that produce the desired structural condition, achievement of these aims is aided by stabilizing and binding soil aggregates with natural or synthetic amendments; crop rotations that include plants with dense, fibrous root systems and that provide a protective surface cover against raindrop impact throughout the period when rainfall is expected; and microtopography that can withstand occasional overland water flow. In the interests of maintaining an open, friable soil structure as well as for economic considerations, it is desirable to accomplish these tasks with the minimum amount of implement traffic over the soil. Some of the above conditions are difficult to achieve simultaneously – for example, a thick mat of crop residues, which is an excellent protective mulch against raindrop impact, makes it difficult to plant seeds to uniform spacing and depth, thus a good seedbed is not achieved. Such a residue mulch might require turning under by plowing before seedbed preparation can be undertaken, and this in turn involves both more implement traffic and exposure of the land to rainfall when it is in a vulnerable state. Desirable as the objectives of conservation tillage appear to be, there are conflicting reports in the literature as to its efficacy. While there seems to be agreement on its ability to conserve soil and water under the proper conditions, there are reports, on the one hand, of enhanced soil water storage and improved yields, but on the other hand there are also reports of consistently lower yields under conservation tillage. In view of the need for special implements and general inconvenience of field operations, some farmers hesitate to undertake this farming method. If conservation tillage is acceptable as an ideal to strive for, there is still need for improved understanding and development of efficient techniques.

7.3.1.1 Contour Cultivation and Strip Cropping

This is one of the measures included under conservation tillage. Undulating landscapes are particularly vulnerable to runoff and erosion, and tillage and planting along the contour are effective in controlling or eliminating them. Under contour tillage, surface roughness is least along the contours, and greatest along the gradient of the land. Where some runoff may be unavoidable, the land is not worked exactly on the contour line, but along a "falling contour", guiding the runoff along a gentle slope to a protected outflow structure. Contour cultivation is often combined with terraces or benches which increase surface detention and guide excess water to the outlet in a controlled manner (Zaongo et al. 1988; English et al. 1990).

To meet the demands of farm management, strips of temporarily bare, fallow land are often included between vegetated strips in contour tillage systems. Even though rainfall during land preparation work may cause some runoff and erosion, this will be intercepted on the next lower cropped strip, thus minimizing the damage. While effective, contour farming requires more labor, is inconvenient, and requires maintenance. There are areas in the USA where contour farming had been established in the 1930s and 1940s, but as family farms were gradually consolidated into large commercial holdings, conservation practices were abandoned in favor of "rectilinear farming", which evidently gave a larger short-term return. A variation of strip cropping has been applied to the ancient practice of runoff farming in semi-arid and arid regions (Evenari 1961). When there is unlimited land but too little rainfall to produce any reliable yield, some of the land is "sacrificed" to act as a donor of runoff. Alternate strips of land are cropped or left bare, with the bare strips being treated to induce more runoff, which is guided to the next lower cropped strip. The ratio of contributing to receiving area must be adjusted to climatic and soil conditions so as to supply the receiving area with enough additional water to produce a crop. While this system may make subsistence farming possible, or provide some supplemental income, by itself this farming system is not economically successful. One of the main problems is that precisely when additional runoff is most needed, e.g., during a drought year, it is least available.

7.3.1.2 Modifying Soil Surface Micro-Relief to Increase Retention

While a rough-tilled, cloddy soil surface increases surface retention, this does not always solve the problem. On the one hand, if this treatment is used for rainy-season fallow, the clods may not be stable and will disintegrate under rainfall impact before the end of the rainy season, and the resulting surface condition will then be very vulnerable to runoff and erosion (Rawitz et al. 1983). On the other hand, if it is intended for the growing season, the cloddy surface makes a poor seedbed, and the clods may be destroyed by essential secondary tillage operations. A very effective alternative solution is to mechanically create a soil surface micro-relief so as to increase surface storage after the soil has been brought to the tilth desired for crop production. In ridge-and-furrow cultivated row crops this technique is variously called basin tillage, furrow-damming, or tied ridges, and consists of constructing small earth dams in the furrows at intervals of approximately 1–5 meters (Lyle and Dixon 1977). This adds several tens of millimeters of surface storage which fills up during excess rainfall, and infiltrates after cessation of rainfall (Rawitz et al. 1983; Morin et al. 1984). It also virtually eliminates runoff and erosion. The furrow dams can be constructed with simple and inexpensive tractor- or animal-drawn implements or even by manual labor, and the system is therefore suitable for a wide range of production systems. In some crops the furrow dams interfere with implement traffic during the season or at harvest and it is not always practical to erase the dams before sensitive operations and rebuild them. In such cases one must either leave untreated furrows where the tractor wheels will pass, or to confine the use of this system to fallow periods if this is relevant. A variation of the basin tillage system is pitting, which can be applied to smooth land surfaces under small grains or pasture. The surface storage added by pitting is appreciably smaller than that attainable by basin tillage. In either case, it is desirable to design the system so as to capture statistically predicted rainfall excess on a soil of given infiltrability and slope. The controllable design parameters are distance between basins and their depth. Steeper slope, lower infiltrability, and larger expected rainfall excess require higher storage capacity, thus deeper and more closely spaced basins. In practice, the range of dimensions is rather limited, with minimal spacing between dams being about 80–100 cm, and maximal depth about 30 cm.

7.3.1.3 Various Unconventional Tillage Principles and Implements

In many areas where crop stubble is used for grazing animals, the final cover consists of root crowns and very short stubble. Instead of disking or plowing, the surface soil layer may be conditioned with special implements such as chisel plows, sweep plows and rod weeders in order to produce an open and friable tilth while preserving whatever vegetative surface mulch there is, without disrupting soil macropores and the binding effect of fibrous root systems. Where these methods are appropriate to local conditions, their net effect will be increased porosity, favorable pore-size distribution, and improved persistence of favorable surface structure, expressed as higher infiltrability and increased water storage capacity. The success of these practices depends on maintaining the lowest possible gradient in the direction of tillage so that piping out will not become a localized, let alone a field-size problem.

7.3.1.4 Major Land Shaping for Runoff Control

There are numerous cases, especially in the less arid parts of the region of interest, where the above measures are not capable of controlling excess rainfall. On the one hand, the problem may be caused by fairly rare but large storms which cause most of the water and soil loss. On the other hand, in the monsoon climate of the semi-arid tropics such runoff-producing storms may be a regular feature of each rainy season. In order to utilize this excess by storing it in the soil, or to avoid erosion damage, there have evolved various farming systems based on terraces which both impound surface water and control its flow. Typical examples are the paddy system of southern Asia, and various conservation terraces of the US Midwest. In the former system water is impounded in level or almost level basins, which have been constructed on slopes ranging from nearly flat land to very steep hillsides. For rice cultivation a constant water depth is maintained in the basins, with overflow guided to the next lower basin. In the dryland system of the American Midwest water is detained by levees constructed along falling contours in such a manner that water is actually impounded only on the lower part of the terrace adjacent to the levee, with the excess being guided along the gentle slope of the levee to protected water ways for disposal by the surface drainage system. Whereas the basic unit areas of the systems described in the previous sections have typical dimensions of perhaps 1–5 m, the typical dimensions of terraces may be 3–15 m wide and several tens to over 100 m long. The construction and maintenance of modern terrace systems may involve both tillage implements (e.g., one-way plows, ridgers) and heavy earth moving machinery.

7.3.1.5 Residue Management

Crop residues covering bare soil reduce impact of rain drops and surface aggregate breakdown, crusting, runoff and erosion, and evaporation, while improving soil water storage by maintaining soil moisture under the mulch for longer periods. They thus improve emergence and establishment of crops (Smika and Unger 1986; Lal 1991). Residues can be left standing while the soil is tilled by sweep plow or rod weeder. This way stubble can retain wind-blown snow, reduce wind erosion and sand blowing. However, it is less effective against raindrop impact and evaporation. If residues are in the form of a mat, evaporation and erosion are better controlled, but sowing prob-lems may arise due to planter clogging. In some instances crop residues host hibernating pests or diseases, and careless management can lead to severe yield losses (Smika and Unger 1986; Unger 1990; Lal 1991).

7.3.2 Crop Rotations Including Fallow Periods

Crops differ in their response to tillage-induced soil conditions and water storage fluctuations caused by erratic rainfall. Farmers have therefore developed crop rotation practices by trial-and-error that aim at water conservation and some degree of weed, pest and disease control. In dry areas (200–350 mm mean annual rainfall) no more than one crop can be grown, and in years with lower than average rainfall, no crop can be produced at all. In these cases fallowing is adopted, and whatever plant material was produced until the decision was made is either grazed by livestock or used as a cover crop or green manure. Crop rotations maintain a change in crop succession to minimize carryover of pests, diseases and weeds, and attempt to maintain soil fertility. Most commonly a field is left in fallow for a year to store rainfall from one year to the next (Arnon 1972, 1980). However, while some increase in fertility seems attainable, the efficacy of dry season fallowing to preserve water is very much in doubt on the one hand, while on the other hand fallowed fields are more susceptible to runoff and erosion (Arnon 1972; Steiner et al. 1988).

It is common to extend the crop rotation and limit periods whenever more rain is anticipated or greater weed, pest and disease carryover is expected. Two to 3-year rotations of small grain, fallow, followed by grain or legume – fallow or grain – legume – fallow repeated in the same order are commonly used under 200–350 mm rainfall in the Mediterranean zone (Arnon 1972). In other regions with similar climate rotations such as grain – pasture (one or more years) – fallow can be used (Steiner et al. 1988).

The extent of fallow periods depends on the stability of the system productivity and on economics. Many factors are involved in choosing a crop rotation and inclusion of fallowing: soil water storage capacity, amount and variability of rainfall, choice of crops and their profitability, pressure of weeds, pests and diseases, requirement and cost of tillage operations.

It is obvious that each crop rotation and fallow combination needs to be tested over a long period in order to evaluate its sustainability. Information pertinent to a given location does not seem to be simply transferable to other locations without additional prolonged testing before adoption. Furthermore, due to low profitability of the system and the costs of developing it, investment and research in dryland farming have been minimal. Only large data bases developed by slow, painstaking dedicated work can alleviate this situation and help to develop new technologies to improve dryland farming, a task in fact adopted by the farmers themselves in their trial-and-error approach.

7.3.3 Conjunctive Use of Rainfall and Irrigation

Under dryland conditions, rainfall is relied upon to provide the crop water needs. Farmers try to time their tillage operations and planting dates to match the crop growth stages and evapotranspiration needs with existing soil water storage and the expected replenishing rainfall. It is obvious that when rainfall combined with residual soil water storage can not support high yields, some supplementary irrigation might become most profitable for the farmers.

Farmers will, and should, time their supplementary irrigation schedule according to: (1) the amount of water available to them; (2) the time "window" when irrigation water is available; and (3) the critical stages of crop development, in order to improve the efficacy of the intercepted precipitation and aim to get the highest production per unit land and unit of water used (English 1990; Bucks et al. 1990). The selection of criteria for irrigation scheduling is a moot question: Whether supplementary irrigation should be aimed to produce maximal yield on a small portion of the total area, or whether it is preferable to accept a lower yield per unit area but apply the limited water to a larger area. The choice is obviously affected by the uncertainty of future rainfall and by the higher costs of farming a larger area (tillage, seeds and fertilizers, herbicides, water application).

Farmers have the option of aiming either toward a schedule that will guarantee the optimal income (Mjelde et al. 1990), or toward a long-term, low profit, low risk practice. The range of optimal return is usually estimated by calculating the water quota where the marginal revenue from the yield, calculated from the crop production function, equals the unit price of water. However, one must always bear in mind the uncertainty involved in using long-term crop production functions and soil water storage data. It is thus difficult to predict water-crop yield relations, which makes each year's decision something of a gamble (English et al. 1990). An ingenious system was devised by Stewart et al. (1981) for supplementary furrow irrigation of row crops in a summer-rainfall area of Texas. The length of a field was divided into three zones of management intensity (fertilization, seeding rate), under basin tillage. Only the upper zone was given full irrigation, assuming zero rainfall. Tailwater resulting from this irrigation plus any rainfall, was intercepted by the two downslope zones. Thus not only were runoff losses eliminated, but the size of the area receiving full irrigation automatically adjusted itself to the (unpredictable) amount of rainfall, thus resolving the question of apportionment of limited irrigation water over an area larger than could be fully irrigated. The graduated management intensity took into account the decreasing probability of adequate water supply with distance from the water source. The overall result was a very efficient conjunctive use of rainfall and irrigation water.

7.3.4 Runoff Farming

In areas where natural rainfall is only marginally sufficient for dryland farming, nonarable areas or hillsides may be used for "water harvesting" to support an economically valuable crop. Such systems were used by ancient communities (Evenari 1961). The basic technique is simple. Rainfall runoff water from uncropped contributing areas is collected and diverted to infiltrate in adjacent cropped recipient areas. The ratio of contributing to recipient areas in the ancient systems ranged from 20:1 to 40:1. This ratio can be considerably reduced by the use of modern methods (Rawitz and Hillel 1973). The water distribution may be through a collecting reservoir or by direct delivery to level basins, with spillways controlling overflow from higher basins to the adjacent lower ones. Another method is to shape the landscape into microcatchments concentrating local runoff towards the low point of a small basin where a tree is planted (Shanan and Tadmor 1976), or to shape the land into alternating runoff contributing and receiving strips.

Since these techniques are totally based on precipitation, they tend to produce inadequate amounts of runoff water precisely when it is most needed – in drought years. While yield of the cropped area will be higher than the yield that would have been harvested had the whole area been cropped, absolute yields may be so low that they may render the system uneconomical unless conjunctive application of supplementary irrigation is possible (Hillel and Rawitz 1972).

7.4 Concluding Remarks

Successful rainfed farming systems require efficient, low risk management of the soil water regime and improved water use efficiency by crops. Various techniques were developed and tested, but no universal practice can be recommended. Specific agricultural systems have been developed for various combinations of crops, climate and soils. These include various tillage practices which promote water storage in the soil at low cost, and minimize evaporation losses and weed competition with the crop for soil water. Cropping practices were developed to time and coordinate sowing date and crop growing season with expected available precipitation and soil water storage.

Since the dominant feature of dryland agriculture is, and will continue to be, the erratic and scarce rainfall, it is implausible to expect any great degree of intensification of the dryland farming system. However, further development and improvement of current and new practices of soil and water conservation while sustaining the income of farmers, may be achieved by using probability analysis and forcasting of precipitation, coupled with versatile, simple yet effective tillage and cropping practices compatible with the sparse, erratic precipitation pattern at the lowest costs and risks for the farmer.

References

Arnon I (1972) Crop production in dry regions. Leonard-Hill, London, 762 pp

Arnon I (1980) Optimizing yields and water use in Mediterranean agriculture. In: Soils in Mediterranean type climates and their yields potential. Proc 14th Coll IPI, Sevilla, Spain, pp 311–317

Bowen H (1981) Alleviating mechanical impedance. In: Arkin GF, Taylor HM (eds) Modifying the root environment to reduce crop stress. ASAE Monogr 4, St Joseph MI, pp 21–60

Brady NC (1988) Scientific and technical challenges in dryland agriculture. In: Proc Int Conf Dryland farming, challenges in dryland agriculture – a global perspective, Aug 1988, Amarillo, TX, pp 6–12

Bresler E, McNeal BL, Carter DL (1982) Saline and sodic soils: Principles-Dynamic-Modeling, vol 10. Adv Ser Agric Sci. Springer, Berlin Heidelberg New York, 236 pp

Bucks DA, Sammis TW, Dickey GL (1990) Irrigation of arid areas. In: Hoffman GJ, Howell TA, Solomon KH (eds) Management of farm irrigation systems. ASAE Mongr St Joseph, MI, pp 499–548

Clempner G, Solomon KH (1987) Accuracy and geographic transferability of crop water production functions. In: Congr Irrigation systems for the twenty-first century, July 1987, Portland OR, pp 285–292

de Wit CT (1958) Transpiration and crop yields. Versl Landbouwk Onderz 64:69–88

Dregne HE (1982) Dryland soil resources. Sci Techonol Agric Rep Ag Int Dev, Washington, DC

Eck HV, Unger PW (1985) Soil profile modification for increasing crop production. In: Stewart BA (ed) Advances in soil science, vol 1. Springer, Berlin Heidelberg New York, pp 66–100

El-Swaify SA, Pathak P, Rego TJ, Singh S (1985) Soil management for optimized productivity under rainfed conditions in the semi-arid tropics. In: Stewart BA (ed) Advances in soil science, vol 1. Springer, Berlin Heidelberg New York, pp 1–65

English MJ (1990) Deficit irrigation: an analytical framework. J Am Soc Civil Eng (Irrig) 116:399–412

English MJ, Musick JT, Murty VVN (1990) Deficit irrigation. In: Hoffman GJ, Howell TA, Solomon KH (eds) Management of farm irrigation systems. ASAE Monogr, St Joseph MI, pp 631–663

Erickson AE (1972) Improving the water properties of sand soils. In: Hillel D (ed) Optimizing the soil physical environment toward greater crop yields. Academic Press, New York, pp 35–42

Evenari M (1961) Ancient desert agriculture in the Negev. Science 133:976–996

Gardner WR, Gardner HR (1983) Principles of water management under drought conditions. Agric Water Manag 7:143–155

Greb BW (1979) Reducing drought effects on croplands in the West Central Great Plains. Inf Bull 420, USDA, Washington, DC, 31 pp

Hadas A (1974) Effects of external evaporative conditions on drying of soils. 10th Int Soil Sci Soc (Moscow) 1:136–142

Hanks RJ (1974) Model for predicting plant yield as influenced by water use. Agron J 66:660–664

Hatfield JL (1990) Agroclimatology of semiarid lands. In: Stewart BA (ed) Advances in soil science, vol 13. Springer, Berlin Heidelberg New York, pp 9–26

Helweg OJ (1991) Functions of crop yield from applied water. Agron J 83:769–773

Hillel D (1971) Soil and water: physical principles and processes. Academic Press, New York London

Hillel D (1972) The field water balance and water use efficiency. In: Hillel D (ed) Optimizing the soil physical environment toward greater crop yields. Academic Press, New York, pp 79–100

Hillel D, Guron Y (1970) The use of radiation techniques in water use efficiency studies. Res Rep IAEA (Vienna) 36 pp

Hillel D, Rawitz E (1972) Soil water conservation. In: Kozlowski TT (ed) Water deficit and plant growth, vol 3. Academic Press, New York, pp 307–338

Kanemasu ET, Stewart JI, van Donk SJ, Virmani SM (1990) Agroclimatic approaches for improving agricultural productivity in semi arid tropics. In: Stewart BA (ed) Advances in soil science, vol 13. Springer, Berlin Heidelberg New York, pp 273–309

Lal R (1991) Conservation tillage for sustainable agriculture tropics versus temperate environment. In: Brady N (ed) Advances in agronomy, vol 42. Academic Press, New York, pp 85–197

Lyle WM, Dixon OR (1977) Basin tillage for rainfall retention. Trans ASAE 20:1013–1018

Marshal TJ (1959) Relation between water and soil. Tech Comm 50, Commonw Bur Soils, Harpenden, UK, 91 pp

Mjelde JW, Lacewell RD, Talpaz H, Taylor CR (1990) Economics of irrigation management. In: Hoffman GJ, Howell TA, Solomon KH (eds) Management of farm irrigation systems. ASAE Monogr, St Joseph MI, pp 461–493

Morin J, Rawitz E, Hoogmoed WB, Benyamini Y (1984) Tillage practices for soil and water conservation in the semiarid zone. III. Runoff modelling as a tool for conservation tillage design. Soil Tillage Res 4:215–224

Proebsting EL, Peretz J (1986) Plant response to methods of irrigation. In: Lasko AN, Lenz F (eds) Regulation of photosynthesis in fruit trees. NY State Agric Exp Stn, Geneva NY, pp 155–161

Radosevich SR, Holt JS (1984) Weed ecology: implication for vegetation management. John Wiley & Sons, New York, 312 pp

Rawitz E (1969) The dependence of growth rate and transpiration rate on plant and soil physical parameters under controlled conditions. Soil Sci 110:172–182

Rawitz E, Hillel D (1969) Comparison of indexes relating plant response to soil moisture status. Agron J 61:231–235

Rawitz E, Hillel D (1973) A runoff farming trial with almonds in the Negev of Israel. In: Hadas A (ed) Ecological studies, analysis and synthesis, vol 4. Springer, Berlin Heidelberg, New York, pp 315–324

Rawitz E, Morin J, Hoogmoed WB, Margolin M, Etkin H (1983) Tillage practices for soil and water conservation in the semi-arid zone. I. Management of fallow during the rainy season preceding cotton. Soil Tillage Res 3:211–231

Rawson MH, Turner NC (1982) Recovery from water stress in five sunflowers (Helianthus annuss L) on leaf area and seed production. Aust J Plant Phys 9:437–448

Reddy SJ (1983a) Agroclimatic classification of the semi-arid tropic. I. A method for the computation of classifactory variables. Agric Meteorol 30:185–200

Reddy SJ (1983b) Agroclimatic classification of the semi-arid tropic. II. Identification of classifactory variables. Agric Meterol 30:293–325

Shainberg I, Letey J (1984) Response of soil to sodic and saline conditions. Hilgardia 52:1–57

Shalhevet J, Mantel A, Bielorai H, Shimshi D (1981) Irrigation of field and orchard crops under semi-arid conditions. 2nd rev edn. Int Irrig Infor Cent Bet-Dagon, 132 pp

Shanan L, Tadmor NH (1976) Micro-catchment systems for arid zone development. Hebrew Univ Jerusalem, Center for International Agricultural Cooperation, Ministry of Agriculture, Rehovot, 129 pp

Sinclair TR, Tanner CB, Bennett JM (1984) Water use efficiency in crop production. Biol Sci 34:36–40

Smika DE, Unger PW (1986) Effect of surface residues on soil water storage. In: Stewart BA (ed) Advances in soil science, vol 5. Springer, Berlin Heidelberg New York, pp 111–138

Steiner JL, Day JC, Papendick RI, Meyer RE, Bertrand AR (1988) Improving and sustaining productivity in dryland regions of developing countries. In: Stewart BA (ed) Advances in soil science, vol 8. Springer Berlin Heidelberg New York, pp 79–122

Stern G, Bresler E (1983) Non-uniform sprinkler irrigation and crop yield. Irrig Sci 4:17–29

Stewart BA, Steiner JL (1990) Water use efficiency. In: Stewart BA (ed) Advances in soil science, vol 13. Springer, Berlin Heidelberg New York, pp 151–174

Stewart BA, Dusek DA, Musick JT (1981) A management system for the conjuctive use of rainfall and limited irrigation of graded furrows. Soil Sci Soc Am J 45:413–419

Stewart JI (1989) Mediterranean type climate wheat production and response farming. In: Proc Worksh Soil water crop/livestock management systems for rainfed agriculture in the Near East region, Amman, Jordan, 1986, pp 5–19

Turner NC, Henson IE (1989) Comparative water relations and gas exchange of wheat and lupins in the field. In: Kreeb KH (ed) Structural and functional response to environmental stresses: water shortage. SPB, The Hague, pp 293–304

UNESCO (ed) (1977) World map of desertification. UN Natl Conf Desertification A/Conf 74/2

Unger PW (1990) Conservation tillage systems. In: Stewart BA (ed) Advances soil science, vol 13. Springer Berlin Heidelberg New York, pp 28–68

van Schilfgaarde J, Rawlins SL (1983) Water resources management in a growing society. In: Taylor HM, Jordan WR, Sinclair TR (eds) Limitation of efficient water use in crop production. ASA, Madison, pp 517–530

Viets FG (1962) Fertilizers and the efficient use of water. Adv Agron 14:223–264

Warrick AW, Yates SR (1987) Crop yield as influenced by irrigation uniformity. Adv Irrig 4:169–180

Whitman CE, Meyer RE (1990) Strategies for increasing the productivity and stability of dryland farming systems. In: Stewart BA (ed) Advances in soil science, vol 13. Springer, Berlin Heidelberg New York, pp 347–358

Zaongo CG, Wendt CW, Hossner LR (1988) Contour strip rainfall harvesting for cereal grains production in the Sahel. In: Proc Int Conf Dryland farming on challenges in dryland agriculture – global prospective, August 1988, Amarillo, pp 242–244

Zaslavski D, Buras N (1967) Crop yield response to nonuniform application of irrigation. Trans ASAE 10:196–198, 200

Part III Problem Water Uses and Treatment

8 Irrigation with Saline Water

S.R. GRATTAN

8.1 Introduction

Crop production in arid and semi-arid regions of the world is dependent upon an adequate supply of suitable-quality water. A supply of water is considered adequate when sufficient quantities are readily available for irrigation throughout the season to meet crop-water needs. This depends not only on the absolute quantity of water available, but on the scale of irrigated agriculture imposed on a region. In areas where irrigated agriculture frequently encounters irrigation water shortages, emphasis is usually placed on methods of increasing water quantity (e.g., utilizing groundwater; Howitt and M'Marete 1991), rather than considering whether water demands placed on such arid regions have been too high to ensure a dependable long-term supply. In some extremely arid countries, however, development of new irrigation water supplies is necessary to maintain a stable food supply. In order to expand its water resource base, the country or region must be able to utilize poorer quality water. The quality of water is considered suitable for crop production providing it can be used alone or in conjunction with other water sources and can sustain economic yields over the long term. The actual quality of water that is suitable for irrigation, as outlined in Chapter 2, depends upon crop salt tolerance, site conditions, and management practices. This chapter emphasizes management practices that utilize saline water as either a sole source of irrigation water or in conjunction with other sources of higher-quality water.

8.2 Potential for Using Saline Water for Irrigation

Salinity problems and their impacts on arid land agriculture have been recognized for centuries. However, only within the past half-century have scientists and engineers devoted their careers to studying salinity in relation to physical, chemical, biological and/or management factors that affect soils, water, and plants. Consequently, much is now known about specific aspects of salinity (e.g., physical and chemical effects on the soil, physiological effects

Adv. Series in Agricultural Sciences, Vol. 22
K.K. Tanji/B. Yaron (Eds.)
© Springer-Verlag Berlin Heidelberg 1994

on the plants, and crop salt tolerance), but predicting the relative yield of a crop given information on water quality, the soil, the climate, and management remains difficult. This is not surprising since plant, soil, and atmospheric factors change over time and crop salt tolerance can vary depending upon crop age, soil conditions, temperature, humidity, and air pollution (Maas 1990).

8.2.1 Historical Evidence

Studies or practices that have been carried out in many arid regions of the world have demonstrated that water, conventionally classified as too saline for agricultural use, has been used successfully to irrigate many annual crops. Based on a worldwide survey of irrigation projects that used saline water, Shalhevet and Kamburov (1976) concluded that the water could be as high as 6000 mg/l TDS. Pillsbury and Blaney (1966) recommended that the upper limit for irrigation water should be an electrical conductivity (EC$_w$) of 7.5 dS/m (about 4800 mg/l TDS). Table 8.1 lists crops that have been grown successfully in different parts of the world using saline water. While crop yields in these cases may not all have been maximal, they provided an economic return. The EC$_w$ of these waters are in many cases far in excess of the maximum EC$_w$ (EC$_w$-threshold), still a particular crop can tolerate and sustain optimal yields, given several assumptions related to water management and crop behavior. The EC$_w$-threshold values reported in Table 8.1 were derived from soil-salinity threshold values (EC$_e$) reported by Maas

Table 8.1. Crops that have been grown successfully in various locations in the world using saline water

Crop	Location	EC$_w$	EC$_w$-threshold	Reference
		..........dS/m..........		
Alfalfa	Bahrain	3.3–5.0	1.3	Ayers and Westcot (1985)
	Colorado, USA	2.3–7.8		Miles (1977)
	Tunisia	1.3–4.7		Ayers and Westcot (1985)
Cabbage	Bahrain	3.3–5.0	1.2	Ayers and Westcot (1985)
Carrots	Bahrain	3.3–5.0	0.7	Ayers and Westcot (1985)
Cauliflower	Bahrain	3.3–5.0	1.9	Ayers and Westcot (1985)
Celery	Bahrain	3.3–5.0	1.2	Ayers and Westcot (1985)
Cotton	Israel	4.6	5.1	Frenkel and Shainberg (1975)
	Uzbekistan, USSR	7.8–9.4		Bressler (1979)
Onions	Bahrain	3.3–5.0	0.8	Ayers and Westcot (1985)
Peppers	Bahrain	3.3–5.0	1.0	Ayers and Westcot (1985)
Sorghum	Colorado, USA	2.3–7.8	4.5	Miles (1977)
	Tunisia	1.3–4.7		Ayers and Westcot (1985)
Tomato	UAE	2.3	1.7	Ayers and Westcot (1985)
Wheat	Colorado, USA	2.3–7.8	4.0	Miles (1977)
	India	15.0		Dhir (1976)

(1990) and from EC_w–EC_e relations reported by Pratt and Suarez (1990) for furrow-irrigated fields with leaching fractions (LF) of 0.15. Exactly how annual crops are able to maintain economic, and in some cases maximal yield utilizing irrigation water with an EC_w in excess of the EC_w-threshold will be discussed later in this chapter.

Less is known about utilizing saline water for irrigation of perennial crops. First, it is difficult to study perennials since their lifetime exceeds 1 year and the effects of saline water applied over 1 to 2 years may influence crop performance in subsequent years. Furthermore, because of this long time frame, it is difficult to assess the response of trees to salinity and account for changing salinity profiles, dormant periods, and seasonal weather changes (temperature, humidity, rainfall, and evaporative demand). Second, many perennial crops, particularly trees and vines, are susceptible to B, Na and Cl toxicities. Although certain rootstocks are effective at controlling the amount of toxic elements into the scion, necrosis can develop on the tips and margins of leaves as well as other tissue after these elements have accumulated over several years. If the problem becomes more acute, leaf drop can occur and reduce crop yields beyond what one would predict due to osmotic effects alone.

8.2.2 Limitations and Potential Long-Term Effects

The cases summarized in Table 8.1 indicate that irrigation water containing salt concentrations in excess of conventional suitability standards can be used successfully on many crops for several years. However, the long-term effects from these practices (one or more decades) are unknown.

The scale of management is a major factor in assessing these long-term effects. Most research is directed toward farm-scale management, but inputs (water, energy, chemicals) and outputs (drainage water, and products) at this level could affect neighboring farms, other water users (fishing, recreation, industry), or the ecology of the receiving water. This suggests the importance of considering both environmental and economic impacts on all scales before introducing irrigation enterprises into a region. Clearly more research is needed to study all interactions, particularly the disposal of agricultural drainage water.

In order to sustain irrigation with saline water indefinitely, it is important to maintain a favorable salt balance on the farm and at the regional level. Achieving a favorable salt balance requires adequate drainage and disposal. In conjunction with soil salinity measurement techniques that monitor changing conditions and irrigation management strategies that maintain the average rootzone salinity within acceptable limits, soils with adequate drainage help to avoid long-term salinization problems.

Other long-term effects on the soil, such as soil crusting, reduced water infiltration capacity, and accumulation of toxic elements, should also be considered. Soils that have a high potential for soil crusting generally have 15-30% clay and less than 1.5% organic carbon (Shainberg and Singer 1990). Crusting, reduced water infiltration, and reduced seedling emergence are of particular concern where soils are poorly structured and the irrigation water's sodium adsorption ratio (SAR_w) exceeds the permeability threshold corresponding to the EC_w. Water infiltration decreases as the exchangeable sodium percentage (ESP) increases and the irrigation water salinity decreases (Pratt and Suarez 1990). Conversely, irrigation with saline water dominated by Na may not affect soil permeability since increased soil salinity compensates for increased sodicity (Thellier et al. 1990). However, problems are most likely to occur during or after rainfall or irrigation with good quality water on a field that was previously irrigated with a saline-sodic water. Such problems were observed at a California experimental site where saline-sodic water (9000 mg/l TDS and $SAR_w = 30$) had been applied for 4 years (Rolston et al. 1988). In some cases, deterioration of soil properties during the rainy season from decreasing EC in the soil surface layer can be irreversible (Shainberg and Singer 1990). For a more in-depth review of this subject, see Emerson (1984) and Shainberg and Singer (1990).

As indicated above, many saline irrigation waters contain chloride and/or boron that can accumulate in plants (particularly trees and vines) to toxic levels that cause foliar injury and reduced yields. Boron can be particularly troublesome if irrigation waters contain more than 1 mg B/l depending upon the plant species and rootstocks. Plants can accumulate more boron when grown on coarse-textured rather than fine-textured soil since the higher adsorptive capacity of fine-textured soil maintains lower soil-water B concentrations (Keren et al. 1984). Furthermore, once in the soil, more leaching is required to remove excess B than to remove excess salts. However, it appears that native boron in the soils is more difficult to leach than B that has accumulated from the irrigation water (Keren and Miyamoto 1990).

Development of foliar injury due to specific ion effects can be accelerated when the foliage is wetted by sprinkler irrigation (Maas 1985). Under sprinkler irrigation ions accumulate in the leaves from foliar absorption as well as root absorption and translocation. Since root absorption is more selective than foliar absorption, any advantage obtained by selecting rootstocks that discriminate in the uptake and translocation of toxic ions can be lost when sprinkled with saline water.

For the reasons described above, use of saline water to irrigate tree and vine crops is not recommended, particularly since financial investments are large and long term.

Some saline waters, especially saline drainage waters that have been collected from irrigated soils derived from certain types of sedimentary rock, may contain trace elements that can be absorbed by the plant, posing a health threat to its consumer. Elements such as Mo and Se are readily absorbed by plants and can be toxic to animals and humans (Page et al. 1990). Trace elements are of particular concern since they can concentrate as they move up the food chain, a process called biomagnification. Some saline drainage waters from the San Joaquin Valley in California contain high concentrations of U and V in addition to Se and Mo (Bradford et al. 1990), but the extent to which these accumulate in forage and crops or in the soil is unknown.

The major limitation with trace elements in drainage water is disposal of the water, especially when the terminal body receiving this water is inland and subsequent evaporation concentrates the elements. This can create a major ecological hazard, as exemplified in the San Joaquin Valley of California (Tanji et al. 1986), and suggests that even though irrigated agriculture may be sustained on the farm level, it can produce adverse effects on other water users or the environment at the regional scale (van Schilfgaarde 1990).

8.3 Factors that Facilitate the Use of Saline Water

Before utilizing saline water for irrigation, it is important to consider such factor as: (1) the availability of irrigation water sources (both saline and non-saline); (2) site conditions and characteristics; (3) crop selection; and (4) management practices. Although these categories seem rather obvious and elementary, historical evidence indicates that they are not always considered in the planning and implementing of an irrigation enterprise. Tanji (1990) cited historical examples of societies either directly or indirectly traumatized or devastated by salinization, such as Mesopotamia (Iraq), the Indus Plain (India and Pakistan), the Viru Valley (Peru) and the Salt River Basin (USA).

8.3.1 Irrigation Water Sources

The extent to which a saline water can be used for crop production depends upon its quality and availability and upon the availability of a good quality water source. A thorough discussion of the water quality parameters that are relevant to crop production is provided in Chapter 2 of this book. Standard water sampling techniques and analyses (Chapman and Pratt 1982) are needed periodically to determine these parameters (e.g., EC_w, SAR_w, and B).

Water availability is equal in importance to water quality, particularly if the water supply (either saline or non-saline) is not available or dependable at certain times of the year. The use of saline water is facilitated when an adequate supply of good quality water is available, at least during the early part of the season, and when saline water of suitable quality is available and accessible during the latter part of the season.

8.3.2 Site Considerations

Site selection is probably the single most important factor determining the
long-term sustainability of irrigated agriculture in a region. Good on-farm
water management is eventually negated if there is no natural or constructed
drainage outlet to the ocean. On the other hand, irrigation that continued
uninterrupted for thousands of years in the Nile delta (van Schilfgaarde
1990) was probably more directly related to location than to on-farm water
management. Before construction of the Aswan Dam, delta soils were
subject to periodic flooding, which not only resupplied nutrients to the soil
from deposited sediments but also naturally leached accumulated salts from
the soil to the Mediterranean Sea. Unfortunately, irrigation enterprises have
historically been developed in countries based on convenience and/or
political factors and rarely, if ever, were potential long-term problems
considered that affect agriculture, other water users, or the environment (van
Schilfgaarde 1990). Irrigation in closed basins, without an outlet to the
ocean, is a classic example of exceptional soil susceptibility to salinization. In
the valley soils within closed basins it is common to find restricting layers
(hard pans, clay pans, etc.), which, if they are close to the soil surface, may
cause waterlogging and subsequent salinization within decades (Tanji 1990).
On the other hand, if there are no restricting layers and the vadose regions
has a large capacity to store salts, it may be centuries before drainage
problems occur.

There are obvious limitations for utilizing saline water for irrigation of
crops grown in soils with a high saline water table. In this case, an increased
leaching fraction will cause the saline water table to advance in the root
zone. In situations where drainage is impossible due to environmental
and/or political factors, fields must be irrigated with good quality water as
efficiently as possible to avoid excess drainage.

Artificial drainage is one method of controlling both the water table and
soil salinization, but this requires transportation of the saline drainage water
from the region. In closed basins, an inland body of water is often the
terminus (e.g., Dead Sea, Israel; Great Salt Lake, Utah; Stillwater Reservoir,
Nevada; the Salton Sea, California; and the Kesterson Reservoir, California).
In these cases, depending upon the inputs and capacity of the terminal body
of water, salts and other constituents will concentrate over time from
evaporation. Eventually, such bodies will become excessively saline and lose
biological and recreational values (van Schilfgaarde 1990).

Installation of drains does not in itself control regional scale salinity. In a
case reported by Wichelns et al. (1988) as cited by Hoffman (1990), growers
in the Broadview Irrigation District in California installed drains in over
80% of the 4000 irrigated hectares within the district. They were installed
over a period of time after good quality water ($EC_w = 0.5$ dS/m) became
available to the district in 1957. Unfortunately no drainage outlet was
available until 1983, when the mean concentration of salts in the drainage

Table 8.2. Change in the area and yield of cotton and tomato in Broadview Water District compared to Fresno County

	1968 to 1972		1978 to 1982		Change (%)	
	Broadview	Fresno County	Broadview	Fresno County	Broadview	Fresno County
Cotton						
Area (ha)	715	76,000	2,130	167,000	+ 200	+ 140
Lint yield (kg/ha)	1,190	1,190	1,305	1,190	+ 10	0
Tomato						
Area (ha)	605	8,150	170	16,300	− 70	+ 100
Fruit yield (Mg/ha)	72	50	59	64	− 18	+ 28

water had increased to 2800 mg/l TDS. Although fields were leached, the salts in the drainage water were mixed with irrigation water and reapplied to the land. As irrigation water salinity increased over the years, the cropping pattern in this district shifted from moderately salt-sensitive tomato to salt-tolerant cotton (Table 8.2).

Recently, concern over the disposal of drainage water has extended beyond salts, nitrates and pesticides to trace elements. The recent discovery of selenium and other trace elements at the Kesterson Reservoir in the San Joaquin Valley in California has increased awareness of the external effects of irrigated agriculture (Tanji et al. 1986). Irrigated areas that naturally contribute large amounts of trace elements to drainage effluents may be forced out of production to reduce adverse environmental impacts on receiving waters.

Ideally, the most conductive site for irrigation with saline water is one where: (1) good quality water is also available for irrigation; (2) the land is not located within a closed basin but near the sea to reduce effects of drainage on receiving waters; and (3) the soils are well drained and do not contain unusually high levels of trace elements.

8.3.3 Crop Selection

Crop selection is an important management decision for maximizing profits. The most desirable characteristics in selecting a crop for irrigation with saline water are: (1) high marketability; (2) high economic value; (3) ease of management; (4) tolerance to salts and specific ions; (5) ability to maintain quality under saline conditions; (6) low potential to accumulate trace elements; and (7) compatibility in a crop rotation (Grattan and Rhoades 1990).

However, no crop is outstanding in all these categories. Economic value per crop area, for example, is negatively correlated with crop salt tolerance

(Grattan and Rhoades 1990), and many high-value crops (e.g., peach, plum, lemon) are sensitive to specific ions. Nevertheless, the enterprise must balance all these factors and select crops with those qualities that are most desirable for a given set of conditions.

Another factor in crop selection is seasonal water requirements. If saline water is being considered for irrigation, this usually implies that there are shortages in good quality water. Therefore, any means of reducing crop water requirements by selecting cultivars with a shorter life cycle or those that can be planted at times of the year when evaporative demands are lower are particularly desirable. Since saline conditions reduce both plant growth and seasonal evapotranspiration (ET), it is important to develop information on crop-water production functions under saline conditions (Letey et al. 1990).

Many factors that facilitate the use of saline water are related to management alone. For example, bed configuration (Bernstein and Fireman 1957; Kruse et al. 1990) can influence where salts accumulate in soil, which can effect young seedlings. Fertilization practices may be modified, particularly if saline drainage water with high levels of NO_5 is used for irrigation.

Irrigation water management factors that facilitate the use of saline water are extremely important, and the following section of this chapter is devoted entirely to that subject.

8.4 Management of Saline Water to Optimize Crop Performance

The extent to which saline water affects crop performance depends not only upon its concentration and composition but upon how it is managed. The degree of concentration of saline water that can be used for a particular crop also depends upon the availability of good quality water.

8.4.1 Crop Salt-Tolerance Expressions

Usually the salt tolerance of a crop is described by plotting relative yield as a function of soil salinity. For most crops, this response function is sigmoidal, and non-linear models (van Genuchten 1983) fit the data better than the two-piece linear response function first introduced by Maas and Hoffman (1977). One non-linear expression described by van Genuchten (1983) takes the following form:

$$Yr = \frac{Ym}{1 + \left(\dfrac{C}{C_{50}}\right)^P},$$

where Yr is relative yield, Ym is maximal yield under non-stressed conditions, C is the average root zone salinity, C_{50} is the average root zone salinity where yields are at 50% Ym, and P is an empirical constant that depends upon the crop and the conditions under which the experiment was conducted. This expression is advantageous in crop simulation modeling since it accurately describes the sigmoidal growth response by plants to salinity (Maas 1990). However, it is not as useful as the two-piece linear response function by Maas and Hoffman (1977) for providing general guidelines on relative salt tolerance and/or for assisting in saline water management decisions. The two-piece linear response function originally introduced by Maas and Hoffman (1977) takes the following form:

$$Yr = 100 - b(EC_e - a),$$

where Yr is relative yield expressed as a percentage, EC_e is the mean electrical conductivity of the saturated-soil extract in the crop root zone expressed in dS/m, "a" is the yield salinity threshold (dS/m) for a given crop, and b is the slope expressed in percent yield per dS/m.

The yield salinity threshold values of "a", reported by Maas (1990), are often used as standards to quantify the maximum salinity in the irrigation water that can be tolerated by a crop without a reduction in yield (Ayers and Westcot 1985). Although this is a logical approach, these threshold values reflect the response of crops to the average root zone salinity from the time seedlings were established until harvest. Most experiments that generated the yield threshold values were conducted using high leaching fractions (e.g., near 50%) to minimize spatial and temporal changes in soil salinity. Experiments such as these create uniform salinity profiles which facilitate data interpretation, but which rarely duplicate field conditions. The salt distribution under field conditions can vary from a uniform salinity profile that remains fairly constant over time to highly irregular profiles where the salinity at the bottom of the root zone is several times that at the soil surface. If there is a saline water table near the soil surface (e.g., less than 1.5 m), the upward movement of saline water by capillary flow will produce an inverted salinity profile.

Attempts have been made to understand how plants respond to non-uniform salinity profiles that change over time. Despite substantial controversy, most evidence indicates that crop growth is more closely related to the average soil-water salinity where most of the soil-water extraction occurs (Hoffman 1990). Given that root depth and density change over time and soil salinity changes spatially and temporally, it is understandable that it is difficult to model or predict plant response to these "real-life" conditions.

For most annual crops, sensitivity to soil salinity often changes from one growth stage to the next (Pearson and Bernstein 1959; Maas et al. 1983; Maas and Poss 1989). Although salinity can delay germination, most crops are tolerant during germination, less so during seedling emergence, and again more tolerant at later growth stages (Maas 1990).

However, caution is advised when concluding that plants are relatively salt-tolerant during the germination stage. Many experiments, on which this conclusion is based, were conducted in the laboratory using petri dishes and filter paper saturated with various salt solutions. In the field, the germinated seed is subjected to additional stresses such as temperature extremes, water deficit, and mechanical impedance. Therefore, seedling emergence is dependent upon how effectively the newly germinated seed can tolerate a combination of these stresses. This is an important research area that has been surprisingly neglected in the past and deserves more attention in the near future.

Based on the discussion above, an annual crop could be expected to tolerate a more saline water if this water were applied later in the season, since: (1) the salt-tolerance of most crops increases as the crop matures; and (2) applying water later in the season reduces the crop's salinity exposure.

8.4.2 Utilizing a Single Source of Water

When only one source of poor-quality water is available for irrigation, crop production is limited by the extent to which rainfall can leach salts from the upper part of the profile and by how the irrigation water is managed.

Leaching is necessary to sustain crop production over time. The amount of leaching required depends upon the crop, the salinity of the irrigation water, soil characteristics, climate, and management (Hoffman 1990). Therefore, depending upon the crop and the water salinity, a fraction of the water that infiltrates the soil surface must drain below the root zone (i.e., leaching fraction). In the simple form the leaching fraction (LF) can be written as:

$$LF = \frac{Vd}{Va},$$

where Va and Vd are volumes of water applied and drained, respectively. In water management, the volume terms are usually replaced by depth terms, and LF is equated to the inverse relationship when quantity terms are replaced by quality terms (i.e., EC or chloride concentrations). Conceptually, the leaching fraction is important; practically, it is difficult to measure. In a grower's field, particularly under surface irrigation, Va is not always easy to measure, and the desired LF is often less than 1-AU (where AU = application uniformity).

The difficulty in measuring Vd is more obvious. One has to estimate Vd based on evapotranspiration estimates, water infiltration, and soil-water retention properties. These estimates alone introduce errors and uncertainties greater than the numerical value of the LF desired.

The leaching fraction concept also ignores time. The time it takes for water to enter the soil surface and exit the root zone depends upon water infiltration characteristics and the depth of the root zone. In some coarse-textured soils with adequate infiltration and drainage, substantial leaching can occur after each irrigation, especially with shallow rooted crops. In many other soils, there is no leaching where water infiltration is low in the middle-to-latter portion of the season and root depths and ET are maximal. For example, in a field experiment with barley, the combination of increased rooting depth and decreased water infiltration over the season decreased the estimated LF as the season progressed (Aragüés et al. 1992). In such cases, a leaching fraction must be considered on an intermittent or seasonal basis. Intermittent leaching can be more effective at leaching salts than continuous leaching, i.e., after each irrigation (Shalhevet 1984).

Since crops respond, for the most part, to the average root zone salinity where most of the water extraction occurs (Hoffman 1990), relationships that predict average root zone salinity based on LF and salinity of the applied water are useful. This relationship, however, depends upon the crop-water extraction pattern in the root zone. One proposed extraction pattern assumes that the plant extracts 40, 30, 20, and 10% of the available water from the upper to lower quarters of the rootzone (Rhoades 1982). Another approach assumes an exponential uptake function (Hoffman and van Genuchten 1983) where the extracting pattern is closer to 78:14:6:2.

The relationships between mean root zone salinity and salinity of the applied water at various LFs are shown in Fig. 8.1. The solid line assumes the 40:30:20:10 extraction pattern (Rhoades 1982) while the broken line assumes the exponential pattern proposed by Hoffman and van Genuchten (1983). Although the root-water extraction pattern influences the relationships, the differences decrease substantially when characterizing only the root zone depth in which 90% of the active roots extract water. This is a reasonable approach in irrigation water management, where the lower half of the profile would be neglected under the exponential pattern since it represents less than 10% of the active roots. The upper half, representing more than 90% of the active roots, closely follows the 40:30:20:10 extraction pattern.

The solid-lines relationships in Fig. 8.1 developed by Rhoades (1982) assumes that the crop is irrigated by conventional methods (e.g., flood or furrow) where considerable drying occurs in between irrigations. Under high frequency irrigation practices, such as those used with drip irrigation, the matric potential is maintained at a higher level, which increases the utility of water at a given EC_w. Therefore, the slopes of the relationships under high-frequency irrigation at a given LF are less than those developed under conventional irrigation practices (Rhoades 1982).

The relationship in Fig. 8.1 can be used as a first approximation in making irrigation management decisions. For example, suppose a grower has an unlimited supply of water with an EC_w of 2.0 dS/m and intends to use this to furrow-irrigate tomatoes. According to Fig. 8.1, the grower would need to maintain a leaching fraction of 0.20 in order to ensure that the root zone does not exceed the salinity threshold for tomatoes (i.e., EC_e = 2.5 dS/m).

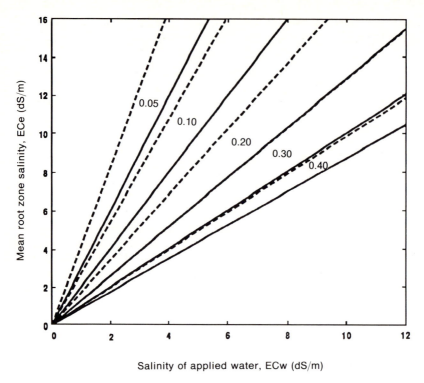

Salinity of applied water, ECw (dS/m)

Fig. 8.1. Relationship between soil salinity (EC_e) and salinity of the applied water (EC_w) using various LFs. The solid and broken lines are relationships developed by Rhoades (1982) and Hoffman and Van Genuchten (1983) assuming a 40:30:20:10 root-water extraction pattern and an exponential root-water pattern, respectively

This can be used as a first approximation only since, under field conditions: (1) LF cannot be readily measured; (2) plant tolerance to salinity can vary depending upon age and environmental conditions; (3) LF throughout the field may vary with soil variability and irrigation non-uniformities; (4) many soils, particularly fine-textured soils, can have low infiltration rates, making it impossible to achieve steady-state conditions or leaching, except during periods of low ET; and (5) many crops are grown over shallow water tables, making excessive leaching undesirable.

Research efforts continue to strive for increased irrigation uniformities, improved estimates of ET and water infiltration, and satisfactory management of highly variable soils. Only by reaching these objectives it is possible to bridge the gap between theoretical LF and achievable LF.

8.4.3 Utilizing Waters From Multiple Sources of Variable Quality

The utility of saline water is substantially increased when a source of good quality water is available for irrigation. In these situations, the salinity of the root zone is not determined solely by the management of a single source of saline water. Transient salinity profiles are intentionally created when multiple sources of water are available. This situation allows the grower to take advantage of crops of varying tolerance within a rotation from season to season (Rhoades 1984; Rhoades et al. 1988, 1989) or of the changing tolerance of a moderately salt-sensitive crop within a season (Pasternak et al. 1986; Grattan et al. 1987; Shennan et al. 1987).

There are two general techniques for utilizing saline water in conjunction with other available supplies (Grattan and Rhoades 1990). The first is to blend irrigation water supplies either before or during irrigation events to dilute the saline fraction of the water. The other is a cyclic technique where irrigation water of different qualities is cycled or alternated during different seasons or different times within a season. Both techniques require that at least two sources of water are available for irrigation. At least one must be low in salts while the others can be saline within limits (i.e., less than 6000 mg/l TDS) (Grattan and Rhoades 1990). Neither technique is mutually exclusive; in fact, both may be used.

8.4.3.1 Blending Irrigation Water Supplies

This is the most common practice for utilizing waters of various qualities. Blending involves mixing two or more supplies of water of different qualities to achieve a water of suitable quality. Dinar et al. (1986) described a method to calculate optimal blending ratios of saline and non-saline water. By adopting this technique to blend waters utilizing the same proportion for each irrigation, the grower would produce a single source of water of intermediate quality. In this case, the grower would follow the same management principles as those used for a single source of saline water.

The blending practice is usually adopted to expand the existing water supply. However, in some cases growers blend water as a temporary means of disposing of drainage water. Wichlens et al. (1988) reported that some growers in the San Joaquin Valley of California were forced to blend their drainage water with good quality irrigation water for over two decades, until a drainage outlet was finally available. Beyond improving irrigation efficiencies, blending drainage water with good quality water was the only option available for managing drainage effluents. The fraction of drainage water utilized increased from zero in the early 1960s to about 0.5 by the early 1980s.

Utilizing the blending method requires a controlled mixing of the water supplies. Two blending processes were discussed by Shalhevet (1984): network dilution and soil dilution. With network dilution, a facility must be constructed that allows waters to be blended in specific proportions within the conveyance system. With soil dilution, the soil acts as a natural medium for blending the water supplies. In the latter case, the different waters are alternated either within an irrigation event or between irrigations.

Meiri et al. (1986) conducted a 3-year study in Israel to compare crop performance under network and soil dilution systems. Utilizing a rotation of potatoes and peanuts under drip irrigation, they concluded that both crops responded to the mean water-uptake-weighted soil salinity, regardless of the blending method.

8.4.4 Cyclic Irrigation Method

The cyclic method of utilizing multiple sources of saline water of different qualities was proposed and tested by Rhoades (1977). His vision was to develop a management scheme utilizing saline water in a rotation of crops with different salt-tolerances. He proposed the non-saline water be used for pre-plant and early irrigations of the salt-tolerant crop and all irrigations of the moderately salt-sensitive crop. Salt-tolerant crops are irrigated with saline water after they have grown out of their salt-sensitive growth stage. After the salt-tolerant crop is grown, good quality water is used to reclaim the upper portion of the soil profile, thus aiding the establishment of the moderately salt-sensitive crop. Subsequent irrigations with good quality water during the remainder of the season move previously accumulated salts farther down the soil profile and, it is hoped, ahead of the advancing roots.

In theory, the cyclic use of high quality and saline water can be repeated year after year, provided that a favorable salt balance is maintained, toxic elements do not accumulate, and physical soil conditions related to tilth and permeability are sustained over the long-term. Field studies conducted by Rhoades (1984); Rhoades et al. (1989); Ayars et al. (1986a, b); Grattan et al. (1987); and Shennan et al. (1987) have shown that the cyclic irrigation technique can be viable.

In an extensive study in the Imperial Valley of California, Rhoades et al. (1988) conducted field experiments to test cyclic irrigation practices on two cropping patterns, one of which consisted of wheat, sugarbeets, and melons in a 2-year rotation. Colorado river water (900 mg/l TDS) was used for irrigating melons and for the preplant and early irrigations of wheat and sugarbeets. Drainage water (3500 mg/l TDS) was used at all other times. The researchers found the drainage water, which supplied 75% of the irrigation water requirements, did not reduce yield for any of the three crops even after two complete rotations. The second crop pattern tested consisted of a dual cropping of cotton and wheat for 2 years, followed by two years of alfalfa. Drainage water was applied only to the cotton (after seedlings were established). Again, saline drainage water did not reduce yields.

Another cyclic irrigation practice was conducted by Ayars et al. (1986a, b) using drip rather than surface irrigation methods. For three consecutive years saline drainage water ($EC_w = 8.0$ dS/m) was applied to cotton after seedlings were established. The subsequent wheat crop was irrigated with good quality water ($EC_w < 0.5$ dS/m). Sugarbeets followed wheat and were irrigated again with saline water after seedlings were established. The researchers found that yields from plots irrigated with drainage water were no different from those irrigated with only good quality water.

Grattan et al. (1987) extended the cyclic irrigation practice to test whether saline drainage water could be used directly to irrigate moderately salt-sensitive crops after they reached a salt-tolerant growth stage. This finding would have particular significance since saline water has reportedly improved fruit quality (Shannon and Francois 1978; Pasternak et al. 1986). They found that saline drainage water (EC = 8.0 dS/m and 6 mgB/l) applied after first flower could be used to irrigate melons and processing tomatoes without reducing yield. Furthermore, drainage water, which supplied up to 65% of the irrigation water requirements, did indeed improve fruit quality.

Shennan et al. (1987) designed a field experiment to test the long-term feasibility of a modified cyclic practice on a crop rotation consisting of two years of cotton followed by one year of processing tomato. In this study, two cyclic irrigation regimes were tested: (1) one year of saline drainage water (EC = 7.3 to 7.7 dS/m and 5 mgB/l) applied to the tomato crop after first flower followed by 2 years of aqueduct water applied to the subsequent cotton crops; and (2) drainage water applied to tomato and the first cotton crop after pre-irrigations with aqueduct water, followed by aqueduct water on the second cotton crop.

Each irrigation treatment began at each of the three points in the cropping sequence. After four years of investigation, saline drainage water, regardless of whether it was applied 1, 2, or 3 of the 4 years, did not reduce the yields of either cotton lint or tomato fruit (Grattan et al. 1991). After the fifth year, however, drainage water reduced tomato fruit yields.

Irrigation with aqueduct water displaced a large proportion of the salts and boron in the upper 60 cm of soil that had accumulated from previous years' irrigation with drainage water (Grattan et al. 1990). At depths below 60 cm, salts and boron increased over time in most plots that received drainage water. Furthermore, yearly changes in soil salinity at lower depths lagged behind those in the upper profile. It was common to find increased salinity at lower depths from irrigation with non-saline water. In this instance, salts that had accumulated in the upper profile from previous years' drainage water applications moved to lower depths.

Such lag-time responses in the soil profile can influence decisions regarding timing of saline water application within a rotation. In well-drained soils, the salinity profile can change more readily than soils with low water infiltration. These latter soils may not be able to leach much of the salts in the upper profile to lower portions ahead of the advancing roots. If this is the case, moderately salt-sensitive plants that follow salt-tolerant crops irrigated with saline water may be adversely affected.

The studies described above provide additional evidence that irrigation water with an EC_w far in excess of the EC_w-threshold reported by Ayers and

Westcot (1985) can be used for several years without reducing yields. Crops are able to withstand growth reductions from saline water in excess of the EC_w-threshold since: (1) the stress experience does not occur during the entire season; and (2) the crop experiences most of the salt-stresses at a salt-tolerant growth stage.

Periodic reclamation of the soil may be required to avoid salt buildup to growth-limiting levels. However, the amount of good quality water needed to periodically reclaim the soil must be taken into account for long-term water economy. Any "good quality" water savings during prereclamation years may be lost during reclamation, particularly if soil B is a constituent that needs to be reduced in the soil profile.

8.4.5 Blending vs. Cyclic Practices

Blending irrigation waters is often proposed as a means to expand the existing water supply and can be particularly attractive when the concentration of the saline water is less than 1500 mg/l TDS (Grattan and Rhoades 1990). Too often, however, growers are tempted to blend water that is too saline. Such practices deprive the plant of the opportunity to fully utilize the high quality fraction of the blended water. The following is an extreme example but illustrates how blending can reduce the "useable water" by the crop. Suppose a grower is producing beans, but realizes there is not a sufficient supply of good quality water to meet crop needs and considers blending this water with saline water to supplement crop requirements. However, the only saline water available is one-half the salinity of seawater. If one liter of saline water is blended with one liter of good quality water, the result is zero liters of usable water since salt-sensitive beans cannot tolerate water containing 25% of the salts in seawater. Although this example is exaggerated, the principle is valid. That is, it cannot be assumed that blending irrigation water supplies will satisfy the objective of increasing water that is usable by the crop.

Although the cyclic practice can utilize waters of higher salinity, it limits the extent to which saline water can be utilized for irrigation. Developing limits or guidelines depends on the intensity and duration of exposure to salinity and the sensitivity of the crop to salinity during the growth stage in which it experiences salinity stress. More research is obviously needed in the area of plant response to changing salinity profiles.

Saline water availability and accessibility can influence which practice is more feasible. The cyclic technique may make more stringent demands on saline water availability than the blending method. The cyclic method requires that all irrigation water requirements be supplied by saline water at a time when the crop is mature and evaporative demands are high. In many cases where drainage water is the source of saline water, its availability is not usually consistent with demands. Very often drains flow in the spring after rainfall and pre-irrigations, yet demands for the water are in the summer. Although water can be pumped later in the season, this practice can operate on only a limited scale.

In general, if water availability is unrestricted, the cyclic technique has many advantages over the blending method: (1) soil salinity can be lowered at certain times to allow more salt-sensitive crops to be included in the rotation; (2) a water blending facility is not required; (3) the use of the saline water supply can be maximized; and (4) water of higher salinities can be used (Grattan and Rhoades 1990).

8.5 Conclusions

The long-term sustainability of irrigation using saline water depends upon the quality and availability of water sources as well as irrigation water management. Irrigation with saline water can be sustained year after year if the soil has a favorable salt-balance and suffers no physical and chemical degradation. Impacts from irrigated agriculture must be assessed at the farm level as well as the regional level. If irrigation sites are selected in closed basins, regional salt balances must be achieved which require off-site disposal of agricultural drainage. Disposal of agricultural drainage into inland bodies of water or into streams can have adverse effects on both other water users and the ecology of the receiving water.

Most of the field research on crop salt-tolerance was conducted using a single source of saline water and achieving high LF's to create salinity profiles with small spatial and temporal changes. These methods facilitated data interpretation. Researchers have also developed relationships to predict average root zone salinity from LF and irrigation water salinity. All of this information has allowed individuals to develop salinity standards for agriculture. However, although this research has been extremely useful, such experiments create unrealistic salinity profiles. Under field conditions these profiles are spatially variable and in constant flux due to rainfall, changes in irrigation water quality, and water infiltration limitations.

There is now a substantial amount of evidence indicating that the water quality standards for agriculture may be too conservative. Recent field experiments have shown that saline water, previously considered unsuitable for irrigation, may be used successfully to produce crops, provided that certain management guidelines are followed.

Water of higher salinity can be used for irrigation if a source of good quality water is also available. Multiple sources of water available for irrigation allow transient salinity profiles to be intentionally created. By adopting a cyclic irrigation method that creates a transient profile, growers can take advantage of crops of different tolerances to salinity between seasons as well as of the changing salt-tolerence of a crop within a season.

Under the cyclic irrigation practice, crops can tolerate waters of high salinity since the crop experiences salt stress for only a fraction of the season and at a relatively salt-tolerant growth stage. However, more research is needed to predict plant performance under salinity profiles that change spatially and temporally.

Depending upon the quality of the saline water, blending irrigation water sources can expand or reduce the supply of water usable by the plant. Very often a high-saline water is blended with good quality water, effectively reducing the plant's usable water supply.

Although saline water has a potential for use in irrigation, the long-term effects of its use must be considered. In well-drained soils where drainage effluents are discharged to the ocean, saline irrigation water may be suitable indefinitely, particularly in conjunction with a suitable supply of good quality water. In poorly drained soils with no off-site drainage disposal options, irrigation with saline water cannot be sustained indefinitely as it causes gradual salinization of the region. Saline water containing trace elements may further limit the extent to which such water can be used. If crops are irrigated with saline water containing high levels of B, large quantities of good quality water will eventually be needed to reduce soil B concentrations to satisfactory levels.

References

Aragüés R, Royo A, Faci J (1992) Evaluation of a triple line source sprinkler system for salinity-crop production studies. Soil Sci Soc Am J 56:377–383

Ayars JE, Hutmacher RB, Schoneman RA, Vail SS, Felleke D (1986a) Drip irrigation of cotton with saline drainage water. Trans ASAE 29(6):1668–1673

Ayars JE, Hutmacher RB, Schoneman RA, Vail SS (1986b) Trickle irrigation of sugar beets with saline drainage water. USDA/ARS Water Manag Lab Annu Rep Fresno CA, pp 5–6

Ayers RS, Westcot DW (1985) Water quality for agriculture. FAO, irrigation and drainage Pap 29. FAO, Rome

Bernstein L, Fireman M (1957) Laboratory studies on salt distribution in furrow-irrigated soils with special reference to the pre-emergence period. Soil Sci 83:249–263

Bradford GR, Bakhtar D, Westcot D (1990) Uranium, vanadium and molybdenum in saline waters of California. J Environ Qual 19:105–108

Bressler MB (1979) The use of saline water for irrigation in the USSR. Joint Commiss Sci Tech Coop, Water Resourc, Bur Reclamat, Eng Res Cent, Denver

Chapman HD, Pratt PF (1982) Methods of analysis for soils, plants and waters. Univ Cal Publ 4034, 309 pp

Dhir RD (1976) Investigations into use of highly saline waters in an arid environment. I. Salinity and alkali hazard conditions in soil under a cyclic management system. In: Proc Int Manag Conf Managing saline water on irrigation, Texas Tech Univ, Lubbock, pp 608–609

Dinar A, Letey J, Vaux HJ (1986) Optimal ratios of saline and non-saline irrigation-waters for crop production. Soil Sci Soc Am J 50:440–443

Emerson WW (1984) Soil structure in saline and sodic soils. In: Shainberg I, J Shalhevet (eds)

Soil salinity under irrigation processes and management. Springer, Berlin Heidelberg New York, pp 65–76

Frenkel H, Shainberg I (1975) Irrigation with brackish water: chemical and hydraulic changes in soils irrigated with brackish water under cotton production. In: Irrigation with brackish water. Int Symp Beer-Sheva, Isr Negev Univ Press, Jerusalem, pp 175–183

Grattan SR, Rhoades JD (1990) Irrigation with saline ground water and drainage water. In: Tanji KK (ed) Agricultural salinity assessment and management. ASCE Manu Rep Eng Practices 71, pp 432–449

Grattan SR, Shennan C, May DM, Mitchell JP, Burau RG (1987) Use of drainage water for irrigation of melons and tomatoes. Cal Agric 41(9/10):27–28

Grattan SR, Shennan C, May D, Roberts B, Hillhouse C, Burau RG (1991) Continuation in the long-term cyclic use of saline drainage water in a rotation of cotton and processing tomato. Tech Prog Rep, UC Salinity/Drainage Task Force, Div Agric Nat Resourc, Univ Cal, Davis

Hoffman GJ (1990) Leaching fraction and rootzone salinity control. In: Tanji KK (ed) Agricultural salinity assessment and management. ASCE Manu Rep Eng Practices 71

Hoffman GJ, van Genuchten M Th (1983) Water management for salinity control. In: Taylor H, Jordan W, Sinclair T (eds) Limitations to efficient water use in crop production. Am Soc Agron Monogr, pp 73–85

Howitt RE, M'Marete M (1991) Well set aside proposal: a scenario for ground water banking. Cal (Agric) 45(3):6–9

Keren R, Miyamoto S (1990) Reclamation of saline, sodic, and boron-affected soils. In: Tanji KK (ed) Agricultural salinity assessment and management. ASCE Manu Rep Eng Practices 71

Keren R, Bingham FT, Rhoades JD (1984) Plant uptake of boron as affected by boron distribution between liquid and solid phases in soils. Soil Sci Soc Am J 48: 297–302

Kruse EG, Willardson L, Ayars J (1990) On-farm irrigation and drainage practices. In: Tanji KK (ed) Agricultural salinity assessment and management. ASCE Manu Rep Eng Practices 71

Letey J, Knapp K, Solomon K (1990) Crop-water production functions under saline conditions. In: Tanji KK (ed) Agricultural salinity assessment and management. ASCE Manu Rep Eng Practices 71

Maas EV (1985) Crop tolerance to saline sprinkling waters. Plant Soil 89:273–284

Maas EV (1990) Crop Salt Tolerance. In: Tanji KK (ed) Agricultural salinity assessment and management. ASCE Manu Rep Eng Practices 71

Maas EV, Hoffman GJ (1977) Crop salt tolerance – current assessment. J Irrig Drain Div, ASCE 103 (IR2): 115–134

Maas EV, Poss JA (1989) Salt sensitivity of wheat at various growth stages. Irrig Sci 10: 29–40

Maas EV, Hoffman GJ, Chaba GD, Poss JA, Shannon MC (1983) Salt sensitivity of corn at various growth stages. Irrig Sci 4:45–57

Meiri A, Shalhevet J, Stozy LH, Sinai G, Steinhardt R (1986) Managing multi-source irrigation water of different salinities for optimum crop production. BARD Tech Rep 1-402-81. Volcani Cent, Bet Dagan, Isr, 172 pp

Miles DL (1977) Salinity in the Arkansas Valley of Colorado. Interagency Agreement Report EPA-AIG-D4-0544. EPA, Denver

Page AL, Chang AC, Adriano DC (1990) Deficiencies and toxicities of trace elements. In: Tanji KK (ed) Agricultural salinity assessment and management. ASCE Manu Rep Eng Practices 71

Pasternak D, DeMalach Y, Borovic J (1986) Irrigation with brackish water under desert condition. VII. Effect of time of application of brackish water on production of processing tomatoes (Lycopersicon esculentum, Mill). Agric Water Manag 12:149–158

Pearson GA, Bernstein L (1959) Salinity effects at several growth stages of rice. Argon J 51:654–657

Pillsbury AF, Blaney HF (1966) Salinity problems and management in river systems. Proc ASCE (IR) 98:77–90

Pratt PR, Suarez DL (1990) Irrigation water quality assessments. In: Tanji KK (ed) Agricultural salinity assessment and management. ASCE Manu Rep Eng Practices 71

Rhoades JD (1977) Potential for using saline agricultural drainage water for irrigation. In: Proc Water management for irrigation and drainage. ASCE, Reno NV, July 1977, pp 85–116

Rhoades JD (1982) Reclamation and management of salt-affected soils after drainage. In: Proc 1st Annu Western Provincial Conf Rationalization of water and soil research and management. Lethbridge, Alberta, Can, Nov 27–Dec 2, pp 123–197

Rhoades JD (1984) Use of saline water for irrigation. Cal Agric 38(10):42–43

Rhoades JD, Bingham FT, Letey J, Dedrick AR, Bean M, Hoffman GJ, Alves WJ, Swain RV, Pacheco PG, LeMert RD (1988) Reuse of drainage water for irrigation: results of Imperial Valley study. I. Hypothesis, experimental procedures and cropping results. Hilgardia 56(5):1–16

Rhoades JD, Bingham FT, Letey J, Hoffman GJ, Dedrick AR, Pinter PJ, Replogle JA (1989) Use of saline drainage water for irrigation: Imperial Valley study. Agric Water Manag 16:25–36

Rolston DE, Rains DW, Biggar JW, Läuchli A (1988) Reuse of saline drain water for irrigation. In: UCD/INIFAP Conf, Guadalajara, Mex, March 1988

Shainberg I, Singer JJ (1990) Soil response to saline and sodic conditions. In: Tanji KK (ed) Agricultural salinity assessment and management. ASCE Manu Rep Eng Practices 71

Shalhevet J (1984) Management of irrigation with brackish water. In: Shainberg I, Shalhevet J (eds) Soil salinity under irrigation. Processes and management. Springer, Berlin Heidelberg New York, pp 298–318

Shalhevet J, Kamburov J (1976) Irrigation and salinity: a world-wide survey. Int Commiss Irrig Drainage, 106 pp

Shannon MC, Francois LE (1978) Salt tolerance of three muskmelon cultivars. J Am Soc Hortic Sci 103:127–130

Shennan C, Grattan S, May D, Burau R, Hanson B (1987) Potential for the long-term cyclic use of saline drainage water for the production of vegetable crops. Tech Prog Rep, UC Salinity/Drainage Task Force, Div Agric Nat Res, Univ Cal, Davis, pp 142–146

Tanji KK (ed) (1990) Nature and extent of agricultural salinity. In: Agricultural salinity assessment and management. ASCE Manu Rep Eng Practices 71

Tanji KK, Läuchli A, Meyer J (1986) Selenium in the San Joaquin Valley. Environment 28-(6):6–11, 34–39

Thellier C, Holtzclaw KM, Rhoades JD, Sposito G (1990) Chemical effects of saline irrigation water on a San Joaquin Valley Soil: II. Field soil samples. J Environ Qual 19:56–60

van Genuchten M Th (1983) Analyzing crop salt tolerance data: model description and user's manual. USDA-ARS-UKSSL Res Rep 120

van Schilfgaarde J (1990) Irrigation agriculture: is it sustainable? In: Tanji KK (ed) Agricultural salinity assessment and management ASCE Manu Rep Practices 71

Wichelns D, Nelson D, Weaver T (1988) Farm-level analyses of irrigated crop production in areas with salinity and drainage problems. San Joaquin Valley Drainage Program. USBR, Sacramento, CA, 79 pp

9 Irrigation with Treated Sewage Effluents

Takashi Asano

9.1 Introduction

Utilization of wastewater for agricultural and landscape irrigation has been practiced in many parts of the world for centuries. With the advent of sewerage systems in the nineteenth century, sewage (domestic wastewater) was used at "sewage farms", and by 1900 there were numerous sewage farms in Europe and in the United States (Reed and Crites 1984; Sterritt and Lester 1988). Although these sewage farms were used primarily for waste disposal, incidental use was made of the water for crop production or other beneficial uses. More recently, a number of planned wastewater reclamation and reuse projects have been developed as a matter of necessity to meet growing water needs in irrigation and other uses. In many industrialized nations, there are growing problems of providing adequate water supply, and municipal and industrial wastewater disposal. In developing countries, particularly those in arid parts of the world, there is a need to develop low cost, low technology methods for acquiring new water supplies, protecting existing water sources from pollution. As the demand for water increases, wastewater reclamation and reuse have become an increasingly important source for meeting some of this demand. Because wastewater reclamation and the planned use of reclaimed water are so closely linked to the freshwater supply of a region, significant wastewater reuse projects are typically implemented in water-short areas of the world.

Wastewater reclamation is the treatment or processing of wastewater to make it reusable, and water reuse is the utilization of this water for a variety of applications. In addition, the direct reuse of water implies existence of a pipe, or other conveyance facilities, for delivering the reclaimed water. Indirect reuse, through discharge of an effluent to a receiving water for assimilation and withdrawl downstream, while important, does not constitute planned reuse. In contrast to reuse, wastewater recycling normally involves only one use or user, and the effluent from the user is captured and redirected back into that use scheme. In this context, wastewater recycling is practiced predominantly in the steam-electric, manufacturing, and mineral industries (Metcalf and Eddy 1991).

The use of water has increased rapidly and significantly in many regions of the world over the past few decades. However, in recent years in the United

Adv. Series in Agricultural Sciences, Vol. 22
K.K. Tanji/B. Yaron (Eds.)
© Springer-Verlag Berlin Heidelberg 1994

Table 9.1. Categories of municipal wastewater reuse and potential constraints

Wastewater reuse categories	Potential constraints
Agricultural irrigation: crop irrigation commercial nurseries	Effect of water quality, particularly salts, on soils and crops
Landscape irrigation: park school yard freeway median golf course cemetery greenbelt residential	Public health concerns related to pathogens (bacteria, viruses, and parasites) Surface and groundwater pollution if not properly managed
Industrial reuse: cooling boiler feed process water heavy construction	Reclaimed wastewater constituents related to scaling, corrosion, biological growth, and fouling Public health concerns, particularly aerosol transmission of organics, and pathogens in cooling water
Groundwater recharge: groundwater replenishment salt water intrusion subsidence control	Organic chemicals in reclaimed wastewater and their toxicological effects Total dissolved solids, metals, and pathogens in reclaimed wastewater
Recreational/environmental uses: lakes and ponds marsh enhancement streamflow augmentation fisheries snowmaking	Health concerns of bacteria and viruses Eutrophication due to N and P
Nonpotable urban uses: fire protection air conditioning toilet flushing	Public health concerns about pathogens transmitted by aerosols Effects of water quality on scaling, corrosion, biological growth, and fouling
Potable reuse: blending in water supply pipe to pipe water supply	Organic chemicals in reclaimed wastewater and their toxicological effects Esthetics and public acceptance Public health concerns on pathogen transmission including viruses

[a] Arranged in descending order of anticipated volume of use.

States, increased water development costs, environmental concerns, and a growing conservation philosophy are key factors accounting for relatively moderate trends in increase in water use. Wastewater potentially available for reuse include discharge from municipalities, industries, and agricultural return flows. Of these, return flows from agricultural irrigation are usually collected and reused without further treatment. The requirements for degree of municipal wastewater treatment and treatment process reliability will depend on the categories of planned reuse, as shown in Table 9.1. These categories of municipal wastewater reuse are arranged in descending order of anticipated future volume of use.

9.2 Need for Preapplication Treatment

Although irrigation with sewage effluents is, in itself, an effective form of wastewater treatment (such as in slow rate land treatment), some degree of treatment must be provided to untreated municipal wastewater (raw sewage) before it can be used for agricultural or landscape irrigation. The degree of preapplication treatment is an important factor in the planning, design, and management of wastewater irrigation systems.

Preapplication treatment of wastewater is practiced for the following reasons (Asano et al. 1985): (1) protection of public health; (2) prevention of nuisance conditions during storage and application; and (3) prevention of damage to crops and soils.

The level of wastewater treatment required for agricultural and landscape irrigation uses depends on the soil characteristics, the crop irrigated, the type of distribution and application systems, and the degree of worker and public exposure.

The level of treatment required by regulatory agencies prior to irrigation of many crops is often not greater than, and is sometimes less than, the level of treatment required for discharge to receiving waters. Additional treatment to remove wastewater constituents that may be toxic or harmful to certain crops is technically possible, but normally is not justified economically in case of agricultural irrigation. To use waters containing such constituents, the crops selected must be tolerant to the wastewater constituents, and systems must be managed to mitigate any harmful effects of these wastewater constituents.

9.3 Municipal Wastewater (Sewage) Characteristics

To discuss wastewater treatment processes and the characteristics of effluent produced by them, it is first necessary to describe the characteristics of untreated municipal wastewater (raw sewage).

9.3.1 Wastewater Sources

Sewage or municipal wastewater is the general term applied to the liquid waste collected in sanitary sewers and treated in a municipal wastewater treatment plant (sewage treatment plant). Municipal wastewater is composed of domestic (sanitary) wastewater (sewage), industrial wastewater, and infiltration inflow. Domestic wastewater is the spent water supply of the community after it has undergone a variety of uses in residences, commercial buildings, and institutions. Industrial wastewater is spent water from manufacturing or food-processing plants. Inflow is storm water that enters the sewer system through manholes and other openings, and infiltration is groundwater that seeps into the sewer through improperly sealed or broken joints or cracks in the pipe.

 The relative quantities of wastewater from each source vary widely among communities and depend on the number and type of commercial and industrial establishments as well as on the age and length of the sewer systems. In a few communities in the USA, storm water runoff is collected in a separate (storm) sewer system with no known domestic or industrial wastewater connections and is conveyed to the nearest watercourse for discharge usually without treatment. Many large cities have a combined sewer system in which both storm water and municipal wastewater are collected in the same sewer. During dry weather, flow in the combined sewers is intercepted and conveyed to the wastewater treatment plant for treatment. During storms, flow in excess of the wastewater treatment plant capacity is either retained within the system and treated subsequently or is bypassed to the point of discharge.

9.3.2 Wastewater Flow Rates

The volume of wastewater generated in a community on a per capita basis varies from 0.15 to $0.6\,m^3/day$ and includes domestic wastewater plus infiltration inflow but excludes industrial wastewaters. The wide range of per capita flows reflects differences in water consumption among communities and is largely a function of climate, the price of water, and reliability of the water supply. An average value of $0.38\,m^3/day$ per capita is often used for planning purposes in the absence of data specific to the community.

 The short-term variations in wastewater flows observed at municipal wastewater treatment plants tend to follow a diurnal pattern. Flow is low during the early morning hours, when water consumption is lowest and when the base flow consists of infiltration inflow and small quantities of sanitary wastewater. The first peak flow generally occurs in the late morning, when wastewater from the peak morning water use reaches the treatment plant. A second peak flow occurs in evening after the dinner hour. The relative magnitude of the peaks and the times at which they occur vary with the size of the community and the length of the sewers. Small communities with small sewer systems have a much higher ratio of peak flow to average flow than do large communities.

Although the magnitude of peaks is depressed as wastewater passes through a treatment plant, the daily variations in flow from a municipal treatment plant make it impractical, in most cases, to irrigate with effluent directly from the treatment plant. Some form of flow equalization or short-term storage of treated effluent is necessary to provide a relatively constant supply of reclaimed water for efficient irrigation. Additional benefits from storage are discussed later in the chapter.

Seasonal variations in wastewater flows are commonly observed at resort areas, in small communities with college campuses, and in communities that have seasonal commercial and industrial wastewater loads. An example is the substantially higher summer flows experienced by communities that receive industrial wastewater from seasonal food-processing industries.

9.3.3 Wastewater Constituents and Compositions

The physical properties and the chemical and biological constituents of wastewater are important parameters in the design and operation of collection, treatment, and disposal facilities and in the engineering management of environmental quality. The constituents of concern in wastewater treatment and wastewater irrigation are listed in Table 9.2. A complete evaluation and classification of water quality criteria for irrigation are presented in the latter part of this chapter.

Composition refers to the actual amounts of physical, chemical and biological constituents present in wastewater. The composition of untreated wastewater and the subsequently treated effluent depends upon the composition of the municipal water supply, the number and type of commercial and industrial establishments, and the nature of the residential community. Consequently, the composition of wastewater often varies widely among different communities. Typical data on the composition of untreated domestic wastewaters in the USA are presented in Table 9.3.

Wastewater quality data routinely measured and reported are mostly in terms of gross, nonspecific pollutional parameters [e.g., biochemical oxygen demand (BOD), suspended solids (SS), and chemical oxygen demand (COD)] that are mostly of interest in water pollution control (see Tables 9.2 and 9.3). In contrast, the water characteristics of importance in agricultural or landscape irrigation are specific chemical elements and compounds that affect plant growth or soil permeability. These characteristics are not often measured or reported by wastewater treatment agencies as part of their routine water quality monitoring program. Consequently, when obtaining data to evaluate or plan a wastewater irrigation system, it is often necessary

Table 9.2. Constituents of concern in wastewater treatment and irrigation with reclaimed wastewater (Pettygrove and Asano 1985)

Constituent	Measured Parameters	Reason for concern
Suspended solids	Suspended solids, including volatile and fixed solids	Suspended solids can lead to the development of sludge deposits and anaerobic conditions when untreated wastewater is discharged in the aquatic environment. Excessive amounts of suspended solids cause plugging in irrigation systems.
Biodegradable organics	Biochemical oxygen demand, chemical oxygen demand	Composed principally of proteins, carbohydrates, and fats. If discharged to the environment, their biological decomposition can lead to the depletion of dissolved oxygen in receiving waters and to the development of septic conditions.
Pathogens	Indicator organisms, total and fecal coliform bacteria	Communicable diseases can be transmitted by the pathogens in wastewater: bacteria, viruses, parasites.
Nutrients	Nitrogen, phosphorus, and potassium	Nitrogen, phosphorus, and potassium are essential nutrients for plant growth, and their presence normally enhances the value of the water for irrigation. When discharged to the aquatic environment, nitrogen and phosphorus can lead to the growth of undesirable aquatic life. When discharged in excessive amounts on land, nitrogen can also lead to the pollution of groundwater.
Stable (refractory) organics	Specific compounds (e.g., phenols, pesticides, chlorinated hydrocarbons)	These organics tend to resist conventional methods of wastewater treatment. Some organic compounds are toxic in the environment, and their presence may limit the suitability of the wastewater for irrigation.
Hydrogen ion activity	pH	The pH of wastewater affects metal solubility as well as alkalinity of soils. Normal range in municipal wastewater is pH = 6.5–8.5, but industrial waste can alter pH significantly.
Heavy metals	Specific elements (e.g., Cd, Zn, Ni, Hg)	Some heavy metals accumulate in the environment and are toxic to plants and animals. Their presence may limit the suitability of the wastewater for irrigation.
Dissolved inorganics	Total dissolved solids, electrical conductivity, specific elements (e.g., Na, Ca, Mg, Cl, B)	Excessive salinity may damage some crops. Specific ions such as chloride, sodium, boron are toxic to some crops. Sodium may pose soil permeability problems.
Residual chlorine	Free and combined chlorine	Excessive amount of free available chlorine (> 1 mg/l Cl_2) may cause leaf-tip burn and damage some sensitive crops. However, most chlorine in reclaimed wastewater is in a combined form, which does not cause crop damage. Some concerns are expressed as to the toxic effects of chlorinated organics in regard to groundwater contamination.

Table 9.3. Typical composition of untreated municipal wastewater[a] (Pettygrove and Asano 1985)

Constituent	Concentration range[b]			US Average[c]
	Strong	Medium	Weak	
Solids, total:				
Dissolved, total[d]	1,200	720	350	–
Fixed	850	500	250	–
Volatile	525	300	145	–
Suspended	350	220	100	192
Fixed	75	55	20	–
Volatile	275	165	80	–
Settleable solids, ml/l	20	10	5	–
Biochemical oxygen demand,				
5-day 20 °C	400	220	110	181
Total organic carbon	290	160	80	102
Chemical oxygen demand	1,000	500	250	417
Nitrogen (total as N)	85	40	20	34
Org-N	35	15	8	13
NH_3-N	50	25	12	20
NO_2-N	0	0	0	–
NO_3-N	0	0	0	0.6
Phosphorus (total as P)	15	8	4	9.4
Organic	5	3	1	2.6
Inorganic	10	5	3	6.8
Chlorides	100	50	30	–
Alkalinity (as $CaCO_3$)[d]	200	100	50	211
Grease	150	100	50	–
Total coliform bacteria,[e]				
MPN/100 ml[f]	–	–	–	22×10^6
Fecal coliform bacteria,[e]				
MPN/100 ml	–	–	–	8×10^6
Viruses, PFU/100 ml[g,h]	–	–	–	3.6

[a] All values are expressed in mg/l, except as noted.
[b] After Metcalf & Eddy, Inc. (1991).
[c] Culp et al. (1979).
[d] Values should be increased by amount in demostic water supply.
[e] Geldreich (1978).
[f] Most probable number/100 ml of water sample.
[g] Berg and Metcalf (1978).
[h] Plaque-forming units.

to sample and analyze the wastewater for those constituents that define the suitability of the water for agricultural or landscape irrigation.

The constituents that largely determine the suitability of wastewater for agricultural or landscape irrigation are the dissolved inorganic solids or minerals. These constituents are not altered substantially in most wastewater treatment processes. In some cases, they may increase in concentration as a result of evaporation in lagoons or storage reservoirs.

Consequently, the composition of dissolved minerals in effluent used for irrigation can be expected to be similar to the composition in the untreated wastewater. The composition of dissolved minerals in untreated wastewater is determined by the composition of incoming domestic water supply plus mineral pickup resulting from water use in the household.

Municipal wastewaters contain pathogens of fecal origin including bacteria, viruses, protozoa, and parasitic worms. In areas where sanitary disposal of human feces is not adequately practiced, diseases caused by these organisms, such as typhoid fever, bacillary dysentery, hepatitis, and poliomyelitis, are common. Because pathogens in water and wastewater are relatively few in number and difficult to isolate, the nonpathogenic coliform group of bacteria, which is more numerous and easily tested for, is used as an indicator of the presence of enteric pathogens in treated effluent and reclaimed wastewater. Coliform bacteria, excreted in large numbers in the feces of humans and other warm-blooded animals, average about 50 million coliform per gram of feces (Geldreich 1978). Untreated domestic wastewater contains millions of coliform per 100 ml. Consequently, the presence in large numbers of coliform bacteria is taken as an indication that pathogens may be present, and the absence of coliform is taken as an indication that the water is free from pathogens.

9.4 Municipal Wastewater Treatment and Effluent Characteristics
(Asano et al. 1985)

Municipal wastewater treatment consists of a combination of physical, chemical, and biological processes and operations to remove solids, organic matter, pathogens, and sometimes nutrients from wastewater. General terms used to describe different degrees of treatment, in order of increasing treatment level, are preliminary, primary, secondary, and tertiary and advanced wastewater treatment. A disinfection step to inactivate pathogens usually follows the last treatment step. The individual processes and operations commonly used in the various wastewater treatment steps are described briefly in this section. A generalized wastewater treatment diagram is shown in Fig. 9.1.

9.4.1 Preliminary Treatment

The objective of preliminary treatment is the removal of coarse solids and other large materials often found in untreated wastewater. Removal of these materials is necessary to enhance the operation and maintenance of subsequent treatment units. Preliminary treatment operations typically include

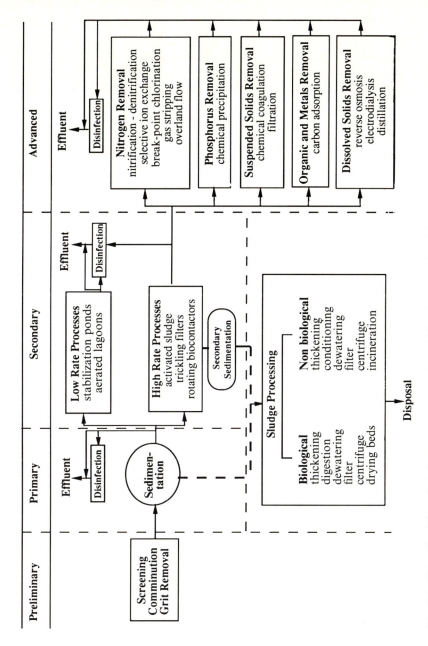

Fig. 9.1. Generalized flow sheet for wastewater treatment. (After Asano et al. 1985)

coarse screening, grit removal by sedimentation, and, in some cases, com-
minution of large objects removed by the screens and grit removal by
sedimentation. In grit chambers, the velocity of the water through the
chamber is maintained sufficiently high to prevent settling of most organic
solids. In most small wastewater treatment plants, grit removal is not
included as a preliminary treatment step.

9.4.2 Primary Treatment

The objective of primary treatment is the removal of settleable organic and
inorganic solids by sedimentation, and the removal of materials that will
float (scum) by skimming. Approximately 25 to 50% of the incoming BOD,
35 to 50% of COD, 50 to 70% of the total suspended solids (TSS), and 55 to
65% of the oil and grease are removed during primary treatment. Some
organic nitrogen, organic phosphorus, and heavy metals associated with
solids are also removed during primary sedimentation, but colloidal and
dissolved constituents are not. The effluent from primary sedimentation
facilities is referred to as primary effluent.

In many industrialized countries, primary treatment is the minimum level
of preapplication treatment required for wastewater irrigation. It is considered sufficient treatment if the wastewater is used to irrigate crops that are
not consumed by humans and may be sufficient treatment for irrigation of
orchards, vineyards, and some processed food crops. However, to prevent
potential nuisance conditions in storage or flow equalizing reservoirs, some
form of secondary treatment will normally be required, even in the case of
nonfood crop irrigation. It may be possible to use at least a portion of
primary effluent for irrigation if off-line storage is provided.

Primary sedimentation tanks or clarifiers may be round or rectangular
basins, typically 3 to 5 m deep. Hydraulic detention times range between 2
and 3 h. Settled solids (primary sludge) are removed from the bottom of
tanks by sludge rakes that scrape the sludge to a central sump from which it
is pumped to sludge processing units. Scum is swept across the tank surface
to a scum skimmer by water jets or mechanical means. Scum is also pumped
to the sludge processing units.

In the USA wastewater treatment plants which are larger than about
7600 m³/d, primary sludge is most commonly processed biologically by
anaerobic digestion. In the digestion process, bacteria metabolize the organic matter in sludge, thereby reducing the volume requiring ultimate
disposal, rendering it stable (nonputrescible) and improving the dewatering
characteristics of the sludge. Digestion is carried out in covered tanks
(anaerobic digesters), typically 7 to 14 m deep. The residence time in a
digester may vary from a minimum of about 10 days for high-rate digesters
(well mixed and heated) to 60 days or more in standard-rate digesters
(unheated and not mixed). Gas containing about 60 to 65% methane is

produced during digestion and can be recovered as an energy source. In plants smaller than $7600\,m^3/d$, sludge is processed in a variety of ways including anaerobic digestion, storage in sludge lagoons, direct application to sludge drying beds, in-process storage (as in stabilization ponds) and land application.

9.4.3 Secondary Treatment

The objective of secondary treatment is the further treatment of the effluent from primary treatment to remove the residual organic and suspended solids. Secondary treatment is the level of preapplication treatment required when the risk of public exposure to wastewater is moderate. Furthermore, secondary treatment of wastewater is required, in most cases, to prevent environmental pollution in the USA and other industrialized countries. The secondary treatment follows primary treatment and involves the removal of biodegradable dissolved and colloidal organic matter using aerobic biological treatment processes. Aerobic biological treatment is performed in the presence of oxygen by aerobic microorganisms (principally bacteria) that metabolize the organic matter in the wastewater, thereby producing more microorganisms and inorganic end products (principally CO_2, NH_3, and H_2O). Several aerobic biological processes are used for secondary treatment. The processes differ primarily in the manner in which oxygen is supplied to the microorganisms and in the rate at which organisms metabolize the organic matter. For the purpose of this discussion, biological wastewater treatment processes are grouped into two processes: high- and low-rate processes (see Fig. 9.1).

9.4.3.1 High-Rate Biological Processes

High-rate biological processes are characterized by relatively small basin volumes and high concentrations of microorganisms compared with the low-rate processes. Consequently, the growth rate of new organisms is much greater in high-rate systems because of a well-controlled environment. The microorganisms must be separated from the treated wastewater by sedimentation to produce the clarified secondary effluent. The sedimentation tanks used in secondary treatment, often referred to as secondary clarifiers, operate in the same basic manner as the primary clarifiers described previously. The biological solids removed during secondary sedimentation, called secondary or biological sludge, are normally combined with primary sludge for sludge processing.

Common high-rate processes include the activated sludge processes, trickling filters or biofilters, oxidation ditches, and rotating biological contactors (RBC). A combination of two of these processes in series (e.g., biofilter followed by activated sludge) is sometimes used to treat municipal wastewater containing a high concentration of organic matter from industrial sources.

In the activated sludge process, the reactor is an aeration tank or basin containing a suspension of the wastewater and microorganisms. The contents of the aeration tank are mixed vigorously by aeration devices that also supply oxygen to the biological suspension. Aeration devices commonly used include submerged diffusers that release compressed air and mechanical surface aerators that introduce air by agitating the liquid surface. Hydraulic detention times in the aeration tanks range from 3 to 8 h. Following the aeration step, the microorganisms are separated from the liquid by sedimentation. The clarified liquid is the secondary effluent. A portion of the biological sludge is recycled to the aeration basin. The remainder is removed from the process and sent to sludge processing to maintain a relatively constant concentration of microorganisms in the system. Several variations of the basic activated sludge process, such as extended aeration, and oxidation ditches, are in common use, but the principles are similar.

A trickling filter or biofilter consists of a basin or tower filled with support media such as stones, plastic shapes, or wooden slats. Wastewater is applied intermittently, or sometimes continuously, over the media. Microorganisms become attached to the media and form a biological film layer. Organic matter in the wastewater diffuses into the film, where it is metabolized. Oxygen is normally supplied to the film by the natural flow of air either up or down through the media, depending on the relative temperatures of the wastewater and air. Forced air can also be supplied by blowers. The thickness of the biofilm increases as new organisms grow. Periodically, portions of the film slough off the media. The sloughed material is separated from the liquid in a secondary clarifier and discharged to sludge processing. Clarified liquid from the secondary clarifier is the secondary effluent. A portion of the effluent is normally recycled to the biofilter to improve hydraulic distribution of the wastewater over the filter.

Rotating biological contactors (RBCs) are similar to biofilters in that organisms are attached to support media. In the case of RBC, the support media are rotating discs that are partially submerged in flowing wastewater. Oxygen is supplied to the attached biofilm from the air when the film is out of the water. Some oxygen is also supplied to the wastewater by the agitation of the disc. Sloughed pieces of biofilm are removed in the same manner described for biofilters.

High-rate biological treatment processes, in combination with primary sedimentation, typically remove 85% of BOD and SS originally present in the wastewater and most of the heavy metals. Activated sludge generally produces an effluent of slightly higher quality, in terms of these constituents, than biofilters or RBSs. When coupled with a disinfection step, these processes provide substantial but not complete removal of pathogens (bacteria and viruses). These processes, however, remove very little phosphorus, nitrogen, nonbiodegradable organics, and dissolved minerals.

9.4.3.2 Low-Rate Biological Processes

These are characterized by lower conversion rates of organic matter contained in wastewater as compared to the high-rate biological systems discussed previously. In most low-rate processes the microorganisms are not usually separated from the liquid in a separate step. In small treatment plants, primary sedimentation prior to low-rate processes is often omitted. In some cases, Imhoff tanks have been used effectively to provide primary treatment. Commonly used low-rate biological processes include aerated lagoons and stabilization ponds. The use of aquatic systems and recirculating sand filters is also increasing because of the high quality of effluent produced.

Aerated lagoons are characterized by hydraulic detention times of 7 to 20 days and water depths of 3 m or more in the basin. Oxygen is usually supplied to the basin by mechanical surface aerators that agitate the water surface, although submerged air diffusion devices have been used. Only the upper layer of the liquid in the basin is normally mixed, and an anaerobic zone develops near the bottom of the lagoon. Organic solids that settle to the bottom of the lagoon are decomposed by anaerobic bacteria.

Stabilization ponds (also called oxidation ponds) use algae to supply oxygen to the basin. The basin is mixed only by periodic wind-generated wave action and thermal currents. Hydraulic detention times range from 20 to 30 days or more, and depths are typically 0.6 to 2.5 m. Only the upper 0.5 to 1.5 m remain aerobic.

Aquatic systems using floating plants such as water hyacinth and emergent plants such as cattail and bulrush have been shown to produce a high-quality effluent often comparable to that produced by the activated sludge process. Water hyacinth treatment systems are typically 0.6 to 1.2 m deep and are operated as modified plug-flow reactors. In some cases, effluent recycle is used to improve performance. Aquatic systems are most suitable for use in temperate climates where the air and water temperatures do not vary dramatically. Aquatic plants can also be used to limit the growth of algae in stabilization ponds.

Recirculating sand filters are a modification of the slow sand filter. The principal modifications include a large increase in the size of the filtering medium and the continuous recirculation of treated effluent mixed with the untreated wastewater. A portion of the treated effluent is discharged continuously. The only drawback of this system is the land area required and the energy needed for the recirculating pump used to apply the wastewater to the filter.

9.4.3.3 Comparison of High- and Low-Rate Biological Processes

Low-rate biological processes are less costly and require less process control than high-rate processes; however, in some of the processes because solids are not separated from the liquid the quality of the effluent is substantially lower than that from high-rate processes. Higher suspended solids in stabilization pond effluent due to algal growth is an example. Aerated lagoons and stabilization ponds are seldom used for preapplication treatment when advanced treatment is required in combination with secondary treatment or when the highest level of disinfection is required in combination with secondary treatment. However, in the San Diego aquaculture demonstration project, the effluent from an aquatic treatment system is to serve as the influent for an advanced treatment process.

In general, low-rate biological processes provide a sufficient degree of preapplication treatment for all other types of irrigation for which secondary treatment is required and also provide sufficient treatment to prevent nuisance conditions in storage reservoirs. Stabilization ponds also provide considerable pathogen and nitrogen removal, depending on the temperature and detention time involved. The low-rate biological processes are particularly important in many wastewater treatment and irrigation applications in developing countries. Some of the considerations on their applicability and costs are further discussed in the latter section of this chapter.

9.4.3.4 Tertiary and Advanced Treatment

Tertiary or advanced treatment is employed when specific wastewater constituents must be removed, but cannot be removed by secondary treatment. As shown in Fig. 9.1, individual treatment processes are necessary to remove nitrogen, phosphorus, additional suspended solids, refractory organics, heavy metals, and dissolved solids. Because advanced treatment usually follows high-rate secondary treatments, it is sometimes referred to as tertiary treatment. However, advanced treatments processes are sometimes combined with primary or secondary treatment (e.g., chemical addition to primary clarifiers or aeration basins to remove phosphorus) or used in place of secondary treatment (e.g., overland flow treatment of primary effluent).

In many situations, where the probability of public exposure to the reclaimed water or residual constituents is high, such as in spray irrigation of parks and golf courses, the intent of the treatment criteria is to minimize the probability of human exposure to pathogens. Effective disinfection of bacteria and viruses is inhibited by suspended and colloidal solids in the water. Thus, in public health sensitive uses, these solids must be removed by advanced treatment before the disinfection step. The sequence of treatment often specified is: secondary treatment followed by chemical coagulation, sedimentation, filtration, and disinfection. This treatment train is often referred as the Title 22 treatment in California, and is assumed to produce an effluent free from detectable pathogens (Asano et al. 1992).

9.4.3.5 Disinfection

The disinfection process normally involves the injection of a chlorine solution at the head end of a chlorine contact basin. The chlorine dosage depends upon the strength of the wastewater and other factors, but dosages of 5 to 15 mg/l are common. Ozone and ultraviolet (UV) irradiation can also be used for disinfection, but these methods of disinfection have not been in common use in the United States. Chlorine contact basins are usually rectangular channels with baffles to prevent short-circuiting, but all are designed to provide a contact time of at least 15 min. However, along with the advanced wastewater treatment requirements, sometimes a chlorine contact time of as long as 120 min is required in the case of spray irrigation of food crops which are consumed uncooked (WPCF 1989).

As mentioned previously, the effectiveness of disinfection is measured in terms of the concentration of indicator organisms (total coliform or fecal coliform bacteria) remaining in the effluent at the end of the chlorine contact basin. The number of coliform bacteria remaining are often expressed in terms of the most probable number of organisms per 100 ml of water sample (MPN/100 ml) and the numbers of coliform bacteria in case of membrane filtration technique.

9.4.3.6 Effluent Storage

Although not considered a step in the treatment process, a storage facility is, in most cases, a critical link between the wastewater treatment plant and the irrigation system. Storage is needed for the following reasons (State of California 1981): (1) to equalize daily variations in flow from the treatment plant and to store excess when average wastewater flow exceeds irrigation demands including winter storage; (2) to meet peak irrigation demands in excess of the average wastewater flow; (3) to minimize disruptions in the operations of the treatment plant and irrigation system. Storage is used to provide insurance against the possibility of unsuitable reclaimed wastewater entering the irrigation system and to provide additional time to resolve temporary water quality problems; and (4) to provide additional treatment. Oxygen demands, suspended solids, nitrogen, and microorganisms are reduced during storage.

9.5 Irrigation with Treated Sewage Effluents

Irrigation water quality is of particular importance in arid zones where extremes of temperature and low humidity result in high rates of evapo-transpiration. The consequence is salt deposition from the applied water which tends to accumulate in the soil profile. The physical and mechanical properties of the soil, such as dispersion of particles, stability of aggregates, soil structure and permeability, are sensitive to the types of exchangeable ions present in irrigation water. Thus, when irrigation with reclaimed wastewater (treated sewage effluents) is being planned, these factors related to soil properties must be taken into consideration. The problems, however, are no different from those caused by salinity or trace elements in freshwater supplies and are of concern only if they restrict the use of the water or require special management to maintain acceptable crop yields (Pettygrove and Asano 1985).

It has been established that the productivity of irrigated land is fundamentally dependent on its internal drainage. No irrigation scheme can succeed unless the soil profile remains permeable. The permeability depends both on the proportions of exchangeable cations, other than sodium, held by the soil and on the total concentration of soluble salt in the percolating water.

9.5.1 Irrigation Water Quality Guidelines

There have been a number of different irrigation water quality guidelines proposed. The water quality guidelines presented in Table 9.4 were developed by the University of California Committee of Consultants and were subsequently expanded by Ayers and Westcot (1985). The guidelines stress the management needed to successfully use water of a certain quality. Four categories of potential water quality problems: (1) salinity; (2) specific ion toxicity; (3) water infiltration rate; and (4) miscellaneous are used for evaluating the suitability of irrigation water. These guidelines emphasize the long-term influence of water quality on crop production, soil conditions, and farm management, and are applicable to both freshwater and reclaimed wastewater.

The health aspect of water quality associated with wastewater reclamation and reuse in California known as the Wastewater Reclamation Criteria (Code of Regulations, Title 22, Division 4: Environmental Health) is summarized in Table 9.5. Note that there are many reuse applications that do not require a high degree of wastewater treatment. These criteria are basically health regulations, such that they do not specifically address the treatment technology or the potential effect of reclaimed water on the crops or soil (cf. Table 9.4 for irrigation water quality). In these regulations, the

Table 9.4. Guidelines for interpretation of water quality for irrigation[a]

Potential irrigation problem	Units	Degree of restriction on use		
		None	Slight to Moderate	Severe
Salinity (affects crop water availability)				
EC_w[b]	dS/m or mmho/cm	< 0.7	0.7–3.0	> 3.0
TDS	mg/l	< 450	450–2000	> 2000
Permeability (affects infiltration rate of water into the soil. Evaluate using EC_w and SAR together)[c,d]				
SAR = 0–3	and EC_w = > 0.7		0.7–0.2	< 0.2
= 3–6	= > 1.2		1.2–0.3	< 0.3
= 6–12	= > 1.9		1.9–0.5	< 0.5
= 12–20	= > 2.9		2.9–1.3	< 1.3
= 20–40	= > 5.0		5.0–2.9	< 2.9
Specific ion toxicity (affects sensitive crops)				
Sodium (Na)[e,f]				
Surface irrigation	SAR	< 3	3–9	> 9
Sprinkler irrigation	mg/l	< 70	> 70	
Chloride (Cl)[e,f]	mg/l	< 140	140–350	> 350
Surface irrigation	mg/l	< 100	> 100	
Boron (B)	mg/l	< 0.7	0.7–3.0	> 3.0
Miscellaneous effects (affects susceptible crops)				
Nitrogen (Total-N)[g]	mg/l	< 5	5–30	> 30
Bicarbonate (HCO_3) (overhead sprinkling only)	mg/l	< 90	90–500	> 500
pH		Normal range 6.5–8.4		
Residual chlorine (overhead sprinkling only)	mg/l	< 1.0	1.0–5.0	> 5.0

[a] Adapted from Ayers and Westcot (1985). The basic assumptions of the guidelines are discussed below.

[b] EC_w means electrical conductivity of the irrigation water, reported in mmho/cm or dS/m. TDS means total dissolved solids, reported in mg/l.

[c] SAR means sodium adsorption ratio. SAR is sometimes reported as R_{Na}. At a given SAR, infiltration rate increases as salinity (EC_w) increases. Evaluate the potential permeability problem by SAR and EC_w in combination (cf. Ayers and Westcot 1985; Pettygrove and Asano 1985).

[d] For wastewaters, it is recommended that the SAR be adjusted to include a more correct estimate of calcium in the soil water following an irrigation. A procedure is given in Ayers and Westcot (1985) and Pettygrove and Asano (1985). The adjusted sodium adsorption ratio (adj R_{Na}) calculated is to be substituted for the SAR value.

[e] Most tree crops and woody ornamentals are sensitive to sodium and chloride; use the values shown. Most annual crops are not sensitive.

[f] With overhead sprinkler irrigation and low humidity (< 30%), sodium or chloride greater than 70 or 100 mg/l, respectively, have resulted in excessive leaf absorption and crop damage to sensitive crops.

[g] Total nitrogen should include nitrate-nitrogen, ammonia-nitrogen, and organic nitrogen. Although forms of nitrogen in wastewater vary, the plant responds to the total nitrogen.

(continuation on page 216)

Footnotes to Table 9.4. Contd.

Assumptions in the Guidelines (Ayers and Westcot 1985). The water quality guidelines in Table 9.4 are intended to cover the wide range of conditions encountered in California's irrigated agriculture. Several basic assumptions have been used to define the range of usability for these guidelines. If the water is used under greatly different conditions, the guidelines may need to be adjusted.

   Wide deviations from the assumptions might result in wrong judgments on the usability of a particular water supply, especially if it is a borderline case. Where sufficient experience, field trials, research, or observations are available, the guidelines may be modified to more closely fit local conditions.

Yield Potential. Full production capability of all crops, without the use of special practices, is assumed when the guidelines indicate no restrictions on use. A "restriction on use" indicates that there may be a limitation such as choice of crop or the need for special management in order to maintain full production capability, but a restriction on use does not indicate that the water is unsuitable for use.

Site Conditions. Soil texture ranges from sandy-loam to clay with good internal drainage. Rainfall is low and does not play a significant role in meeting crop water demand or leaching. In the Sierra and extreme north coast areas of California where precipitation is high for part or all of the year, the guideline restrictions are too severe. Drainage is assumed to be good, with no uncontrolled shallow water table present.

Methods and Timing of Irrigations. Normal surface and sprinkler irrigation methods are used. Water is applied infrequently as needed, and the crop utilizes a considerable portion of the available stored soil water (50% or more) before the next irrigation. At least 15% of the applied water percolates below the root zone (leaching fraction [LF] > 15%). The guidelines are too restrictive for specialized irrigation methods, such as drip irrigation, which result in near daily or frequent irrigations. The guidelines are not applicable for subsurface irrigation.

Water Uptake by Crops. Different crops have different water uptake patterns, but all take water from wherever it is most readily available within the root zone. Each irrigation leaches the upper root zone and maintains it at a relatively low salinity. Salinity increases with depth and is greatest in the lower part of the root zone. The average salinity of the soil solution is about three times that of the applied water.

   Salts leached from the upper root zone accumulate to some extent in the lower part but eventually are moved below the root zone by sufficient leaching. The crop responds to average salinity of the root zone. The higher salinity in the lower root zone becomes less important if adequate moisture is maintained in the upper, "more active" part of the root zone.

median number of total coliform count and turbidity are used for the assessment of treatment reliability of wastewater reclamation plant.

9.5.2 Irrigation of Vegetable Crops –
a 5-Year Field Demonstration Project in Monterey, California

To demonstrate the concepts discussed in the previous sections, the Monterey Wastewater Reclamation Study for Agriculture (Engineering-Science 1987) is summarized in this section. During the early 1970s it became evident that northern Monterey County's groundwater supply was decreasing because of extensive withdrawl of groundwater for irrigation. This overdraft lowered the water tables and created an increasing problem of saltwater intrusion from the Pacific Ocean. At the same time, wastewater treatment

Table 9.5. California wastewater reclamation criteria for irrigation and recreational impoundments

Reclaimed wastewater applications	Description of minimum wastewater characteristics			
	Primary effluent[a]	Secondary and disinfected	Secondary, coagulated, filtered[b] and disinfected	Median total caliform/ 100 ml
Crop irrigation:				
Fodder crops	X			NR[c]
Fiber	X			NR
Seed crops	X			NR
Produce eaten raw, surface irrigated		X		2.2
Produce eaten raw, spray irrigated			X	2.2[d]
Processed produce, spray irrigated		X		23[e]
Landscape irrigation:				
Golf courses, freeways		X		23
Parks, playgrounds			X	2.2
Recreational impoundments:				
No public contact		X		23
Boating and fishing only		X		2.2
Body contact (bathing)			X	2.2

[a] Effluent not containing more than 0.5 ml/l/h of settleable solids. No primary effluent is used presently in California and this category will be deleted in the future regulation.
[b] This requirement is often referred to as the "Title 22 requirement" or the "Title 22 treatment process". Effluent does not exceed an average of 2 turbidity units (NTU) and does not exceed 5 NTU more than 5% of the time during any 24-h period.
[c] No requirement.
[d] The median number of coliform organisms in the effluent does not exceed 2.2/100 ml and the number of coliform organisms does not exceed 23/100 ml in more than one sample within any 30-day period.
[e] The median number of coliform organisms in the effluent does not exceed 23/100 ml in 7 consecutive days and does not exceed 240/100 ml in any two consecutive samples.

facilities were reaching full capacity, requiring expansion to meet the growing needs of the region. The water quality management plan recommendations recognized that wastewater reclamation and reuse had to be proven safe before regional implementation of agricultural irrigation using reclaimed wastewater could be considered. This provided the impetus for the Monterey Wastewater Reclamation Study for Agriculture (MWRSA), which was conceived as a pilot project designed to assess the safety and feasibility of agricultural irrigation with reclaimed municipal wastewater.

The combination of fertile soils and a long growing season makes the lower Salinas Valley in northern Monterey County a rich agricultural region. Artichokes are a major crop, but a variety of annual crops is also grown: broccoli, cauliflower, celery, and lettuce are grown throughout the region.

Planning for the project begun in 1976 by the Monterey Regional Water Pollution Control Agency, the regional agency responsible for wastewater collection, treatment, and disposal. Full-scale field studies began in 1980 and continued through May of 1985. During these five years, a perennial crop of artichokes was grown along with rotating annual crops of celery, broccoli, lettuce, and cauliflower. Extensive sampling and analysis of waters, soils, and plant tissues were conducted throughout the 5 years by Engineering-Science, a consulting engineering firm; University of California; and State agencies (Sheikh et al. 1990).

9.5.2.1 Description of the Project

The site for the MWRSA field operations was a farm in Castroville, California. The existing 1500 m^3/d Castroville Wastewater Treatment Plant was selected for modification and upgrading to be used as the pilot tertiary reclamation plan for MWRSA. A portion of the secondary effluent was diverted to a new pilot tertiary treatment plant which consisted of two parallel treatment process trains. The first process, the Title-22 process (T-22) (see Sect. 9.4.3.4), conformed strictly to the requirements of the California Wastewater Reclamation Criteria for irrigating food crops that may be consumed without cooking. The second process produced a treated wastewater designated as filtered effluent (FE). This is a wastewater treated less extensively than T-22 effluent via direct filtration of secondary effluent. Well water from local wells was the control for the study (Kirkpatrick and Asano 1986).

The 12-ha field site was divided into two parts, demonstration fields and experimental plots. Large demonstration fields were established because farm-scale feasibility of using reclaimed water is of special importance to the growers, farm managers, and operators responsible for day-to-day farming practices. To investigate large-scale feasibility of using reclaimed waste-water, two 5-ha plots were dedicated to reclaimed water irrigation, using the FE flow stream. On one plot, artichokes were grown; on the other lot, a succession of broccoli, cauliflower, lettuce, and celery plants were raised during the first three years of the field investigation. The crops were observed carefully for appearance and vigor. Normal farming practices of local growers were duplicated on these fields with the exception of harvest, which was not carried out. Because of its experimental nature, the produce from these plots was not marketed. Six field observation days were held, and the local growers and the news media were invited to acquaint the agricul-tural community with the ongoing MWRSA activities and to obtain feed-back regarding their perceptions, questions, and concerns.

A split-plot design was chosen for the experimental plots. This design allowed the use of two treatment variables: water type and fertilization rate. Four replicates of three types of main plots were irrigated with T-22 effluent, FE, or well water. These three water types were assigned randomly to main plots within each block or replicate to achieve a randomized complete block (i.e., each block contains all three of the main water type treatments).

Each main plot was then divided into four subplots, each of which was randomly assigned a different fertilization rate treatment: the full amount of nitrogen fertilizer used by local farmers (3/3), two-thirds the full rate (2/3), and one-third the full rate (1/3), and no fertilizer (0/3). The full design thus had 48 plots. This process was performed for artichokes and repeated for annual row crops, for a total of 96 plots which occupied 1.2 ha. This experimental design allowed comparison of both irrigation with different water types and the effect of varying fertilization rates. The fertilization rates were designed to elucidate the value of the two effluents as a supplement to fertilization (Sheikh et al. 1990).

Five years of field data were collected and analyzed. Table 9.6 lists physical and chemical properties of irrigation waters which were used in MWRSA. The following results and conclusions were extracted from the Monterey Wastewater Reclamation Study for Agriculture – Final Report (Engineering-Science 1987).

9.5.2.2 Results of Public Health Studies: Virus Survival

Monitoring for the presence of naturally occurring animal viruses showed that the influent to the two pilot processes (Castroville unchlorinated secondary effluent) contained measurable viruses in 53 of the 67 samples taken. The median concentration of virus was 2 plaque-forming units per liter (PFU/l): 90% of the samples contained less than 28 PFU/l. During the approximate 5-year period, no in situ viruses were recovered from the chlorinated tertiary effluent of either process. No viruses were recovered from any of the crop samples. This was also the case for the soil irrigated with the reclaimed water.

Virus Seeding of Plants and Soil. Although no in situ viruses were recovered from irrigated plants and soil, it was important that an estimate be made of the ability of virus to survive under these conditions. Virus survival measurements were made in the laboratory and under field conditions. In the laboratory, the times required for a 99% die-off of the viruses (T_{99}) ranged from 7.8 days for broccoli to 15.1 days for lettuce. In field studies in Castroville, the T_{99} values were 5.4 days for artichokes, 5.9 days for romaine lettuce, 7.8 days for butter lettuce.

The survival of virus in Castroville soil was determined both under environmental chamber conditions and under field conditions. The T_{99}

Table 9.6. Chemical characteristics of irrigation waters used in the experimental fields (in mg/l unless otherwise noted (After Sheikh et al. 1990)

Parameter[a]	Well water		Title-22 water		Filtered effluent	
	Range	Median	Range	Median	Range	Median
pH[a]	6.9–8.1	7.8	6.6–8.0	7.2	6.8–7.9	7.3
Electrical conductivity[b]	400–1344	700	517–2452	1256	484–2650	1400
Calcium	18–71	48	17–61.1	52	21–66.8	53
Magnesium	12.6–36	18.8	16.2–40	20.9	13.2–57	22
Sodium	29.5–75.3	60	77.5–415	166	82.5–526	192
Potassium	1.6–5.2	2.8	5.4–26.3	15.2	13–31.2	18
Carbonate, as $CaCO_3$	0.0–0.0	0.0	0.0–0.0	0.0	0.0–0.0	0.0
Bicarbonate, as $CaCO_3$	136–316	167	56.1–248	159	129–337	199.5
Hardness, as $CaCO_3$	154–246	202.5	187–416	217.5	171–435	226.5
Nitrate as N	0.085–0.64	0.44	0.18–61.55	8.0	0.08–20.6	6.5
Ammonia as N	ND[e]–1.04	–	0.02–30.8	1.2	0.02–32.7	4.3
Total phosphorus	ND–0.6	0.02	0.2–6.11	2.7	3.8–14.6	8.0
Chloride	52.2–140	104.4	145.7–7.841	221.1	145.7–620	249.5
Sulfate	6.4–55	16.1	30–256	107	55–216.7	84.8
Boron	ND–9	0.08	ND–0.81	0.36	0.11–0.9	0.4
Total dissolved solids	244–570	413	643–1547	778	611–1621	842
Biochemical oxygen demand (BOD)	ND–33	1.35	ND–102	13.9	ND–315	19
Adjusted SAR[c]	1.5–4.2	3.1	3.1–18.7	8.0	3.9–24.5	9.9
MBAS[d]	–	–	0.095–0.25	0.136	0.05–0.585	0.15

[a] Standard pH units.

[b] Decisiemens per meter (dS/m).

[c] Adjusted sodium adsorption ratio. adj. $SAR = \dfrac{Na}{\sqrt{\dfrac{Ca + Mg}{2}}}[1 + (8.4 - pH_c)]$.

[d] Methylene-blue active substance (MBAS).

[e] ND = Chemical concentration below detection limit. Detection limits are as follows: NH_3-N = 0.02 mg/l; P = 0.01 mg/l; B = 0.02 mg/l; BOD = 1 mg/l; and MBAS = 0.05 mg/l.

values for the decay of virus under environmental chamber conditions were respectively, 5.4, 9.7, and 20.8 days for 60, 70, and 80% relative humidity. In the field the T_{99s} were 5.2 and 4.8 days for runs one and two, respectively. Thus, the rate of virus removal under chamber and field conditions was quite similar. No viruses were recovered from any soil section after 12 to 14 days of exposure.

Bacteria and Parasites. During the 5 years of the study, the quality of irrigation waters improved because of the continued improvement in treatment plant operations and storage procedures. All three types of waters, including the well water control, periodically exhibited high coliform levels. No Salmonellae, Shigellae, *Ascaris lumbricoides, Entamoeba histolytica,* or other parasites were ever detected in any of the irrigation waters.

The levels of total and fecal coliform in soils and plant tissue irrigated with all three types of water were generally comparable. No significant difference attributable to water type was observed. No parasites were ever detected in soil samples. Parasites were detected in plant tissue only in Year One, and there were no differences in level of contamination between effluent and well water-irrigated crops.

Sampling of neighboring fields detected no relationship between bacteriological levels and the distance from the field site. The aerosol transmission of bacteria was thus deemed unlikely.

Groundwater Protection. No discernible relationship existed between the quality of the shallow ground water underlying the site and the type of applied irrigation water. An examination of all water quality data collected suggests that the ground water quality trends were associated with trends generally applicable in irrigated areas such as increased TDS and nitrate.

Aerosols. It was concluded early in the field operations of MWRSA that aerosol-carried microorganisms from FE sprinklers were not significantly different from those generated by well-water sprinklers. This finding was verified through replications both in daytime and nighttime operations to account for die-offs of organisms caused by ultraviolet rays of the sun. Subsequently reported studies by others have corroborated these findings and established the safety of aerosols from an FE spray.

Health of Field Workers. In addition to these studies, the health status of each person assigned to the field tasks in MWRSA was monitored regularly through frequent questionnaires and through initial and exit medical examinations administered by qualified medical professionals. One hundred questionnaires were completed by personnel during the 5 years. No complaints could be related by personnel during the 5 years. No complaints could be related to contact with treated wastewater effluents. No formal epidemiological investigation was deemed appropriate or necessary for the purposes of MWRSA.

9.5.2.3 Results of Agricultural Studies: Irrigation Water Quality

As one would expect, the two treated effluents had higher levels of most chemical and metal constituents than did well water. The nutrient value of both effluents was substantial. The salt content of irrigation waters was important because of the potential for deleterious effects on crops and soils. Sodium content of irrigation waters was of particular concern because high levels of sodium along with low salinity can create poor soil physical conditions, which reduce permeability.

Salinity of irrigation waters was determined by measuring electrical conductivity (EC) and total dissolved solids (TDS), as well as the concentration of boron, chloride, sodium, bicarbonate, calcium, and magnesium. Concentrations of TDS less than 480 mg/l are recommended for irrigation waters, and levels above 1920 mg/l are considered to be a severe problem (Ayers and Westcot 1985).

Levels of EC, TDS, boron, chloride, and sodium in the two treated effluents were comparable and were higher than those in well-water. Concentrations of TDS in all three types were below the "severe problem" range, but effluent TDS fell into the range of "increasing problems". Levels of magnesium and calcium were similar in all three water types. Bicarbonate levels were higher in filtered effluent than in the other two water types, which showed similar concentrations. Irrigation water data indicate that the reclaimed water is generally in the favorable range for irrigation, because high SAR is accompanied by similarly high salinity.

Heavy Metals in Soils. None of the nine heavy metals studied (cadmium, zinc, iron, manganese, copper, nickel, cobalt, chromium, or lead) manifested any consistently significant difference in concentration among plots irrigated with different water types. Furthermore, except in the case of copper, no increasing trends with time over the 5 years were observed. The gradual increase observed for copper occurred equally for all water types, and at the end of the 5 years, copper concentrations were still below the average for California soils. Iron was generally measured at higher concentrations in the well water than in either effluent. Zinc, however, was higher in both effluents than in well water, although the actual concentrations were on the order of 0.1 mg/l in the two effluents. At these levels, uptake by plants would be faster than accumulation from irrigation input.

Input of zinc and other heavy metals, from the commercial chemical fertilizer impurities, is far greater and accounts for the large concentration differences observed at the three soil depth samples throughout the 5 years. The differences have occurred over many decades of continuous farming with regular applications of chemical fertilizers.

Heavy Metals in Plant Tissues. The same nine metals studied in the soils were also investigated in samples of the edible tissues of plants collected at harvest at each of the 96 subplots. The most important of the many results is that no consistently significant difference in heavy metal concentrations was observed in plants irrigated with either effluent and with well water in any of the 16 samplings over the five-year field trials.

Analysis of cadmium and zinc in residual tissue produced results very similar to those from edible tissues, i.e., no consistent, significant differences were observed between plants irrigated with well water and with either of the two reclaimed waters. However, consistent differences in the accumulation of zinc and cadmium were observed between edible and residual tissues (higher cadmium in residual tissues and higher zinc in edible tissues for all vegetables studied). This difference in accumulation is in fact fortuitous, because it results in relatively higher zinc to cadmium ratios in the edible portion of the crops, belileved to be a safeguard against cadmium bioaccumulation and the resultant health hazards (Engineering-Science 1987).

Soil Permeability. Infiltration rates in lettuce fields were highest in plots irrigated with well water, but these levels were not significantly different because of the great variation of infiltration rates within each water type. Infiltration rates in the artichoke field were higher than in the lettuce field. This is probably due to the fact that the artichoke field received less irrigation water and was less frequently compacted by equipment used for field preparation.

Crop Yields. Artichoke yields were similar for all three water types; in the first 2 years, the different fertilization rates had no effect on yield. In the last 3 years, a significant effect of fertilization became apparent. All three fertilization rates showed significantly higher yields than did the unfertilized plots. There were, however, no significant differences in yield among the 1/3, 2/3, and 3/3 rates. The typical full fertilization rate may thus be in excess of the artichoke plants requirements. The lack of fertilization effect in the first 2 years may have been due to the presence of residual fertilizer left by previous overfertilization.

For most vegetables, yield was somewhat higher with irrigation with FE and Title-22 than with well water, and increases in yield with increasing fertilizer tended to level off at the 2/3 fertilizer rate. Yields of all seven lettuce crops were similar for the three different water types. Increases in lettuce yield tended to level off at the 2/3 rate.

Crop Quality. Field quality assessments and shelf-life measurements uncovered no differences between produce irrigated with reclaimed water and that irrigated with well water. Visual inspection of artichoke plants in the field showed no differences in appearance or vigor of plants irrigated with different water types. Occasional problems with mouse damage were not related to water type.

Shelf life and quality of row crops were similar for all water type treatments. No problems with increased spoilage of produce irrigated with effluents were encountered.

9.5.2.4 MWRSA Findings

Based on virological, bacteriological and chemical results from sampled vegetable tissues, irrigation with filtered effluent or T-22 appears to be as safe as with well water.

After 5 years of field experimentation, results show few statistically significant differences in measured soil or plant parameters attributable to the different water types. None of these differences has important implications for public health. Yield of annual crops is often significantly higher with reclaimed water.

No enteric virus was detected in any of the reclaimed waters sampled although it is often detected in the secondary effluent.

The T-22 process is somewhat more efficient than the FE process in removing viruses when influent is artificially inoculated (seeded) at extremely high rates. Both treatment systems removed more than five logs of poliovirus (i.e., 99.999% removal).

Marketability of produce is not expected to be a problem.

The cost of producing filtered effluent (after secondary treatment) is estimated to be $0.06/m^3, excluding conveyance and pumping costs.

9.6 Health and Safety of Using Reclaimed Wastewater for Irrigation

In every wastewater reclamation and reuse operation, there is some risk of human exposure to infectious agents. The contaminants in reclaimed wastewater that are of health significance may be classified as biological and chemical agents. For most of the uses of reclaimed wastewater, pathogenic organisms pose the greatest health risks, which include bacterial pathogens, helminths, protozoa, and viruses.

To protect public health without unnecessarily discouraging wastewater reclamation and reuse, there have been many efforts to set up conditions and regulations that would allow for safe use of reclaimed wastewater for irrigation. Although there is no uniform set of federal standards for wastewater reclamation and reuse in the United States, several states have developed wastewater reclamation regulations, often in conjunction with regulations on land treatment and disposal of wastewater. Reclaimed wastewater regulations for a specific irrigation use are based on the expected degree of contact with the reclaimed wastewater and the intended use of the irrigated crops. For example, the State of California requires that reclaimed water used for landscape irrigation of areas with unlimited public access must be "adequately oxidized, filtered, and disinfected prior to use", with median total coliform count of no more than 2.2/100 ml (State of California 1978, 1981). To achieve these requirements, it requires the wastewater treatment processes consisting of biological secondary treatment and terti-ary treatment with filtration followed by disinfection (cf. Table 9.5).

Wastewater reuse regulations adopted by the State of Arizona contain enteric virus limits; for example, not to exceed 1 plaque-forming unit (PFU) per 40 l for the most stringent reclaimed water applications such as spray irrigation of food crops. In the similarly stringent reuse applications, the State of Florida requires tertiary treatment to maximize disinfection effect-iveness, resulting in no detectable fecal coliform per 100 ml by maintaining 1.0 mg/l total chlorine residual for 30 min contact time at average daily flow. Although these wastewater reuse standards lack explicit epidemiological evidence to assess health risks, they have been adopted as the attainable and enforceable regulations in planning and implementation of wastewater reclamation and reuse in the United States.

Further safety measures for nonpotable water reuse applications include: (1) installation of separate storage and distribution systems of potable water; (2) use of color coded labels to distinguish potable and nonpotable installa-tion of the pipes; (3) cross-connection and backflow prevention devices; (4) periodic use of tracer dyes to detect the occurrence of cross contamina-tion in potable supply lines; and (5) irrigation during off hours to further minimize the potential for human contacts.

9.6.1 Wastewater Reclamation Criteria in Other Countries

Reclaimed water quality criteria for protecting health in developing coun-tries must be established in relation to the limited resources available for public works and other health delivery systems that may yield greater health benefits for the funds spent. Confined sewage collection systems and waste-water treatment are often non-existent and reclaimed wastewater often provides an essential water resource and fertilizer source. Thus, for most developing countries, the greatest concern for the use of wastewater for irrigation are caused by raw sewage or inadequately treated wastewater which contain the enteric helminths such as hookworm, ascaris, trichuris, and under certain circumstances, the beef tapeworm. These infectious agents as well as bacterial pathogens can damage the health of both the general public consuming the crops irrigated with wastewater and sewage farm workers and their families (Feachem et al. 1983; Shuval et al. 1986).

The World Health Organization (WHO) recommended that crops eaten raw should be irrigated with treated wastewater only after biological treatment and disinfection to achieve a coliform level of not more than 100/100 ml in 80% of the samples (Shuval et al. 1986). These WHO wastewater reuse standards have generally been accepted as a reasonable basis for design of wastewater treatment systems for reuse in many Medi-terranean countries. However, in those Middle East countries which have recently developed facilities for wastewater reuse, the tendency has been to adopt more stringent health criteria similar to the California regulations. This arises out of the desire to protect an already high standard of public health by preventing at any expense the introduction of pathogens into the human food chain (Pescod and Arar 1988).

9.6.2 Wastewater Irrigation in Developing Countries

Wastewater irrigation practices vary widely throughout the world. Some of the major differences are related to the quality of the reclaimed water applied, the rates and total volume applied, the degree of treatment before application, storage requirements, irrigation methods used, and the type of crops grown. Climate is important as it influences such requirements as the amount of water applied, storage volumes, the type of crops and, possibly, the method of irrigation. Although in the past it was common practice to use untreated municipal wastewater to irrigate crops (a practice followed even today in some regions of the world), the need for some preapplication treatment of the wastewater is now recognized in most regions.

As reported elsewhere, because of the strict water quality regulations in the United States that apply to the direct irrigation of food crops, secondary effluent needs to undergo tertiary filtration and disinfection. In other countries, less rigid wastewater reclamation criteria are used. In fact, in many parts of the world, irrigation water may be drawn from rivers and streams that are often heavily polluted with inadequately treated or un-treated wastewater. It can, therefore, be assumed in such situations that a substantial health risk would result from the unrestricted irrigation of agricultural crops with untreated or inadequately treated wastewater. An-other problem is to protect the health of agricultural workers when waste-water is used. It has been reported from India that hookworm and other enteric infections are much more common among workers on sewage farms using untreated wastewater than among the farming population in general (WHO 1981). On the other hand, a follow-up study of the health of workers at municipal wastewater treatment plants in the USA did not reveal any excessive risk of disease or disability in this group (Pahren and Jakubowski 1980).

9.7 Appropriate Technologies for Wastewater Treatment

Appropriate technologies are derived from a variety of sources. Most, though not all, technologies are appropriate to the specific time and place in which they are developed. Some can be transferred to other times and places and many can be improved in stages as additional resources become available (Gunnerson and Kalbermatten 1979).

Technology options for wastewater treatment and disposal are limited by the resources available and the cost considerations of conventional and unconventional alternatives. Existing or traditional wastewater treatment facilities are often cost-effective and frequently can be transferred or upgraded in stages as funds become available. In many situations, when the intent of wastewater treatment is to minimize the probability of human exposure to pathogenic organisms as exemplified in irrigation with wastewater, storage ponds and waste stabilization ponds are often the appropriate technologies. For small communities and/or developing countries, waste stabilization ponds, aquatic systems, and similar low-rate biological processes are usually the most cost-effective method of pretreatment for agricultural irrigation. Where a high quality effluent is required a recirculating sand filter may be appropriate.

9.7.1 Selection of Appropriate Technologies

Important issues in the selection of appropriate technologies for developing countries include: (1) local health concerns; (2) required effluent quality; (3) required treatment plant capacity; (4) initial capital cost; (5) operation and maintenance costs; and (6) required energy for treatment.

The order of importance of the above factors will vary with each reuse application. For example, aerated lagoons and stabilization ponds can be used to treat municipal wastewater adequately for most irrigation purposes. Pond systems also have the advantage of acting as a storage reservoir for nonirrigating seasons. A major factor to consider when deciding whether to construct stabilization ponds is the amount of land they require. If little land is available near a wastewater source, untreated or treated wastewater will have to be pumped to stabilization and/or storage ponds in the closest agricultural area.

From the data and discussion presented in this section, and other parts of the chapter, it can be concluded that low-rate biological processes offer significant economic advantages, especially for small communities. Further, the operation of the low-rate systems is not dependent on the availability of highly skilled operating personnel. Also, because significant reductions in pathogenic organisms can be achieved in pond systems, these systems are well suited for many developing countries where water is short and resources are limited. Where higher levels of treatment are needed, aquatic and recirculating sand filter systems and other energy intensive systems may be more feasible options.

References

Asano T, Smith RG, Tchobanoglous G (1985) Municipal wastewater: treatment and reclaimed water characteristics. Pettygrove GS, Asano T (eds) Irrigation with reclaimed municipal wastewater – a guidance manual. Lewis, Chelsea MI

Asano T, Richard D, Crites RW, Tchobanoglous G (1992) Evolution of tertiary treatment requirements for wastewater reuse in California. J Water Environ Technol 4, 2:36–41

Ayers RS, Westcot DW (1985) Water quality for agriculture. FAO Irrigation and drainage paper 29, rev 1. FAO, Rome

Berg G, Metcalf TB (1978) Indicators of viruses in waters. Berg G (ed) Indicators of viruses in water and food. Ann Arbor Science Publishers, Ann Arbor

Culp/Wesner/Culp (1979) Water reuse and recycling, vol 2: evaluation of treatment technology. US Dep Interior, Off Water Res Technol, Washington, DC

Engineering-Science Inc (1987) Monterey wastewater reclamation study for agriculture. Monterey Regional Water Pollution Agency, Berkeley

Feachem RG, Bradley DJ, Garelick H, Mara DD (1983) Sanitation and disease – health aspects of excreta and wastewater management. World Bank studies in water supply and sanitation 3. John Wiley & Sons, Chichester

Geldreich EE (1978) Bacterial populations and indicator concepts in feces, sewage, stormwater and solid wastes. In: Berg G (ed) Indicators of viruses in water and food. Ann Arbor Science Publishers, Ann Arbor

Gunnerson CG, Kalbermatten JM (eds) (1979) Appropriate technology in water supply and water disposal, ASCE, New York

Kirkpatrick WR, Asano T (1986) Evaluation of tertiary treatment systems for wastewater reclamation and reuse. Water Sci Technol 19, 10:83–95

Metcalf & Eddy Inc (1991) Wastewater engineering: treatment, disposal, and reuse, 3rd edn. McGraw-Hill, New York

Pahren H, Jakubowski W (eds) (1980) Wastewater aerosols and disease. In: Proc Symp, Sept 19–21, 1979, US Environ Protect Ag, Cincinnati

Pescod MB, Arar A (eds) (1988) Treatment and use of sewage effluent for irrigation. Butterworths, London

Pettygrove GS, Asano T (eds) (1985) Irrigation with reclaimed municipal wastewater – a guidance manual. Lewis, Chelsea MI

Reed SC, Crites RW (1984) Handbook of land treatment systems for industrial and municipal wastes, Noyes, Park Ridge NJ

Sheikh B, Cort RP, Kirkpatrick WR, Jaques RS, Asano T (1990) Monterey wastewater reclamation study for agriculture. Res J Water Pollut Control Fed 2, 3:216–226

Shuval HI, Adin A, Fattal B, Rawitz E, Yekutiel P (1986) Wastewater irrigation in developing countries – health effects and technical solutions. World Bank Tech Pap 51, Washington, DC

State of California (1978) Wastewater reclamation criteria, an excerpt from the California Code of Regulations, Title 22, Div. 4, Environmental Health, Department of Health Services, Berkeley, CA

State of California (ed) (1981) Evaluation of agricultural irrigation projects using reclaimed water. Prepared by Boyle Engineering Corporation, Office of Water Recycling, State Water Resources Control Board, Sacramento CA

Sterritt RM, Lester JN (1988) Microbiology for environmental and public health engineers. Spon, London

WHO-World Health Organization (1981) Health aspects of treated sewage reuse. In: Rep WHO Sem, Algiers 1–5 June 1980. EURO Rep Stud 42. WHO Reg Off Eur, Copenhagen

WPCF (1989) Water reuse. Manual of practice SM-3, 2nd edn. Water Pollution Control Federation, Washington, DC

10 Drainage Water Treatment and Disposal

E.W. Lee

10.1 Introduction

Effective management of water resources demands the protection of both surface and ground waters from the impacts of water quality deterioration from irrigated lands. It is a common agricultural practice to reuse surface return water or to discharge runoff to local water courses, usually for downstream reuse. However, this practice has limitations for the disposal of subsurface drainage, which contains not only accretions of dissolved salts but also trace elements that may be of environmental concern. Irrigation technologies which improve the efficiency of water conservation practices will reduce the quantity of subsurface drainage and thus provide some mediation of deleterious impacts. However, in many locations, irrigation improvements are not adequate to meet regulatory requirements for water quality standards in receiving waters. In such cases, irrigation practices must be supplemented by drainage water treatment. In recent years the growing concern for the protection and the preservation of wildlife habitat and water quality has presented a new challenge to water resource managers.

This section will discuss state-of-the-art drainage water treatment, reuse, and disposal technologies as well as current research, potential applications, and future directions.

10.2 Treatment Technology

10.2.1 Objectives

Although ancient civilizations have been ruined by salinization of productive agricultural lands (FAO 1973), the simple expedient of drainage disposal into local water courses has sustained agriculture throughout history. Today, however, water resource managers are faced with the technical challenge of halting or slowing the deterioration of water quality to levels that can seriously impact both agriculture and the environment over time.

Gappa (1990) reported that steadily increasing salt loads in the Colorado River Basin have demonstrated the need for salinity control. These salt

Adv. Series in Agricultural Sciences, Vol. 22
K.K. Tanji/B. Yaron (Eds.)
© Springer-Verlag Berlin Heidelberg 1994

imbalances are largely attributed to irrigation return water and subsurface drainage throughout the basin. While salinity and associated boron and sodium accumulations have long concerned growers, toxic trace elements are emerging as the latest threat to sustainable agriculture and environmental quality.

Ohlendorf et al. (1988) studied the impact on aquatic birds of selenium in subsurface drainage at the Kesterson Reservoir in San Joaquin Valley, California. Such reports on selenium toxicosis have heightened the public's awareness of the health and environmental hazards of toxic trace elements (Klasing et al. 1990). Since selenium tends to bioaccumulate in the land and aquatic food chains, there is a growing urgency to manage drainage disposal in an ecologically sound manner.

The primary objective of drainage treatment technologies is to reduce salt and trace element concentrations to protect the beneficial uses of local water resources. Broader goals include protecting public health and environmental values. The challenge in achieving these goals is to make applied technologies both affordable for the agricultural economy and sufficiently effective to meet the water quality objectives established by regulatory agencies.

The chemical composition of drainage, the complex physiochemical and biochemical interactions of the major constituents, and the presence of toxic trace elements create major difficulties in managing drain water. Table 10.1 illustrates the chemical and physical composition of subsurface drainage in the San Joaquin Valley, California, while Table 10.2 lists regulatory water quality objectives. A comparison of these tables shows that total dissolved solids (TDS), a measure of drainage salt content (average 9820 mg/l), must be greatly reduced to meet the receiving water standards of 1 dS/m (approx. 600 mg/l) for San Joaquin River water. The selenium content of drainage into the San Joaquin (average 325 µg/l) must be reduced by more than 99% to meet the regulatory objective of 5 µg/l. Boron in drainage water (14 400 mg/l) must be reduced by more than 99% to 700 µg/l. Molybdenum in drainage water (88 µg/l) must be reduced by at least 88% to 10 µg/l. These are stringent demands for technical development to meet environmental objectives.

10.2.2 Available Technologies

Conventional municipal and industrial wastewater treatment technologies can be applied to agricultural drainage, as reviewed by Hanna et al. (1990) and Lee (1991) for a drain located in the San Joaquin Valley, California. The literature indicates that, although research in desalting is widely documented, there is very limited experience in the removal of trace toxic substances from a large volume of drainage. Numerous technologies have

Table 10.1. Drainage water analysis: San Luis Drain at Mendota, San Joaquin Valley, California (Lee et al. 1988)

Constituent	Units	Average	Maximum
Sodium	mg/l	2,230	2,820
Potassium	mg/l	6	12
Calcium	mg/l	554	714
Magnesium	mg/l	270	326
Alkalinity (as $CaCO_3$)	mg/l	196	213
Sulfate	mg/l	4,730	6,500
Chloride	mg/l	1,480	2,000
Nitrate/nitrite (as N)	mg/l	48	60
Silica	mg/l	37	48
TDS	mg/l	9,820	11,600
Suspended solids	mg/l	11	20
Total organic carbon	mg/l	10.2	16
COD	mg/l	32	80
BOD	mg/l	3.2	5.8
Temperature[a]	°C	19	29
pH	-	8.2	8.7
Boron	µg/l	14,400	18,000
Selenium	µg/l	325	420
Strontium	µg/l	6,400	7,200
Iron	µg/l	110	210
Aluminum	µg/l	< 1	< 1
Arsenic	µg/l	1	1
Cadmium	µg/l	< 1	20
Chromium (total)	µg/l	19	30
Copper	µg/l	4	5
Lead	µg/l	3	6
Manganese	µg/l	25	50
Mercury	µg/l	< 0.1	< 0.2
Molybdenum	µg/l	88	120
Nickel	µg/l	14	26
Silver	µg/l	< 1	< 1
Zinc	µg/l	33	240

[a] Temperature varied from 23–25 °C (summer) to 12–15 °C (winter).

Table 10.2. Water quality objectives for San Joaquin River, California (Lee et al. 1988)

Constituent	Objective
Selenium (wetland use)	2 µg/l
Selenium (in river)	5 µg/l
Electrical conductivity	1.0 dS/m
Boron	700 µg/l
Molybdenum	10 µg/l

been identified with potentials, but there are few viable processes that can achieve the dual challenges of technical effectiveness and economic feasibility.

Early studies in drainage management concentrated on reducing salts, but this emphasis gradually shifted to controlling nutrients, pesticides and herbicides, and currently focuses on managing toxic trace elements. Select technologies have been under research and development ranging from laboratory to field level pilot plants. While advances in desalination have gained global recognition, major research in the removal of nutrients and toxic trace elements has been confined to developed countries, mainly the western USA. The following sections review these developments and their potential application to other locations throughout the world.

10.2.2.1 Desalination

Desalting technology has been used in many areas of the world, mainly for converting marine and brackish water to drinking water. In most situations, cost is not a major constraint in using desalting technology to meet basic human needs.

Reverse osmosis (RO) purportedly has the greatest potential for desalting agricultural drainage. Distillation methods can also separate salts but are not considered affordable for drainage waters because of their high energy requirements.

The practical application of RO has been amply documented in the technical literature and demonstrated in many facilities (Culp/Wesner/Culp 1979; PCR Toups 1982). RO involves the use of semipermeable membranes to separate salts. Saline feed water is pressurized so that product water is forced through membranes while salts are left behind in a brine stream. In many situations, pretreatment of the feed water is necessary to mitigate the effects of fouling and deterioration to prolong the useful life of the membranes.

The RO process was evaluated by the consultant CH2M Hill (1986) for desalting drainage water and removing toxic trace elements in the San Joaquin Valley. This conceptual study was based on existing off-the-shelf RO technologies for a $0.4 \text{ m}^3/\text{s}$ (10 mg/d) plant, and on drainage collected at the San Luis Drain at Mendota, California (see Table 10.1). The consultants concluded that the TDS concentration could be reduced to a range of 550 to 650 mg/l and selenium to about 10 to 20 µg/l in the product water. About 7 to 8 mg/l of boron would remain in the finished water. For the drainage under evaluation, the RO treatment would cost about $1344 per 1000 m^3 ($1090/ac-ft). These costs included capital and operating and maintenance expenses, but added costs would be incurred for the collection of drainage and the disposal of residual brine. Cost recovery for product water was not considered.

Development of RO in the San Joaquin Valley has been initiated by the California Department of Water Resources (1986) with a pilot demonstration project at Los Banos. The objectives were to study pretreatment processes and to reduce treatment costs to affordable levels. Further studies are scheduled to determine the efficiency of RO for treatment of subsurface drainage.

Under a 1972 agreement with Mexico, the USA developed plans for salinity control in the Lower Colorado River (USBR 1985), where agricultural drainage is a major contributor to salts in the basin. The US Department of Interior, Bureau of Reclamation, has been testing RO systems and performing pilot demonstration studies at Yuma, Arizona since 1974. It has designed a full-scale, 2.88 m^3/s (72 mg/d)-capacity plant, scheduled to begin operating in 1992, to meet the international obligations for water quality of the Colorado River flowing into Mexico. Based on 1986 prices, the costs for desalting at this plant are estimated at $759 per 1000 m^3 ($615/ac-ft). These include costs for capital investment, maintenance, and operation, but do not include drainage collection or brine disposal costs.

Although promising, RO technologies are considered costly for the desalting and removal of toxic trace elements from agricultural drainage, but further developments may reduce present cost estimates. The integration of RO technologies with the reclamation of salts of commercial value and the sale of product water can offset these costs. As discussed later, these integrated systems are under evaluation.

10.2.2.2 Biological Processes

Conventional municipal wastewater technologies include biological processes with a potential for application to the treatment of drainage water. Further research and development is needed to realize this potential. Anaerobic treatment with bacteria is being used to treat drainage water from the San Joaquin Valley, California (EPOC 1987; Binnie California and California Department of Water Resources 1988). In these pilot plant studies, selenium in the oxidized state is reduced by a bacterial-mediated process to its insoluble elemental form and removed by filtration or gravity separation from the waste stream. Methanol provides energy for the buildup of a dense bacterial mass in biological reactors, but other carbon materials can be used. In these studies, the oxidized forms of selenium in drainage are reduced from 300 to 550 µg/l to about 16–50 µg/l in reactors within one hour. Filtration can remove colloidal particles to further refine effluent selenium to levels ranging from 10 to 40 µg/l. With the use of ion exchange resins, selenium can be further reduced to less than 10 µg/l and, in the same process, boron can be reduced to less than 1 mg/l. The estimated costs for this anaerobic process range from $276 per 1000 m^3 ($224/ac-ft) for a 0.04 m^3/s (1 mg/d) plant to $179–$201 per 1000 m^3 ($145 to $163/ac-ft) for a

0.4 m³/s (10 mg/d) plant. These include capital investment, maintenance, and operating costs but do not include drainage collection or waste stream disposal costs.

The anaerobic process in these pilot plant studies involved sludge blanket and fluidized bed reactors, which were found to be unstable because of product gases uplifting the sludge blankets and solid buildup in the fluidized bed media (Schroeder et al. 1990). For stable performance, sequential batch reactors were tested under laboratory conditions using selenium-laden drainage from the San Joaquin Valley. In these bench studies, selenium was reduced to detectable levels (1 µg/l) and preliminary estimates indicated that costs would be less than $123 per 1000 m³ ($100/ac-ft). Further pilot plant studies are scheduled at a field location for this biological process.

An innovative microalgal bacterial process for the reduction of selenium in drainage was studied in the San Joaquin Valley (Oswald and Gerhardt 1990). Microalgae are grown in open ponds at high rates and harvested by a low-cost bioflocculation process for fermentation in anaerobic digesters. Drainage is introduced into the digesters to reduce oxidized selenium to less soluble forms for separation. Methane gas is generated by bacteria in the digesters and this can be used to power the plant operation. The combustion gases are introduced to the high-rate algal pond, as a supplemental carbon source and by-pass heat is used to raise the temperature of the digesters to increase the rate of bacterial conversion of methane.

In laboratory bench experiments and pilot plant studies in the field, digested algal material was able to reduce soluble selenium in drainage from 367 µg/l down to 20 µg/l, an about 95% removal. The remaining selenium is in soluble and colloidal forms. Further reduction can be achieved by sand filtration or dissolved air flotation with the addition of ferric chloride. Reduction of total selenium level down to about 7 µg/l can be achieved in laboratory studies. This process can also reduce to some extent other trace elements such as barium, boron, chromium, cobalt, copper, lead, manganese, nickel, vanadium and zinc.

Nitrate in drainage in excess of 10 µg/l as N can inhibit the reduction of selenate but selenate reduction can still be effective in the presence of low (2–5 mg/l) nitrate, but at slower reaction rates. However, nitrate can be denitrified and removed in the digesters but this can consume portions of algal organics that are needed for selenate reduction. Unless excessively high nitrates (greater than 50 mg/l as N) are present, this is not considered a major problem for drainage treatment. Atmospheric oxygen also can inhibit the selenium reduction and nitrate denitrification processes. Therefore, anaerobic conditions must be maintained in the digesters to provide the necessary environment leading to selenate reduction. Further pilot plant studies are needed to optimize the microalgal bacterial process for proto-type development but findings to date are promising for this innovative technology.

Preliminary cost estimate for this process range from $125 per 1000 m³ ($102/ac-ft) in a 0.4 m³/s (10 mg/d) plant to $84 per 1000 m³ ($68/ac-ft) in a

$4 \text{ m}^3/\text{s}$ (100 mg/d) plant. These include capital investment, maintenance and operation costs but do not include drainage collection or disposal of waste streams and process residuals.

Toxic trace substances from drainage water have contaminated soil systems. In these locations, studies have been made in an in-situ bio-remediation process (Frankenberger 1990) for the microbial methylation of selenium to gas forms for volatilization, dilution and dispersion into the atmosphere. Although these are naturally occurring microbial processes, methylation rates can be greatly accelerated with the addition of organic amendments to the soil system. In addition, optimum conditions include the maintenance of moisture, aeration and metal activators. These activators include zinc, nickel, and cobalt but high nitrates inhibit the reaction. The obligate aerobic fungi require adequate moisture and good aeration for enhancing methylation and evolution of selenate. Laboratory studies demonstrated that specific fungi can detoxify soil and sediments contaminated with high concentrations of selenium by converting selenium into methylated species, mostly dimethylselenides.

Contaminated sediments in Kesterson Reservoir in the San Joaquin Valley were studied with the in-situ bio-remediation process. A dome capture device was employed to capture the volatilized selenium to determine emission rates. Casein (milk) and citrus (orange) pulp were used to stimulate high rates of volatilization. During summer conditions, emission rates of more than $800 \text{ µg/m}^2/\text{h}$ were recorded. Other organic materials were tested but lower volatilization rates were recorded. In these field studies, the half-life of selenium in the sediments ranged from 1.5 to 5.5 years. Currently, studies are in progress at Kesterson Reservoir to develop this process further before initiating full scale operations to decontaminate the soil to acceptable environmental levels.

In inland areas where discharge to receiving water is not possible, drainage water is stored in evaporation ponds for final disposal. Through surface evaporation, selenium concentrations are elevated to high levels. A process for microbial methylation of selenium from water bodies was studied by Frankenberger (1988). Laboratory and field studies in evaporation ponds in the San Joaquin Valley indicate that selenium can be methylated and volatilized from surface waters. Organic amendments to the water can stimulate these emissions. Initial studies show potentials but further studies are needed to determine the feasibility of this in-situ bio-remediation method to detoxify contaminated ponds.

10.2.2.3 Chemical Treatment

A chemical process for the treatment of selenium-laden drainage water with freshly made ferrous hydroxide was developed by chemists working at the laboratory of the US Bureau of Reclamation (USBR) at Denver, Colorado. The chemical reactions are described in an article by Murphy (1988).

Ferrous hydroxide reacts with selenates to produce insoluble elemental selenium, which can be separated from the aqueous solution. In laboratory experiments with synthetic drainage, ferrous hydroxide reduced and precipitated selenium (Rowley et al. 1989). About 99% can be removed within 30 min of contact time. The process can also remove chromium and nickel at the same time. However, field tests with San Joaquin Valley drainage water indicated slower reactions but reduction down to 1 µg/l was achievable with longer contact time. Optimum conditions require a pH range of 8–10 and are inhibited by the presence of free oxygen, nitrates and bicarbonates.

Pretreatment studies were conducted to reduce these inhibitary effects (Rowley et al. 1989). It was found that acidification followed with aeration could remove bicarbonates. Lime precipitation can also be effective. Removal of free oxygen could be accomplished with sulfur dioxide. The USBR chemists experimented with a promising method to reduce and to remove nitrates with powdered aluminium (Murphy 1991). This innovative process could overcome one of the major obstacles for the successful application of the ferrous hydroxide process.

Preliminary estimates for this treatment process is about \$86 per 1000 m^3 (\$70/ac-ft) for a 2 m^3/s (50 mg/d) plant but this can increase to over \$185 per 1000 m^3 (\$150/ac-ft), if nitrates, oxygen and bicarbonates in drainage water are excessively high. The USBR is continuing studies for possible application in drainage management plans in several locations in the western US, where water quality has been deteriorated by salts and trace elements from agricultural developments.

10.2.2.4 Physical Treatment

Many natural and synthetic resin substances can adsorb ions but the ability to adsorb specific trace substances such as selenium narrows the range of useful materials that can be used effectively for drainage water. The presence of other major ions with similar chemical characteristics as sulfates can compete for adsorption sites and thus can be serious drawbacks.

Iron filings were used in laboratory and mini-pilot plants in the San Joaquin Valley for the treatment of drainage, utilizing a process patented by Mayenkar (1986). The Harza Engineering Company (1986) conducted these studies. Iron filings were activated with oxygen so that a hydroxide adsorption surface is formed from aeration and hydration. The iron filings beds must be kept submerged after activation to prevent the formation of ferric oxide. Studies indicate that the optimum activation occurs when the feed water has a pH between 6 and 10. As the process depends on adsorption, sludge is not produced but spent iron filings are replaced for disposal and the contact beds are refilled with fresh iron filings.

Field studies in a micro-pilot plant indicated that the iron filings developed agglomerations that clog the beds and thus shorten life expectancies. The causes for the cloggings were investigated by Anderson (1989), who postulated that the formation of ferromagnetic species bonded the filings under magnetic influence. However, more studies would be needed to investigate the clogging phenomenon to determine causes and to determine if any mitigation measure could be developed to prolong the life of the iron filing beds.

Preliminary estimates indicated a wide range of costs, $86 to $352 per 1000 m^3 ($70 to $285/ac-ft), depending on the expected bed life and removal efficiency required. Currently further pilot plant studies are underway in the San Joaquin Valley to determine the cause for the iron filing clogging and to evaluate the process to develop an effective system for removing selenium from agricultural drainage.

Selective ion exchange resins have been studied for removal of selenium from drainage. A review of the literature was made by Herrmann (1985) for selenium removal with ion exchange resins. Subsequently, ion exchange processes for the treatment of drainage from the San Joaquin Valley was evaluated by Boyle Engineering Corporation (1988). The study was to determine if selenium-selective resins could be developed. Laboratory isotherm tests indicated that ion resins with large alkyl groups can have increasing selectivity for selenate over sulfate, a major competitive ion in the local drainage water. By increasing the size of the alkyl groups, it may be possible to develop greater affinity for selenium. A computer simulation model confirmed the experimental tests. Further development with selected resins with increasing alkyl groups is needed to demonstrate the feasibility of an effective ion exchange process. Development costs can be reduced if alkyl groups can be attached to existing commercial resins in lieu of synthesizing new resins.

10.3 Disposal of Residual Products

The treatment of agricultural subsurface drainage water for removal of dissolved salts and reduction of trace elements still leave residuals composed of brines, salts and waste solids. Nishimura et al. (1988) completed a review of disposal options in the inland basin of the San Joaquin Valley and identified disposal and reclamation and reuse as possible alternatives. Water resource managers must consider these options in the management of water in agriculture.

10.3.1 Disposal

There are many methods for the disposal of treatment waste streams but in many locations these are limited to: (1) discharge to local surface waters; (2) evaporation ponds; and (3) deep well injection.

10.3.1.1 Surface Waters

Disposal of surface and subsurface agricultural drainage water into receiving waters in oceans, bays, deltas, rivers and small streams is a common practice throughout the world. However, where receiving water quality standards have been established these existing practices will be subject to enforcement by regulatory agencies, which would severely restrict these open disposal practices. Many regulatory agencies have established water quality plans for coastal and inland waters to maintain salinity standards as well as toxic substances control. Enforcement actions will require consideration by water resource managers for the conveyance and disposal facilities, as well as treatment systems to meet environmental protection needs. Depending on specific locations, disposal to surface water may be the most economical approach, providing that dilution or treatment technology can be effective in maintaining established water quality standards.

10.3.1.2 Evaporation Ponds

In many inland locations, disposal of drainage is limited to evaporation ponds because hydrological confinement of basin outlets or constraints by regulatory agencies in restricting discharge to surface water foreclose other options. Evaporation ponds pose many risks to the environment because toxic substances can be concentrated to dangerous proportions or percolation from ponds can threaten ground waters.

The concentration of toxic substances can create dangerous habitats for terrestrial and aquatic animals and birds attracted to evaporation ponds. For example, selenium can bioconcentrate in the aquatic food chain to more than 35 000 times (SJVDP 1990) and therefore can threaten fish and wildlife in evaporation ponds. These pond products can be threatening to public health if consumed in excess. It was also found in these studies that high salinity in evaporation ponds can also affect the health of aquatic birds by encrustations in feathers that impact flying and diving. Drainage water can be treated to remove toxic substances before discharging into evaporation ponds. However, the concentration of toxic constituents can be increased by evaporation to nullify the reduction by treatment.

Mitigation of these potential impacts can be effected by special construction and operation of ponds. In many locations, wastes in ponds are classified for regulatory purposes according to concentrations of toxic substances. If toxic limits are exceeded, special construction on operation are needed for environmental safeguard. Special design and construction are required and these can include double lining of containment ponds, leachate collection, and rainfall and peripheral drainage controls. Operational controls may require clearing banks from weeds, hazing of aquatic birds and monitoring of toxic ponds and local ground water for buildup of toxic constituents. To meet these stringent requirements, the construction of evaporation ponds can exceed $500 000/ha ($200 000/ac), while ordinary evaporation ponds can cost about $50 000/ha ($20 000/ac). Monitoring of chemical and biological constituents in compliance with regulatory requirements can add major costs to operation.

Land requirement for evaporation ponds is an important factor in drainage management. Ponds must be suitably sited and plans must be considered for the disposal of accumulated salts. Local evaporation rates, wind conditions, precipitation, geology, and salinity levels determine land requirements and accordingly ponds must be designed for specific locations. Based on intensive field investigation of ponds in the San Joaquin Valley, Tanji et al. (1990a) estimate that each acre of tiled drained field would require 0.5 ha (0.125 ac) of ponds and that about 0.74×1000 m^3 (0.6 ac-ft) of drainage is collected annually from 0.4 ha (1 ac). Design for evaporation ponds at other geographical locations should be guided by local climatic, physical and chemical factors.

10.3.1.3 Deep-Well Injection

In some locations the disposal of waste brine in the oil industry is by injection into deep geological structures, where oil is extracted or to other underlying strata. This method has potential application for the disposal of agricultural subsurface drainage. In the San Joaquin Valley, a study was made for the application of this technology (URS 1986). The study reviewed the technical and institutional factors in deep well injection.

The geohydrological conditions of the injection sites must be suitable to receive drainage. The geology must be porous with low permeability and isolated by impermeable rock strata to protect existing usable groundwater. The potential for induced seismic events must be considered also in the injection process. Existing seismic and water and oil drilling data can provide useful information in any evaluation plan.

The chemistry of geologic formations in the injection zone must be compatible with the characteristics of injected drainage water since induced chemical reactions can clog geologic formations and reduce or terminate injection programs. In addition, biological activities in drainage can cause clogging and early failures.

In most locations, underground injection programs must meet strict regulatory requirements. Toxic substances are under rigid control in these projects to protect existing underground aquifers. Also, there are rigid regulatory procedures to gain approval for the installation of deep-well injection facilities.

The expected life of facilities is determined by geohydrological and biochemical factors, which influence the buildup of aquifer pressure during injection. Pretreatment of drainage water prior to injection may minimize physical clogging and biological fouling. The premature failure of a system can be the result of several complicating factors. In the oil industry, the life expectancy of injection wells can extend to about 35 years but an estimate for agricultural drainage water would be about 25 years. Based on these assumptions, the estimated cost for deep well injection in the San Joaquin Valley can range from \$201 to \$263 per 1000 m^3 (\$163 to \$213/ac-ft) depending on pretreatment needs.

A test well was drilled in the San Joaquin Valley down to more than 2438 m (8000 ft) in 1989 and the initiation of an injection program has been delayed pending decision to proceed with the needed capital investment by the local irrigation and drainage management agency.

10.3.2 Reclamation and Reuse

Although disposal of drainage and residual drainage products from treatment processes may meet with success, there are many locations where reclamation and reuse alternatives may be possible. For the effective management of agricultural water, good conservation practices with reclamation and reuse strategies should be incorporated. There are several approaches that should be considered.

Drainage water can be reclaimed for agricultural, municipal and industrial purposes. In practice, agricultural drainage is recycled at field level and thus reuse is part of most farming operations particularly in water short areas. However, present recycling practice has limitations in terms of eventual buildup of salts. Drainage water has limitations, depending on the purpose of reuse and the quality of the treated product water. For agricultural purposes, the reused water must meet acceptable standards for total dissolved solids, boron and molybdenum. For drinking water purposes toxic trace elements are important constituents in public health considerations.

The Yuma Desalting Project in Arizona (USBR 1985) is an example where product water from the reverse osmosis process is discharged into the Colorado River to maintain salinity standards for downstream municipal and agricultural users.

Reuse of drainage water on salt-tolerant crops is another possibility. An agroforestry program was demonstrated in the San Joaquin Valley by Cervinka (1990). In 1985 eucalyptus, casuarina, poplar, mesquite and elderica pines were planted. During the first year, freshwater was used and then subsurface drainage water was used to irrigate the trees. Harvesting of trees was envisaged to provide biomass for electrical power and pulp production. Latest progress reports on these studies indicated that soil salinity and boron buildup may have constraints on the growth of eucalyptus trees (Tanji et al. 1990b). Increase in leaching fraction need to be considered to gain a sustaining biomass yield.

There is potential to use drainage for cooling power plants in inland areas. A study was made by the California Department of Water Resources and

University of California (1978) to evaluate the use of subsurface drainage in the San Joaquin Valley and a review on the subject was made by Nishimura (1986) to update findings. Drainage water has great potentials for corrosion and scale formation in cooling system and pretreatment would be needed. The DWR study evaluated the cooling water needs for a 100 MW power-plant which required a flow of 43.5×10^3 m^3/day (11.5 mg/d). Treatment was provided by ion exchange resins and in the regeneration of resins, concentrated cooling tower blowdown was used. Brine disposal was in evaporation ponds. The total cost for treatment and brine disposal was about \$334 per 1000 m^3 (\$270/ac-ft). If toxic substances were to be concentrated to hazardous levels, special containment ponds for evaporation would be needed and this would substantially increase the total cost.

The US Bureau of Reclamation (1986) studied the use of agricultural drainage water from the Palo Verde Irrigation Outfall in the Colorado River Basin for power plant cooling. A hypothetical power plant sited near Las Vegas, Nevada, was evaluated. Pretreatment was determined to be required. The study concluded that agricultural drainage was a feasible cooling option at this inland location.

Agricultural drainage water contains a mix of salts that can be recovered from evaporation ponds or by-pass streams from treatment processes. Salt recovery can provide a repayment potential to recover costs for drainage management. The US Bureau of Reclamation (1979) studied this possibility as part of plans for drainage management in the San Joaquin Valley. More recently, Tanji et al. (1990a) provided detailed information on the magnitude and composition of salts and mineralogy of evaporation ponds, including trace elements in evaporites from agricultural drainage ponds.

Mini salt ponds were studied by EPOC Agriculture (1987) and it was concluded that commercial grade sodium sulfate could be obtained. The marketing of harvested salt is dependent on global markets and proximity to these markets. These are important factors in determining the favorable economics needed to export salts out of high salinity areas. However, further studies are needed to determine the feasibility of salt harvesting before large-scale operation can be initiated.

Other reclamation and reuse options have been proposed, including aquaculture, wetlands and salinity repulsion in surface and ground water. Most of these proposals have been conceptual and have not been evaluated for practical applications. A major problem is associated with trace toxic substances in drainage and related regulatory constraints.

10.4 Integrated Systems

Although technologies for treatment and disposal for subsurface drainage may be effective, the challenge to develop an affordable system remains difficult. Integrating systems of treatment and disposal technologies can provide cost optimization that may not be apparent from individual process.

10.4.1 Cogeneration-Desalination

Cogeneration-desalination systems can treat drainage, remove toxic trace elements, recover by-products, generate energy and produce a useable supply of water for agriculture, industry or drinking purposes. Conceptually these integrated systems have potentials to reduce costs.

One of these integrated systems was studied by URS Corporation (1987) for application in the San Joaquin Valley. The concept includes the use of gas turbines for the generation of electrical power. Heat from the combustion is used for drainage desalination for recovery of usable water and salts. Electrical power in excess of plant operational needs can be sold to help in recovering cost.

The concept was further evaluated by Resource Management International (1989) for possible application in the San Joaquin Valley. For a 100 MW project, revenue during the first 5 years is estimated to produce a positive net revenue of approximately \$370 per 1000 m^3 (\$300/ac-ft) of drainage water treated. However, as natural gas cost is expected to increase and over the next 20 years, the estimated cost will rise to approximately \$228 to \$330 per 1000 m^3 (\$185 to \$270/ac-ft). If natural gas prices do not increase, the long-term cost for drainage management would be reduced. The potential of this integrated system is sufficiently promising for the design plan for a pilot plant to be completed and construction is waiting for funding support (LPA 1990).

10.4.2 Solar Gradient Ponds

Another integrated system involves the use of solar gradient ponds developed with brines or salts from drainage residuals to extract heat, to generate electrical power and to operate desalination processes. The solar pond technology was based on development by Ormat Turbines, Ltd (1981). Using specially designed turbines and generators, Ormat demonstrated that low-temperature energy could be extracted from solar ponds and could be converted to electrical energy. The US Bureau of Reclamation (1987) demonstrated this technology in El Paso, Texas for electrical power generation and heat distillation of saline water. The application of this technology was demonstrated by the California Department of Water Resources in the San Joaquin Valley (1987).

These demonstrations showed that heat could be extracted from solar gradient ponds by heat exchangers placed in the heat storage zone in the pond or by external heat exchangers, with cooled brine returned to the pond.

  While integrated system concepts show great promise in using present state-of-the-art technologies to provide potential lower costs, practical field demonstration of prototypes will be needed to gain experience and to prove affordability of these innovative systems for application in subsurface drainage management.

10.5 Summary and Conclusions

Promising treatment, reuse and disposal technologies have been investigated in the search for viable alternatives for the management of drainage water with heavy salt loads and toxic trace elements.

Treatment studies include bench-scale models in laboratory and mini-pilot plants at the field level. It has been reported that promising treatment technologies can reduce salts and selenium to low levels in laboratory experiments. In mini-pilot plants, selenium can be reduced to 10–40 mg/l in field locations. Treatment costs range from a low of $55 per 10^3 m^3 to a high of $880 per 10^3 m^3 ($68 to $1090/ac-ft). These costs do not include provisions for the collection of drainage and the disposal of effluent, residual waste streams and other by-products.

Although many of the treatment methods have demonstrated technical promise, there remains the further need to research basic mechanisms of processes. Pilot-prototype studies are needed to determine process feasibility under field conditions and to evaluate real costs, operational problems, and environmental impacts.

Studies of treatment technologies have centered on selenium as a primary trace element of concern. Removal of other trace elements from drainage, such as boron, molybdenum, chromium, and arsenic, should be investigated in a broadly based drainage management program.

Disposal of treated effluent and residual by-products presents technical and environmental problems. Disposal of wastewater into evaporation ponds, deep geological formations and toxic pits has long-term limitations. Disposal of residual solids from treatment processes needs to be evaluated. Environmental concerns of disposal must be considered carefully in all situations since many of the salt products are toxic at elevated concentrations.

Unless adequate surface water dilution is available to meet regulatory requirements, a reclamation and reuse strategy may be the most promising long-term approach. This can involve irrigation of agroforestry and salt-tolerant plants.

Reclamation of salts from drainage is technically possible, but the economics of recovery systems depends on prevailing prices and market outlets. At the present time, salt reclamation for the commercial market can be successful only under favorable conditions of sales and transport costs.

Innovative technologies involving the integration of treatment, reclamation, reuse and residual disposal are promising alternatives. These include cogeneration-desalination and solar gradient ponds for electrical power and the reclamation of by-products. However, studies to date have used state-of-the-art technologies which have not been applied to agricultural drainage water. Further studies are needed to demonstrate the economics of integrating these promising innovative technologies.

References

Anderson MA (1989) Fundamental aspects of selenium removal by Harza process. Rep San Joaquin Valley Drainage Program, US Dep Interior, Sacramento

Binnie California and California Department of Water Resources (eds) (1988) Performance evaluation of research pilot plant for selenium removal. Fresno, CA

Boyle Engineering Corporation (ed) (1988) Report on selenium selectivity in ion exchange resins. Rep San Joaquin Valley Drainage Program, US Dep Interior, Sacramento

California Department of Water Resources (ed) (1986) Los Banos demonstration desalting facilities, status of operations. Sacramento

California Department of Water Resources (ed) (1987) Technical information record on the salt-gradient solar pond system at the Los Banos demonstration desalting facility. Fresno, CA

California Department of Water Resources and University of California (eds) (1978) Agricultural wastewater for powerplant cooling development and testing of treatment processes. Sacramento

Cervinka V (1990) A farming system for the management of salt and selenium on irrigated land (agroforestry). Cal Dep Food Agric, Sacramento

CH2M Hill (ed) (1986) Reverse osmosis desalting of San Luis Drain, conceptual level study. Rep San Joaquin Valley Drainage Program, US Dep Interior, Sacramento

Culp/Wesner/Culp (1979) Water reuse and recycling evaluation of technology. US Dep Interior, Off Water Resour and Technol, Washington, DC

EPOC Agriculture (ed) (1987) Removal of selenium from subsurface agricultural drainage by an anaerobic bacterial process. Rep Cal Dep Water Resour, Sacramento

FAO-Food and Agriculture Organization (1973) Irrigation, drainage and salinity. FAO, Rome

Frankenberger WT (1988) In-situ volatilization of selenium from evaporation ponds. Rep San Joaquin Valley Drainage Program, US Dep Interior, Sacramento

Frankenberger WT (1990) Dissipation of soil selenium by microbial volatilization at Kesterson Reservoir. Final Rep US Bur Reclamation, Sacramento

Gappa WS (1990) The Colorado River salinity program, an overall perspective. In: Proc 1990 Natl Conf Irrigation and drainage. Am Soc Civil Eng, 11–13 July 1990, Durango, CO

Hanna GP, Kipps JA, Owens LP (1990) Agricultural drainage treatment technology review. Mem Rep San Joaquin Valley Drainage Program, US Dep Interior, Sacramento

Harza Engineering Co (1986) Selenium removal study. Report to Panoche Drainage District, Firebaugh, CA

Herrmann CC (1985) Removal of ionic selenium from water by ion exchange – review of literature and brief analysis. Rep San Joaquin Valley Drainage Program, US Dep Interior, Sacramento

Klasing SA, Shull LR, Peterson RV, Rosetta TN (1990) Public health evaluation of agricultural drainage water contamination. Rep San Joaquin Valley Drainage Program, US Dep Interior, Sacramento

Lee EW (1991) Treatment, reuse and disposal of drain waters. Water Sci Technol 24:183–188
Lee EW, Nishimura GH, Hansen HL (1988) Agricultural drainage water treatment, reuse and disposal in the San Joaquin Valley of California, pt 1. Treatment. San Joaquin Valley Drainage Program, US Dep Interior, Sacramento
LPA – Land Preservation Association (ed) (1990) Cogeneration plant moves closer to reality. W Valley J, July 1990, LPA, Fresno, CA
Mayenkar A (1986) Removal of dissolved heavy metals from aqueous waste effluents. US Pat 4,565,644, Chicago
Murphy AP (1988) Removal of selenate from water by chemical reduction. Ind Eng Chem Res 27:181–191
Murphy AP (1991) Chemical removal of nitrate from water. Nature 350:21
Nishimura GH (1986) Use of agricultural drainage water for power plant cooling. Rep San Joaquin Valley Drainage Program, US Dep Interior, Sacramento
Nishimura GH, Lee EW, Hansen HL (1988) Agricultural drainage water treatment, reuse and disposal in the San Joaquin Valley of California, pt 2. Reuse and disposal. San Joaquin Valley Drainage Program, US Dep Interior, Sacramento
Ohlendorf HM, Kilness HW, Simmons JL, Stroud RK, Hoffman DJ, Moore JF (1988) Selenium toxicosis in wild aquatic birds. J Toxicol Environ Health 24:67–92
Ormat Turbines Ltd (ed) (1981) A study of the feasibility of solar pond generating facility in the State of California. Final Rep S Cal Edison Co, Los Angeles
Oswald WJ, Gerhardt MB (1990) Microalgal – bacterial treatment for selenium removal from San Joaquin Valley drainage waters. Rep San Joaquin Valley Drainage Program, US Dep Interior, Sacramento
PRC Toups (1982) Evaluation of desalination technology for wastewater reuse. Rep US Dep Interior, Off Water Res Technol, Washington
Resource Management International (ed) (1989) Concept evaluation report, selenium removal/cogeneration project. Prepared for Westlands Water District, Sacramento
Rowley LH, Moody CD, Murphy AP (1989) Selenium removal with ferrous hydroxide. US Bur of Reclamation, Denver
Schroeder E, Ergas S. Lawyer R, Pfeiffer W (1990) Microbial process for removal of selenium from agricultural drainage water. Rep San Joaquin Valley Drainage Program. US Dep Interior, Sacramento
SJVDP – San Joaquin Valley Drainage Program (1990) Fish and wildlife resources and agricultural drainage in the San Joaquin Valley. Rep of SJVDP, US Dep Interior, Sacramento
Tanji K, Dahlgren A, Quek A, Smith G, Ong C, Karajeh F, Peters D, Yoshimoto J, Otani K, Herbel M (1990a) Efficacy of evaporation ponds for disposal of saline drainage waters. Final report to Department of Water Resources, Sacramento
Tanji K, Karajeh F, Quek A (1990b) Agroforestry demonstration project: water and salt balance study. A progress report to California Department of Food and Agriculture, Sacramento
URS Corporation (ed) (1986) Deep-well injection of agricultural drain waters. Rep San Joaquin Valley Drainage Program, US Dep Interior, Sacramento
URS Corporation (ed) (1987) Agricultural drainage salt disposal. Rep Westland Water District, Sacramento
USBR – US Bureau of Reclamation (ed) (1979) Agricultural drainage and salt management in the San Joaquin Valley. Interagency Drainage Program, US Dep Interior, Sacramento
USBR – US Bureau of Reclamation (ed) (1985) Colorado River basin salinity control project, back-ground, plan and status report. USBR, Denver
USBR – US Bureau of Reclamation (ed) (1986) Study of saline water use at the Etiwanda generating station. USBR, Denver
USBR – US Bureau of Reclamation (ed) (1987) Installation and operation of the first 1000-KW solar pond power plant in the United States. USBR, Denver

Part IV Policy and Management Evaluations

11 Economics of Non-Uniform Water Infiltration in Irrigated Fields

E. FEINERMAN

11.1 Introduction

In irrigated agriculture, the unevenness with which water infiltrates the root zone is determined mainly by the spatial variability (non-uniformity) of irrigation water application, on the one hand, and on the variability of hydrologic soil properties, on the other (Warrick and Gardner 1983; Dagan and Bresler 1988). The yield of a given crop, grown during a specific season in a certain field and under certain management and cultivation conditions, is also spatially variable as it is assumed to be directly dependent on the spatially variable water infiltration (Stern and Bresler 1983; Warrick and Yates 1987; Bresler and Laufer 1988). If water infiltration is non-uniform, there will be under-irrigated and over-irrigated areas and, assuming a concave crop-water response function, total yield per unit of land area will be smaller as compared to conditions of uniform infiltration. Therefore, uniformity is considered desirable in yield production processes. It has been noted that the variation of water infiltration as well as of the yield is not completely disordered in space but can be analyzed within the frame of stochastic modeling, i.e., regarding the infiltered water and the resulting yield of a given field as random functions of space coordinates characterized by their probability density functions (pdf) and correlation structure, rather than by their deterministic values.

There is substantial literature focused on the description and modeling of spatially variable processes, which can be roughly classified into two approaches. The first is the unconditional approach, which assumes that the stationary probability density function of the variable of interest has been identified with certainty, without making direct use of the measurements on the spatial variables which serve only for the inference of the pdf. The second is the conditional approach, known also as the geostatistical approach, which was developed by Matheron (1971). This approach makes explicit use of available measurements in order to reduce the uncertainty of the random variables (or "regionalized variables" in the geostatistical terminology) at any point of the field. The geostatistical school has mainly investigated the problem of stochastic interpolation by using the kriging method (Delhomme 1978). An alternative but similar approach which relies on the classical

Adv. Series in Agricultural Sciences, Vol. 22
K.K. Tanji/B. Yaron (Eds.)
© Springer-Verlag Berlin Heidelberg 1994

statistical concept of conditional probability (assuming multivariate normal or log-normal pdf) has been employed by Dagan (1985) and Feinerman et al. (1986). The fact that the parameters characterizing the pdf inferred from a finite sample are themselves random variables has been generally disregarded in most leading studies, with a few exceptions (Kitanidis and Vomvoris 1983; Feinerman et al. 1989a).

The above-mentioned studies are not concerned, however, with the economics of spatially variable resources. Most economic studies of efficient water use were based on the assumption that field soils are perfectly homogeneous and their properties are deterministic (Yaron and Olian 1973; Moore et al. 1974; Yaron and Dinar 1982). The economic implications of non-uniform irrigation application on crop yield and profit maximization have been analyzed by Seginer (1978), Feinerman et al. (1983) and Letey et al. (1984). However, these researchers assumed that the distribution of applied water over the field is known with certainty. The impact of uncertainty was emphasized by Feinerman et al. (1985) who presented a methodology for evaluation of optimal water use in-field which depends on a spatially random function, taking into account the decision-maker's attitude to risk. This analysis was (implicitly) carried out along the lines of the above-mentioned unconditional approach. The statistical moments of interest were not conditioned on some prior information, and the randomness of the estimated parameters characterizing the pdf was not considered. Feinerman et al. (1989a) have developed a stochastic optimization model of a spatially variable irrigated field which addresses the issues of statistical inference, uncertainty of estimation, and efficient use of a limited body of available information. They compared the unconditional with the conditional approaches and concluded that the conditional analysis has the potential to increase the farmer's welfare substantially more than the unconditional approach.

Although water application uniformity can be partly controlled by the farmer via the choice of irrigation method and, e.g., sprinkler lines and spacing, most economic studies of efficient water use under non-uniform irrigation focus on optimization with respect to the quantity of irrigation water and tend to ignore the optimization with respect to the uniformity level (for example, Seginer 1978; Feinerman et al. 1983, 1985). In a few previous economic studies (for example, Hill and Keller 1980; Chen and Wallender 1984; Gohring and Wallender 1987), the joint effects of uniformity and quantity of applied water on irrigation system selection or on economical sprinkler spacing were investigated. The dependence of the irrigation system cost on uniformity was derived by varying the sprinkler spacing and calculating the resulting changes in uniformity and cost. However, the spatial variation of the irrigation water was regarded as deterministic, and uncertainty was not accounted for in the economic application. Feinerman et al. (1989b) have considered a similar problem assuming stochastic spatial variation and taking into account the farmer's attitude toward risk. Seginer (1987) presented a comprehensive review describing a general approach to economic optimization of sprinkler irrigation systems considering the quantity of applied water and application uniformity under spatially variable water application. A methodology for comparing the profitability of various irrigation methods taking into account their application uniformity performance and their associated investment and operating costs is presented in Feinerman et al. (1989c). In addition to application uniformity, this study takes explicit account of the random variability of soil properties which determines how much water infiltrates into the root zone. All the studies on joint optimization of water quantities and application uniformity were carried out using the unconditional approach, made no explicit use of available measurements to reduce uncertainty, and ignored estimation risk.

11.2 Distribution of Average Yield

This section presents a conceptual framework to calculate the statistical structure of the space-dependent crop yield. For convenience, the presentation is based on the commonly used unconditional approach. The detailed mathematical derivations of yield distribution under the conditional approach, which are much more complex, can be found in Feinerman et al. (1989a).

Consider a field of area A hectares (ha). Let $\chi \in A$ be the coordinate vector of a point in the field. To avoid unnecessary complications, it is assumed that there is no run-off or run-on and that precipitation is negligible, so that crop yield responds to the amount of water which reaches point χ. The depth of infiltered water, $Q(\chi)$, at any single point χ in the field is related to the spatial field average depth of applied water, $\bar{Q}$, by:

$$Q(\chi) = \bar{Q}\beta(\chi), \tag{11.1}$$

where $\bar{Q}$ is defined by:

$$\bar{Q} = (1/A) \int_A Q(\chi) \, d\chi, \tag{11.2}$$

and $\beta(\chi)$ is a space-dependent spatial random function, representing the degree of water infiltration uniformity. The uncertainty of $\beta(\chi)$ is related to the uniformity of water application and the variability of soil properties such as hydraulic conductivity which affect water infiltration.

The statistical structure of β is defined completely by the joint probability density function $g(\beta_1, \beta_2, \ldots, \beta_M)$ of its values at an arbitrary set of M points $\chi_1, \chi_2, \ldots, \chi_M$ within the domain definition of A. It is commonly assumed that the vector $(\beta_1, \ldots, \beta_M)$ is multivariate normal (MVN), or has been rendered such by a suitable transformation (e.g., Lumley and Panofsky

1964; Journel and Huijbregts 1978). It is also commonly assumed (the ergodic assumption) that this vector is stationary in the ordinary sense, i.e., that g is invariant under a translation in space of the points χ_i. Under this condition, the expected value:

$$E\,\beta(\chi) = \int \beta g(\beta_1, \ldots, \beta_M)\,d\beta \equiv \bar{\beta} \tag{11.3}$$

is constant, whereas the two-point covariance:

$$\mathrm{Cov}(\beta(\chi_I), \beta(\chi_{II})) = \int\int (\beta(\chi_I) - \bar{\beta})(\beta(\chi_{II}) - \bar{\beta})\,g(\beta_I, \beta_{II})\,d\beta_I d\beta_{II} \tag{11.4}$$

depends only on the lag $|\chi_I - \chi_{II}|$. The variability (or uniformity) of water infiltration can be depicted by the variance of $\beta(\chi)$, σ_β^2, which is given by:

$$\sigma_\beta^2 = \mathrm{Cov}(\beta(\chi), \beta(\chi)) \,. \tag{11.5}$$

Note that the variance σ_β^2 partly depends on the irrigation method and, for a given method, on its performance (for example, spacing between sprinklers, emitters, etc). Hence σ_β^2 is, in addition to $\bar{Q}$, a decision or man-control variable of the farmer and is not exogenous. An increase in the value of σ_β^2 describes a situation in which water application uniformity decreases and the probability moves from the center towards the tails of the pdf of β, while the mean ($\bar{\beta}$) remains unchanged. This holds true regardless of the form of the pdf of β. A change of this type is known as a "mean preserving spread" (MPS) of the distribution under consideration (for example, Sandmo 1971).

The physical model describing the spatial non-uniformity of water infiltration is linked to economic optimization models (see below) by crop-water production function. Obviously, in addition to infiltered water, yield depends on many other inputs such as fertilizers, labor, machinery, etc. Focusing on water, it is assumed hereafter that the farmer uses "best management practices" for inputs other than water. That is, these non-water inputs are assumed to be fixed and are suppressed hereafter. Thus, for K commercial yield components (for example, kernels and total dry matter in corn, lint and seeds in cotton, etc.),

$$Y_j(\chi) = f_j[Q(\chi)] = f_j[\bar{Q}\beta(\chi)] \quad j = 1, \ldots, K \,, \tag{11.6}$$

where $Y_j(\chi)$ is the yield per unit area of component j at the point χ in the field, and f_j is its associated production function.

The quantities of interest in the economic optimization are the spatial field averages over the field of all the relevant commercial yield components, which are given by:

$$\bar{Y}_j = (1/A) \int_A f_j[\bar{Q}\beta(\chi)]\,d\chi \quad j = 1, \ldots, K \,. \tag{11.7}$$

Note that $\bar{Y}_j (j = 1, \ldots, K)$ is a random variable which depends on the random function $\beta(\chi)$ and on the decision (or man-control) variable $\bar{Q}$, and its expectation is given by:

$$E(\bar{Y}_j) = (1/A) \int_A E\{f_j[\bar{Q}\beta(\chi)]\} d\chi$$

$$= (1/A) \int_A \int_{-\infty}^{\infty} f_j[\bar{Q}\beta(\chi)] g[\beta(\chi)/\sigma_\beta^2] d\chi d\beta . \tag{11.8}$$

Here, $g[\beta(\chi)/\sigma_\beta^2]$ is the probability density function of β which is conditioned on the variable σ_β^2 which, in turn, can be partly controlled.

The approximate relationships between the average yield expectation $E\bar{Y}_j$ and the expectation $(\bar{\beta})$ and the variance (σ_β^2) of β can be obtained by employing a second-order Taylor expansion of $Y_j(\chi)$ about $f_j(\bar{\beta}\bar{Q})$, which yields:

$$E(\bar{Y}_j) \approx f_j(\overline{\beta Q}) + (\bar{Q}^2 \sigma_\beta^2) \frac{\partial^2 f_j(\cdot)}{\partial(\beta\bar{Q})^2}\bigg|\beta = \bar{\beta} \quad j = 1, \ldots, K . \tag{11.9}$$

Hence, as long as the yield is a concave function of applied water, that is, $\dfrac{\partial^2 f_j(\cdot)}{\partial(\beta\bar{Q})^2} < 0$, expected yield will increase as the variance of β decreases. The highest value of $E(Y_j)$ for a given level of Q will be achieved under completely uniform water application conditions, that is, when $\sigma_\beta^2 = 0$.

The variance $\sigma_{\bar{Y}_j}^2$, of $\bar{Y}_j$ is given by:

$$\sigma_{\bar{Y}_j}^2 = E(\bar{Y}_j - E(\bar{Y}_j)]^2 = \int_{-\infty}^{\infty} [\bar{Y}_j - E(\bar{Y}_j)]^2 g[\beta(\chi)/\sigma_\beta^2] d\beta . \tag{11.10}$$

Note that the estimation of the statistical moments of $\beta(\chi)$ is based on a given sample of N available field observations $\beta_1, \ldots, \beta_N$ within the domain definition of A. The derivation of the unconditional estimates is based on a traditional approach in modeling processes which take place in heterogeneous formation. This approach is called here "unconditional" because once the parameters are estimated (by the maximum likelihood estimation procedure for multivariate normal pdf, for example), the estimates of the pdf g(·), of the expectation $E(\beta)$ [Eq. (11.3)], and of the covariance COV(·) [Eq. (11.4)] are all assumed to be independent of the values and position in space of the measurements $(\beta_1, \ldots, \beta_N)$.

The conditional analysis (which is not detailed here), unlike the unconditional one, uses the information which is embodied in the finite measured sample $(\beta_1, \ldots, \beta_N)$, not only for the inference of the pdf g(·), but to also reduce the uncertainty of the inferred random function $\beta(\chi)$ at any point in the field. The reduction is achieved by explicit conditioning of the statistical moments of interest on the positions in space of the measurement points and their values. This reduction in the uncertainty is the major difference between the unconditional and the conditional approaches.

11.3 Economic Optimization

For the economic evaluation of the optimal level of the control variable $\bar{Q}$ and of σ_β^2, a profit function II (in \$ per unit area) is defined:

$$II(\bar{Q}, \sigma_\beta^2) = \sum_{j=1}^{K} P_j \bar{Y}_j(\bar{Q}, \sigma_\beta^2) - P_Q \bar{Q} - C(\sigma_\beta^2), \qquad (11.11)$$

where P_j is the price per unit of the j-th crop yield's component, P_Q is the price per unit of water, and $C(\sigma_\beta^2)$ is a cost function, the level of which depends on the variance of β (application uniformity) such that $\partial C/\partial \sigma_\beta^2 \equiv C' < 0$. Being dependent on the random $\bar{Y}_{j's}$, the profit itself is a random variable.

Before formulating the optimization problem and for the sake of completeness, a few concepts of utility theory and attitude to risk will be given herein (an extensive discussion on these concepts can be found in Keeney and Raiffa 1976, Chap. 4 and many other sources). Utility theory, which is central to decision-making under uncertainty, provides a way of encapsulating the decision-maker's (DM) attitude to risk (expressed via the effect of risk on his behavior) in terms of a utility function U. The bulk of the research material concerned with applications of utility theory has been in a univariate framework where utility is defined on a single (random) variable, which is, in our case, the profit, II. The utility function U(II) assigns an appropriate utility to each possible II, and the optimum level of the decision-maker's control variables are the ones which maximize the expected utility $E[U(II)]$ [rather than the expected profit $E(II)$].

An individual is "risk neutral" if he is indifferent to the certainty of the expectation of random profit and to the uncertain profit itself, i.e., $E[U(II)] = U[E(II)]$. An individual is "risk-averse" if he prefers the certainty of $E(II)$ to the uncertain II itself (another way of putting this is that such an individual will pay to avoid uncertainty), i.e., $E[U(II)] < U[E(II)]$. The utility function of a risk-averse individual is concave, and that of a risk-neutral individual is linear.

To select optimal values for the two decision variables – $\bar{Q}$ and σ_β^2 –, a risk-neutral farmer has to maximize expected profits:

$$\underset{\bar{Q}, \sigma_\beta^2}{\text{maximum}} \left\{ \sum_{j=1}^{K} P_j E[\bar{Y}_j(\bar{Q}, \sigma_\beta^2)] - P_Q \bar{Q} - C(\sigma_\beta^2) \right\}, \qquad (11.12)$$

with first-order conditions for optimum:

$$\sum_{j=1}^{K} P_j E(\partial \bar{Y}_j/\partial \bar{Q}) - P_Q = 0, \qquad (11.12a)$$

$$\sum_{j=1}^{K} P_j E(\partial \bar{Y}_j/\partial \sigma_\beta^2) - C' = 0. \qquad (11.12b)$$

The optimization problem of a risk-averse farmer is given by:

$$\text{maximum}_{\bar{Q},\,\sigma_\beta^2} \; E\{U[\Pi(\bar{Q}, \sigma_\beta^2)]\}, \tag{11.13}$$

with first-order conditions for optimum:

$$E\left\{U'(\Pi)\left(\sum_{j=1}^{K} P_j(\partial\bar{Y}_j/\partial\bar{Q}) - P_Q\right)\right\} = 0 \,, \tag{11.13a}$$

$$E\left\{U'(\Pi)\left(\sum_{j=1}^{K} P_j(\partial\bar{Y}_j/\partial\sigma_\beta^2) - C'\right)\right\} = 0 \,. \tag{11.13b}$$

It is interesting to compare now the optimal amounts of irrigation water for the risk-neutral farmer with the risk-averse farmer. With the ergodic and stationarity assumptions, the average yield function in Eq. (11.7) can be rewritten (after omitting the index j for convenience) as:

$$\bar{Y} = h(\bar{Q}\beta) \,. \tag{11.14}$$

For a given level of σ_β^2 the optimization problem for the risk-neutral farmer is now:

$$\max_{\bar{Q}} \; \{PE[h(\bar{Q}\beta)] - P_Q\bar{Q} - C\} \rightarrow \bar{Q} = \bar{Q}_n \,, \tag{11.15}$$

where $\bar{Q}_n$ is the optimal water application of the risk-neutral farmer. The first-order condition is:

$$PE[\partial h/\partial\bar{Q}] = P_Q \,. \tag{11.15a}$$

The optimization problem for the risk-averse farmer is:

$$\max_{\bar{Q}} \; E\{U[Ph(\bar{Q}\beta) - P_Q\bar{Q} - C]\} \rightarrow \bar{Q} = \bar{Q}_a \,, \tag{11.16}$$

where $\bar{Q}_a$ is optimal water application of the risk-averse farmer. The first-order condition for Eq. (11.16) may be derived by a method first used by Horowitz (1970):

$$PE[\partial U/\partial\Pi]\,E[\partial h(\bar{Q}\beta)/\partial\bar{Q}] + P\,\text{cov}\,[\partial U/\partial\Pi, \partial h(\bar{Q}\beta)/\partial\bar{Q}]$$
$$= P_Q E[\partial U/\partial\Pi] \,. \tag{11.16a}$$

Dividing both sides of Eq. (11.16a) by $E[\partial U/\partial\Pi]$ yield:

$$PE[\partial h/\partial\bar{Q}] - P_Q = -P/E[\partial U/\partial\Pi]\,\text{cov}\,[\partial U/\partial\Pi, \partial h/\partial\bar{Q}] \,. \tag{11.16b}$$

Now, $\partial(\partial U/\partial\Pi)/\partial\beta = (\partial^2 U/\partial\Pi^2)\,(\partial\Pi/\partial\beta)$; consequently, $\partial^2 U/\partial\Pi^2 < 0$ (risk aversion) and $\partial\Pi/\partial\beta = P\bar{Q}\,\partial h/\partial(\bar{Q}\beta) > 0$ imply that $\partial(\partial U/\partial\Pi)/\partial\beta < 0$. Additionally, assuming concavity of the relevant part of the production function, $\partial(\partial h/\partial\bar{Q})/\partial\beta = \bar{Q}\partial^2 h/\partial(\bar{Q}\beta)^2 < 0$. The work of Lehman (1966) allows us to conclude that cov $(\partial U/\partial\Pi, \partial h/\partial\bar{Q}) > 0$. Hence, the right-hand side of

Eq. (11.16b) is negative, permitting us to write:

$$PE[\partial h/\partial \bar{Q}] < P_Q .\tag{11.17}$$

Thus [cf. Eq. (11.15a)] the risk-averse farmer will demand more water than the risk-neutral one (that is, $\bar{Q}_a > \bar{Q}_n$) and water, then, can be characterized as a marginally risk-reducing input (see Fig. 11.1). This is so because the risk-averse farmer utilizes more water than the risk-neutral farmer when other input conditions are fixed (Pope 1979).

11.3.1 Cost Estimates of Water Distribution Patterns

To quantify water distribution patterns and the degree of application non-uniformity, a number of characterizing indices have been suggested. The most common one was proposed by Christiansen (1942) as the Christiansen uniformity coefficient (CUC).

$$CUC = \left\{ 1 - \sum_{i=1}^{n} |Q_i - \bar{Q}| / n\bar{Q} \right\},\tag{11.18}$$

where Q_i is the quantity of water measured in the i-th collection can, n is the number of collection cans in the arrangement, and $\bar{Q}$ is the average defined by $\bar{Q} = \Sigma Q_i/n$.

An additional common uniformity coefficient is the statistical uniformity coefficient (SUC):

$$SUC = 1 - CV_Q ,\tag{11.19}$$

where $CV_Q = \sigma_Q/\bar{Q}$ is the coefficient of variation of Q and σ_Q is the

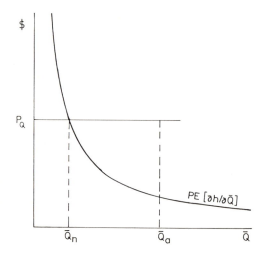

Fig. 11.1. Optimal quantities applied irrigation water for the risk-neutral ($\bar{Q}_n$) and the risk-averse ($\bar{Q}_a$) farmer

standard deviation estimate defined by:

$$\sigma_Q = \left\{ \sum_{i=1}^{n} (Q_i - \overline{Q})^2 / (n-1) \right\}^{\frac{1}{2}}. \tag{11.20}$$

By inspecting (Eq. 11.1) it can be easily verified that $CV_Q = \sigma_\beta$. From this identity and for a known pdf (cf., for example, Warrick 1983) the relationship between CUC and σ_β for normal, lognormal and uniform pdf of β can be obtained, respectively, from:

$$CUC = 1 - 0.789\,\sigma_\beta, \tag{11.20a}$$

$$CUC = 3 - 4Z\{0.5[\ln(1 + \sigma_\beta^2)]^{0.5}\}, \tag{11.20b}$$

$$CUC = 0.866\,\sigma_\beta, \tag{11.20c}$$

where $Z\{\cdot\}$ represents the probability of a standard normal variable.

As mentioned, for a given irrigation technology σ_β^2 (or CUC or SUC) can be partly controlled by the farmer via sprinklers (or emitters) and line spacing. Therefore, the cost C in (11.11) as a function of σ_β (or CUC) can be estimated by taking into account the equipment, labor, and operating costs associated with various spacing between adjacent lines and sprinklers (or emitters).

11.4 Empirical Findings and Their Implications

This section is aimed at summarizing a few empirical results and discussing their associated managerial and economic significance. Most of the results presented below were borrowed from the studies of Letey et al. (1984) – hereafter study 1 – and Feinerman et al. (1989a, b, c) – hereafter studies 2, 3 and 4, respectively – who emphasized managerial and economic aspects. These results display a wide variation in their properties and can be generalized for various crops, soils and irrigation methods. While studies 1 and 2 investigated the impacts of the uncertainty of the random function β without attempting to explicitly relate it to its origin, they were directly related to application uniformity in study 3 and to the variability in both water application and saturated soil hydraulic conductivity in study 4.

In study 3, Feinerman et al. (1989b) have developed and applied a stochastic economic optimization problem by which optimal levels of applied water and sprinkler spacing are determined. Two objective functions were considered: (1) maximizing profit expectation [Eq. (11.12)] of a risk-neutral farmer; and (2) maximizing utility expectation [Eq. (11.13)] of a risk-averse farmer. For the latter, a negative exponential utility function was assumed. Data for the empirical analysis were taken from well-controlled sprinkler irrigation plots of sweet corn and were utilized to estimate a sigmoid-type crop water production function and its associated statistical moments $E(\overline{Y}_j)$ – Eq. (11.8); and $\sigma_{\overline{Y}_j}^2$ – Eq. (11.10).

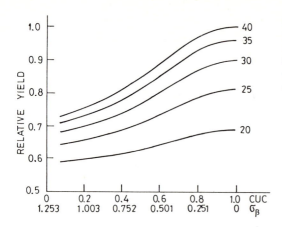

Fig. 11.2. Expected dry matter yield relative to a yield of $\bar{Q} = 40$ cm and $CUC = 1$ as a function of CUC (*numbers labeling the lines* indicate the average depth of applied water $\bar{Q}$)

Two yield components were considered: total dry matter and marketable kernels. The relative expectation for dry matter as a function of CUC and $\bar{Q}$ is presented in Fig. 11.2. This figure demonstrates the adverse impact of non-uniformity application on the level of expected yield as well as the fact that increased water application ($\bar{Q}$) can be substituted for uniformity (CUC) with decreasing marginal rate of technical substitution between ($\bar{Q}$) and CUC (i.e., $d\bar{Q}/dCUC < 0$). Similar findings for cotton and corn were reported in study 1.

Examination of Fig. 11.2 also shows that expected yield depression resulting from decreasing CUC at small water quantity (for example, $\bar{Q} = 20$ cm) is more moderate than the depression at high $\bar{Q}$ (for example, $\bar{Q} = 40$ cm). The reason is that at low $\bar{Q}$ water stress has a dominant effect on yield, while at high $\bar{Q}$ the effect of water is small and uniformity of application is dominant.

It was also found that for a given CUC, the variance of average yield (kernels and dry matter) decreases as $\bar{Q}$ increases and that the positive contribution of high CUC in reducing the variance of average yield is more effective as $\bar{Q}$ is smaller.

The principal findings of the applied stochastic economic optimization model in study 3 can be summarized as follows: the optimal level of applied water $\bar{Q}$ for both risk-neutral and risk-averse farmers decreases as application uniformity and water prices increase and the rate of the decrease is higher for the lower range of water prices. It was also found that $\bar{Q}$ increases as the degree or risk aversion increases. A risk-averse farmer "pays" for his aversion by using more water. This suggests that a savings in irrigation water in areas where water is scarce can be achieved not only by raising water prices but also by subsidizing irrigation systems with higher field application uniformity. This subsidy will be more effective at lower water prices and for the risk-averse farmer than for the risk-neutral one. The higher the degree of risk aversion, the more effective the impact of the subsidization. The optimal values of uniformity of application (CUC) were higher for the risk-averse farmer than for the risk-neutral one and were increased at an increasing rate with the increase in water prices. Obviously, the increase in water prices resulted in a decrease of profit expectations.

The impacts of CUC (or σ_β) on the expectation and variance of average yield were estimated in study 3 via the commonly used unconditional approach. The contribution of the conditional analysis (as compared to the unconditional one) in reducing the uncertainty of the random function $\beta(\chi)$, and as a result of $Y(\chi)$, was investigated in study 2. The methodology of study 2 was illustrated for irrigated corn assuming a Mitscherlich type crop-water production function: $Y(\chi) = Y_{max}[1 - \exp(-\bar{Q}\beta(\chi))]$, where Y_{max} is the maximum possible yield. To illustrate, a random set of 400 β points, $\beta_1, \ldots, \beta_{400}$ within a square field of 1 ha has been simulated for a multivariate normal distribution $g(\beta_1, \ldots, \beta_{400})$ with a constant mean and exponential covariance.

The values of $\beta_1, \ldots, \beta_{400}$ simulate the "true" field (or the "full-information") situation. Subsequently, a finite sample of 25 "measurement" points (out of the 400 points), with their associated β values, has been randomly selected and utilized to derive the best attainable unconditional and conditional estimates of the expectation and the variance of average yield. The results presented in Table 11.1 demonstrate the crucial contribution of the conditional analysis in reducing the variance of the average yield by about 8 times ($\tilde{\sigma}_{\bar{Y}}^{2, UC}$ is, on the average 7.8 times higher than $\tilde{\sigma}_{\bar{Y}}^{2, C}$). It is also observed in Table 11.1 that the estimated conditional average yield is much closer to the "true" average yield than the unconditional one (the "true" average yield is 11–38% higher than $\tilde{Y}^{UC}$ and only 1.4–1.8% higher than $\tilde{Y}^C$).

The economic optimization problem in study 2 assumes a risk-averse farmer with a negative exponential utility function in the form $U(II) = -\exp(-\gamma II)$, which is widely used, where γ is the measure of absolute risk aversion. The optimal value of applied irrigation water, $\bar{Q}$ (a single control-variable), and its associated estimated average yield statistical parameters

Table 11.1. Best attainable unconditional and conditional estimates of the expectation and variance of average yield for various water quantities

$\bar{Q}$ (m³/ha)	Yield expectation (t/ha)			Yield variance (t/ha)²	
	$\tilde{Y}$ UC Unconditional	$\tilde{Y}$ C Conditional	"True"	$\tilde{\sigma}_{\bar{Y}}^{2, UC}$ Unconditional	$\tilde{\sigma}_{\bar{Y}}^{2, C}$ Conditional
3000	8.31	11.27	11.47	5.70	0.825
4000	11.23	13.53	13.75	7.01	0.950
5000	13.51	15.32	15.55	7.80	0.975
6000	15.31	16.74	16.98	8.22	0.936

Table 11.2. Optimal quantities of applied water and its associated estimated and "true" average yields, parameters, profit and utility

Case	Estimated values					"True" values		
	Optimal Q (m^3/ha)	$\hat{E}\bar{Y}$ (t/ha)	$\hat{\sigma}^2_{\bar{Y}}$ (t/ha)^2	$\hat{E}$II ($/ha)	$\hat{E}$U	$\bar{Y}$	II	U
Unconditional	4982	13.46	7.65	209.8	-0.39	15.49	314.6	-0.0430
Conditional	4187	13.90	0.72	312.4	-0.05	14.11	323.7	-0.0393
Full information	4250	—	—	—	—	14.24	324.0	-0.0391

$\hat{E}\bar{Y}$ and $\tilde{\sigma}^{2,\,UC}_{\bar{Y}}$, the estimated expected profit $\hat{E}$II and the estimated expected utility, $\hat{E}$U, are summarized in Table 11.2. The "true" (and deterministic) values of average yield, profit and utility are also included in Table 11.2.

Scrutiny of Table 11.2 shows that the levels of applied water under the uncertain conditional ($\bar{Q}_C$) and unconditional ($\bar{Q}_{UC}$) cases are suboptimal (or second best) as compared with optimal $\bar{Q}$ under the deterministic full information case ($\bar{Q}_T$). In other words, $II(\bar{Q}_T) = 324.0$ and $U[II(\bar{Q}_T)] = -0.0391$ are the highest possible levels of profit and utility, and they should be regarded as reference results. The results in Table 11.2 enable to compare quantitatively the impacts of the unconditional and conditional approaches on the welfare of the farmer. In general, it can be stated that the farmer's welfare increases as the amount of uncertainty present in his decision problem decreases, i.e., as the deviation of his decisions from the optimal full information decisions decreases. Under the unconditional (conditional) analysis, water application $\bar{Q}_{UC}$ ($\bar{Q}_C$) is about 17% higher (only 1.5% lower) than $\bar{Q}_T$. Indeed, the true level of utility under the conditional analysis is only 0.5% lower than the full information utility level, while the true utility level under the unconditional analysis is 10% lower than $U[II(\bar{Q}_T)]$. As expected, the uncertainty reduction due to the conditioning by the measurement points increases the farmer's welfare.

The results presented in Table 11.2 also show that the true average yield, profit, and utility are higher than the estimated ones by, respectively, 1.5, 3.6, and 32% under the conditional analysis, and by 15, 50, and 900% under the unconditional analysis. In other words, the conditional approach is a better way to use the given body of information which is embodied in the finite measured sample $\beta_1, \ldots, \beta_{25}$.

One question related to the problem of choosing the best way to employ a given body of information is, what is the optimal amount of information to collect? As the number of observations (field measurements) increases, $\beta(\chi)$ can be described with less uncertainty and thus with a smaller deviation between the planned quantity of applied water and Q_T, i.e., with an increased value of the utility function. However, the effect of additional observations on the farmer's welfare also depends on the cost of their acquisition. Determining the optimal number of observations and their optimal spread over the field area is a complicated statistical problem, which was not investigated in study 2. However, in examining the effect of the number of measurement points, β_i on the level of $\bar{Q}_C$ – quantity of applied water under the conditional analysis – it was found that the calculated values of Q_i for N = 5, 10, . . . , 35 measurement points were, respectively, 3616, 3857, 3977, 4124, 4187, 4189, and 4193. Although the additional "measurements" were not selected in an optimal manner but were drawn from regular grids, the decrease of the deviations $Q_T - Q_C$ with N is clear.

The economic analyses in studies 1, 3, and 4 were carried out along the lines of the unconditional approach.

Study 1 optimizes the average applied irrigation water quantity of a risk-neutral cotton grower. The results for cotton are summarized in Figs. 11.3 and 11.4 for various infiltration uniformity levels. Several conclusions can be drawn from inspecting these figures. Increasing water price caused optimal $\bar{Q}$ to decline in every case, but it is less sensitive to water price when uniformities are relatively high. For a given level of water price, the optimal level of $\bar{Q}$ decreases when uniformity increases (Fig. 11.3), especially for the lower water prices. This demonstrates that conventional economic analyses which fail to account for the variations in infiltration uniformity may substantially underestimate optimal levels of water application.

The impacts of non-uniform infiltration on profits are depicted in Fig. 11.4 for various levels of water prices. As illustrated in this figure, low infiltration uniformities entail significant losses in profits that are not sensitive to water prices over the entire water price range. This suggests that increasing water prices would provide no incentive for the cotton grower to modify irrigation systems or management practices to improve uniformity. The incentive for switching to more uniform systems or practices is provided

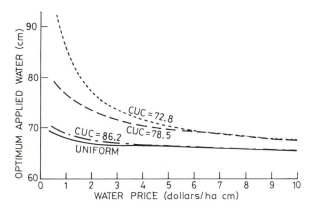

Fig. 11.3. Relationship between optimum $\bar{Q}$ and water price for various values of infiltration uniformity for cotton

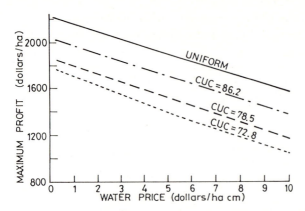

Fig. 11.4. Relationships between maximum profits (revenues net of water costs) and water price for various levels of infiltered water uniformity for cotton

by profit losses which occur irrespective of the price of water. For the empirical data used in study 1 for irrigated cotton in California, an investment of up to $280/ha would be economically justified to improve the CUC from 72.8 to 86.2.

Under uniform conditions the optimal quantities of applied irrigation water in study 1 approximated cotton evapotranspiration – ET – (consumptive use). Data presented in Fig. 11.3 illustrate that under non-uniform infiltration, an amount of water greater than consumptive use is optimal. The applied water in excess of crop ET percolates beyond the rootzone, whereas irrigation that leads to leaching is commonly considered wasteful. Indeed, under uniform conditions leaching would be wasteful. However, under non-uniform infiltration considerable leaching in some parts of the field is justified with optimal water management. The amount justified is dependent upon the level of non-uniformity, the shape of the crop-water production function and economic factors such as water price and revenues from the crop. The important dependency between the impact of infiltration uniformity (or lack thereof) on optimum levels of applied water and the shape and form of the water production function is discussed in Feinerman et al. (1983).

The analysis of study 1 was extended by Feinerman et al. (1984) who investigated the interactive effects of water salinity and uniformity of water infiltration on average crop yield (corn), optimal water application and expected profits. Salt accumulation in the rootzone was determined under both steady-state and transient conditions. The combined effects of salinity and non-uniformity determined crop yield as a function of the quantity and the quality of the irrigation water. Since the transient state salinity converges to the steady-state conditions after sufficient time, the transient state salinity is always less than or equal to steady-state salinity. In other words, since salinity adversely affect crop yield, the steady-state analysis represents a worse case analysis. It was found that expected profit-maximizing water applications, $\bar{Q}^*$, increase under conditions of increased irrigation water salinity, decreased uniformity of infiltered water, and decreased water price (see Figs. 11.5 and 11.6 for the transient case).

The calculated maximum net revenue from irrigating with water of various salinities are presented in Fig. 11.6 for transient and steady-state

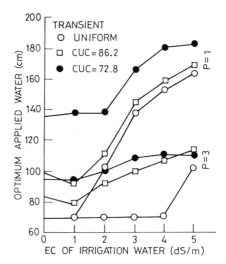

Fig. 11.5. Optimal water application when irrigation with water of various salinity levels (measured by *EC*) under transient conditions for various levels of infiltration uniformity and two water prices (*P*)

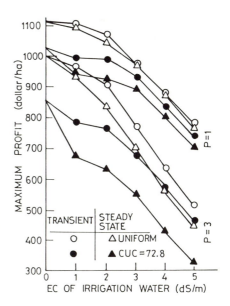

Fig. 11.6. Maximum profits when irrigating with water of various salinity levels (*EC*) under steady-state and transient conditions, two levels of infiltration uniformity and two water prices (*P*)

conditions, two values of water price P, and two infiltered water uniformities. Increasing water salinity decreases gross profits under all conditions. Except for P equal to 1 and uniform water application, gross profits are less under steady-state than under transient conditions for any level of electrical conductivity (EC). Decreasing uniformity of infiltered water results in lower gross profits for all values of EC for both the transient and steady-state conditions.

Note that higher profits are possible when irrigating with water of EC equal to 2 than with nonsaline water if the price for the nonsaline water is $3/ha-cm and the price for the saline water is $1/ha-cm. This result occurs because relatively high yields can be achieved with saline water if large amounts of water are applied for leaching. The costs associated with slight yield decrement and additional $\bar{Q}^*$ resulting from saline water are offset by the less expensive water. Water prices therefore significantly affect potential use of brackish water in a profitable agricultural operation. To conclude: economically optimal water applications and expected profits can vary by a factor of 3 or more depending on water prices, salt concentrations, and uniformity of infiltered water. Therefore, accurate determination of these variables is important for decision-making at the farm level.

Study 4 focused on the development and application of an economic optimization model aimed at determining the optimal level of applied irrigation water under variable soil conditions (non-uniform saturated hydraulic conductivity) and non-uniform water distribution, which maximizes the expected profits of a (risk-neutral) farmer. The model results have subsequently been used to evaluate and compare the profitability of two distinct irrigation methods for cotton, taking into account their uniformity performance and their associated investment and operating costs.

The principal findings and conclusions of study 4 can be summarized as follows: lower uniformity of water application is associated with a lower expected average yield (and obviously with a lower expected profits) and, generally, with a higher optimal water level. The increase in the expected average yield under the more uniform irrigation method is higher when the uniformity of water infiltration is affected by the spatial variability of soil properties. This occurs in soils with a low water application rate, with low variability of hydraulic conductivity, or both. In this case, the relative profitability of the more uniform irrigation method (A), which involves higher investment and operation costs than the less uniform and less costly irrigation method (B), is increased. Sensitivity analyses of water prices conducted in study 4 showed that an increase in water price implied a decrease in optimal water level as well as in expected profits. It was found that the difference between the expected profits from the two irrigation methods is relatively unaffected by water prices.

The results of study 4 were very sensitive to the form of the yield function. Given a concave yield function, the level of average yield increases with the uniformity level of water infiltration. Two cotton yield functions – "more concave" and "almost linear" – were examined and compared. It was found as expected, that the sensitivity of the average yield to non-uniformity in water infiltration is higher the more concave the function. Under the more concave yield function, irrigation method A was more profitable than B, whereas under the almost linear function the opposite was the case. These findings demonstrate the substantial sensitivity of the results to the shape and form of the yield function and the importance of the identification and the estimation of the right yield function: two yield functions of the same crop yielded opposite conclusions.

Finally, Hill and Keller (1980) have presented a computerized mathematical model that relates the selection of irrigation design parameters to crop production profitability, water application uniformity, and irrigation system capital and operating costs. The profit-maximizing quantity of applied water is the model's decision variable. Field-based crop production and economic data for sugar cane were used to illustrate the technique under sprinkle, trickle, and furrow irrigation systems. The major contribution of this study is the inclusion of the irrigation system's physical characteristics, water and crop production costs and returns, crop growth patterns, soil properties, crop response to non-uniform water applications, fertility, and on-farm management practices in one integrated system. The authors of this paper concluded that ". . . increased attention to improved irrigation system uniformity, at reasonable costs, may be mandatory to maintain a favorable profit picture".

11.5 Conclusions and Recommendations for Further Research

Complete uniformity of irrigation water infiltration is unattainable and in general economically undesirable. However, poor infiltration uniformity may entail significant losses in crop yield and profit or utility. A principal conclusion that can be drawn from the studies reported here is that economic prescriptions for optimal use of irrigation water which are based on the conventional assumption that soils and water application are perfectly homogeneous have the potential of being substantially biased or suboptimal. In general, increased infiltration uniformity can be substituted for water application with decreasing marginal rate of technical substitution. Under certain conditions, controlling irrigation uniformity may be as important as controlling irrigation water quantities. This suggests that a savings in irrigation water in areas where water is scarce can be achieved by encouraging irrigation systems with higher field application uniformity.

Another important conclusion is that the impact of infiltration uniformity (or the lack thereof) on optimum levels of water application and on the relative profitability of various irrigation methods depends primarily on the shape and form of the crop-water production function. This implied that the formulation and estimation of the production function should be done with considerable care. In the absence of a biological theory that may enable us to choose the "right" production function, a few functional relationships should be estimated and compared in relation to their fit to the available empirical data, and the best one should be chosen based on statistical criteria.

Another important finding is that the inference of the statistical structure and the stochastic economic optimization of irrigation in a spatially variable field that is achieved via the conditional analysis has the potential to increase the irrigator's welfare significantly more than with the commonly used unconditional approach. This increase results because the conditioning makes a crucial contribution to reducing the variance of the average yield, thus increasing the expected profit or expected utility and decreasing the discrepancy between the predicted and "true" results. The implication is that although the mathematical derivations and the empirical application of the unconditional analysis are much simpler (and hence more attractive) than those associated with the conditional one, the advantage belongs to the latter and the extra effort of performing it is worthwhile.

The analyses and findings reported here represent only a portion of the comprehensive assessment of the economic implications of the spatial variability of applied water, soils and yields. The directions into which the analysis can be profitably extended include: (1) examining the case where several relevant soil properties which affect the rate of water infiltration are simultaneously spatially variable. Such an examination requires data on the multivariate joint distribution of these properties and estimation of a crop-water production function in which they are explicitly included as arguments; (2) investigating the impacts of various agronomic technologies on the soil's spatial variation characteristics and including these technologies in the stochastic optimization model as control variables; (3) considering the impact of excessive water infiltration (in a few portions of the spatially variable field) on the leaching of nutrients below the root zone, particularly nitrogen (N).

The extra N which is lost through leaching should be compensated by additional N application, or decreased yield would be expected. The costs associated with increased fertilizers or revenues lost from declines in yield would have to be included in the analysis. Social costs associated with degraded ground water from nitrate leaching should also be considered; (4) improving the design of empirical experiments aimed at inferring the statistical structure of the random spatially variable water application and relevant soil properties. The improvement should be performed by determination of the optimal number of the relevant observations and their location over the field space; and (5) extending the analyzed economic unit to a multi-output (multi-crop) farm which has at its disposal various land plots and irrigation systems differing in their relevant spatially variable properties.

The studies reported here can be used as important building blocks in such extended analyses.

References

Bresler E, Laufer A (1988) Statistical inferences of soil properties and crop yields as spatial random functions. Soil Sci Soc Am J 52:1234–1244

Chen D, Wallender WW (1984) Economic sprinkler selection, spacing and orientation. Trans ASAE 27:737–743

Christiansen JE (1942) Irrigation by sprinkling. Univ Cal, Berkeley. Agric Exp Stn Bull 670

Dagan G (1985) Stochastic modeling of ground water flow by unconditional and conditional probabilities: the inverse problem. Water Resour Res 21:65–73

Dagan G, Bresler E (1988) Variability of yield of an integrated crop and its causes. III. Numerical simulation and field results. Water Resour Res 24:395–401

Delhomme JP (1978) Kriging in the hydrosciences. Adv Water Resour 7:251–266

Feinerman E, Letey J, Vaux JH Jr (1983) The economics of irrigation with non-uniform infiltration. Water Resour Res 19:410–1414

Feinerman E, Knapp KC, Letey J (1984) Salinity and uniformity of water infiltration as factors in yield and economically optimal water application. Soil Sci Soc Am J 48:477–481

Feinerman E, Bresler E, Dagan G (1985) Optimization of a spatially variable resource: an illustration for irrigated crops. Water Resour Res 6:793–800

Feinerman E, Dagan G, Bresler E (1986) Statistical inference of spatial random function. Water Resour 22:935–942

Feinerman E, Bresler E, Dagan G (1989a) Optimization of inputs in a spatially variable natural resource: unconditional vs conditional analysis. J Environ Econ Manag 17:140–154

Feinerman E, Shani Y, Bresler E (1989b) Economic optimization of sprinkler irrigation considering uncertainty of spatial water distribution. Aust J Agric Econ 33:88–107

Feinerman E, Bresler E, Achrish H (1989c) Economics of irrigation technology under conditions of spatially variable soils and non-uniform water distribution. Agronomie 9:819–826

Gohring, TR, Wallender WW (1987) Economics of sprinkler irrigation systems. Trans ASAE 30:1083–1090

Hill RW, Keller J (1980) Irrigation systems selection for maximum crop profit. Trans ASAE 3: 366–373

Horowitz I (1970) Decision making and the theory of the firm. Holt Rinehard & Winson, New York

Journel AG, Huijbregts CT (1978) Mining geostatistics. Academic Press, New York

Keeney RL, Raiffa H (1976) Decisions with multiple objectives: preferences and value tradeoffs. John Wiley & Sons, New York

Kitanidis PK, Vomvoris EG (1983) A geostatistical approach to the inverse problem in ground water modeling (steady state) and one-dimensional simulations. Water Resour Res 19:677–690

Lehman EL (1966) Some concepts of dependence. Ann Math Statist 103:115–134

Letey J, Vaux HJ Jr, Feinerman E (1984) Optimum crop water application as affected by uniformity of water infiltration. Agron J 76:435–441

Lumley JL, Panofsky A (1964) The structure of atmosphere turbulence. John Wiley & Sons, New York

Matheron G (1971) The theory of regionalized variables and its applications. Ecole de Mines, Fountainbleau, Fr

Moore CV, Synder HJ, Sun P (1974) Effects of Colorado River water quality and supply on irrigated agriculture. Water Resour Res 10:137–144

Pope RD (1979) The effects of production uncertainty on input demands. In: Yaron D, Tapiero CS (eds) Operation research in agriculture and water resources. North-Holland Biomedical Press, New York, pp 123–136

Sandmo A (1971) On the theory of the competitive firm under price uncertainty. Am Econ Rev 61:65–73

Seginer I (1978) A note on the economic significance of uniform water application. Irrig Sci 1:19–25

Seginer I (1987) Spatial water distribution in sprinkle irrigation. Adv Irrig 4:119–168

Stern J, Bresler E (1983) Non-uniform sprinkler irrigation and crop yield. Irrig Sci 4:17–29

Theil H (1971) Principle of econometrics. John Wiley & Sons, New York

Warrick AM (1983) Interrelationships of irrigation uniformity terms. J Irrig Drainage ASAE 109:317–332

Warrick AW, Gardner WD (1983) Crop yield as affected by spatial variations of soil and irrigation. Water Resour Res 19:181–186

Warrick AW, Yates SR (1987) Crop yields as influenced by irrigation uniformity. Adv Irrig 4:169–180

Yaron D, Dinar A (1982) Optimal allocation of farm irrigation water during peak seasons. Am J Agric Econ 64:681–689

Yaron D, Olian A (1973) Application of dynamic programming in Markov chains to the evaluation of water quality in irrigation. Am J Agric Econ 55:467–471

12 Irrigation Management Under Drought Conditions

N.K. Whittlesey, J. Hamilton, D. Bernardo, and R. Adams

12.1 Introduction

Benjamin Franklin is credited with saying "When the well's dry, we know the worth of water" (Schwinden 1984). Since the beginning of civilization, humans have attempted to harness nature's water supply, but have struggled with quantity, quality, or timing that do not match their needs. Food has always come first where water is an essential and often limiting input, but water also meets may other needs of civilization, including municipal, industrial, navigation, power, recreation, waste disposal, and esthetics. Water demand for these competing uses frequently conflicts with agricultural water supplies.

Irrigated agriculture is the largest consumer of water in the western US and throughout the world. This chapter discusses water management in irrigated agriculture – in particular, the management of water during periods of drought.

12.1.1 Defining Drought

What is drought? Felch (1978) distinguishes drought from aridity, which is a permanent climatic feature of a region resulting from low average rainfall. Drought, by contrast, is a short-term lack of precipitation, a temporary feature of climate. However, the literature uses many other definitions of drought as illustrated by Tannehill (1947).

Drought is generally extensive in space and time (Huschke 1959). For irrigated agriculture, drought does not necessarily begin with the cessation of rain, but rather when available irrigation water falls below normal for a period of time. Depending upon how irrigation water is supplied, the lag between rainfall changes and the effects on irrigation water supplies can be quite extensive. Riefler (1978) claims that the largest challenge is understanding the individual farmer's response to drought situations. Farmers tend to plan their irrigation water demand based on average year conditions. Drought may require both managerial and technological adjustments in farming operations, depending upon its length and severity.

Adv. Series in Agricultural Sciences, Vol. 22
K.K. Tanji/B. Yaron (Eds.)
© Springer-Verlag Berlin Heidelberg 1994

12.1.2 Drought and Water Supply

Drought is a problem of water supply, which is not sufficient to meet normal demands for its use. Since irrigation water may be obtained from either surface or subsurface sources, inadequate rainfall will have differential impacts on irrigated agriculture. Periods of drought can have immediate impacts on agriculture which is heavily dependent on surface storage or streamflow from recent rainfall or snowpack. The severity of these impacts will depend on the farmer's ability to adjust to water availability.

Irrigated agriculture dependent on groundwater for irrigation may be less vulnerable to drought, or the time lag may be more extended and tenuous between the drought period and the irrigation water supply changes. The connection between natural rainfall and groundwater irrigation water supplies is site-specific. Shallow aquifers dependent on annual recharge can be very responsive to drought. On the other hand, deep aquifers with minimal annual recharge can be mined for irrigation with little regard for the cyclic nature of rainfall.

12.1.3 Drought and Water Demand

Drought may also affect agriculture's demand for irrigation water. Most irrigated regions use irrigation to supplement natural rainfall for crop production. Irrigation management should maximize capture and use of natural rainfall on crops, but drought may increase irrigation water demand by reducing the rainfall available for crop production. As rainfall diminishes the complementary amount of irrigation must be increased. Hence, the impacts of drought can be compounded by both the demand and the supply for irrigation water.

12.1.4 Drought and Water Quality

Drought can affect water quality in several ways, although these interactions are quite site-specific. Much irrigation operates with a deliberate level of inefficiency. That is, water is leached through the soil profile to maintain acceptable soil salinity, as in the upper Colorado River Basin, or farmers may consciously use low efficiency application systems for economic reasons. Excess water may recharge aquifers, enter surface drains, or percolate through the soil to be recovered through subsurface drains. Surface streams receiving concentrated saline irrigation drainage can rely only on dilution to mitigate the effects. As surface stream flows diminish in a drought, they lose their ability to assimilate polluted water and maintain water quality.

Where it is common to use both surface flows and groundwater on the same lands, drought may encourage use from lower quality water sources. In some areas of northwestern and southwestern US, groundwater can be quite saline, which can be detrimental to both crops and soils.

Drought can also have positive impacts on the quality of water affected by irrigation return flows. Drought, by definition, means less water is available for irrigation. As irrigation efficiency improves to combat drought, both the amount of water leached through the soil and the amount of field runoff will decrease. This will in turn reduce leaching of nutrients and chemicals into groundwater and transport of sediment and nutrients into streams and lakes.

12.2 On-Farm Strategies for Drought Management

Crops, management strategies, and irrigation technologies for a particular region are usually based on expected regional weather patterns. As a deviation from the expected pattern, drought means that normal irrigation practices will often be inadequate to cope with water shortage. Depending upon the severity and length of a drought, farmers may react in several ways to the reduced supplies and increased demands for water.

Haas (1978) described some general strategies for water use during drought, though they are not specific to irrigated agriculture. The first is to augment water supplies. Cloud seeding has been extensively used in some areas to improve rainfall or snowpack, with mixed results. However, there is evidence that seeding sometimes increases precipitation. Construction of water storage facilities and further exploitation of groundwater supplies are other long-run strategies for drought management. Farmers can also resort to alternative cultivation practices such as summer fallow, trashy fallow, strip cropping, contouring and terracing, land leveling, and windbreaks. These techniques generally help make more efficient use of rainfall in arid regions. They are viable strategies for drought management if not already in common use under normal weather conditions.

Since drought is infrequent and unpredictable, Yevjevich et al. (1978) observe that society has a tendency to follow the "leaky roof" philosophy, wherein it cannot be fixed during a rain and when it is not raining there is no need to fix it (p. 34). In any case, we shall proceed to discuss farmer reactions to past droughts and suggest management practices that could be useful in such periods.

Yevjevich et al. (1978) suggest that the standard for effective water use in periods of drought should be maximization of social benefits (p. 36). The decisions of individuals acting in self-interest may not always achieve this objective. Most individual decisions are based on cost effectiveness with little concern for community impacts. For example, pro rata use of available water supplies during drought may result in crop failure for all producers, whereas community sharing of water could produce at least some crops.

Although increased efficiency is an attractive drought strategy for individual producers, in a river basin where upstream users have first right of diversion, this can have the effect of shifting the burden of drought from upstream to downstream users. Legal and institutional frameworks for water allocation and management must be sensitive to such problems.

The remainder of this chapter will focus on individual responses to water shortage without concern for the community effects of individual actions. Drought is a non-event, unlike a hurricane, earthquake, or flood. Since management and investment decisions have been developed for more favorable conditions, management and technical options for combating drought will depend upon the time available to prepare for the drought and the expectation of its potential length. Time can be valuable for adapting to changing water supplies. Moreover, reactions to drought will be tempered by the risks or costs of enduring without adaptive tactics. Options available to the farmer will be influenced by the ruling level of technology and management during normal periods. Reactions to drought can be divided into short- and long-term adjustments.

12.2.1 Short-Term

We will define the short term as a period too short to change the existing irrigation technology, water supply facilities, or investment in perennial crops. This section will describe strategies of irrigation management (application efficiency, deficit irrigation, etc.) and crop selection (land use) for drought management. In all cases, these actions result in temporary improvements in water use efficiency or the productivity of water use.

In an ideal irrigation system, water could be stored, delivered and applied without loss. However, in the real world this ideal is seldom met, for technical and economic reasons. Wade (1986) describes each of these as a potential source of water savings and improved water use efficiency.

To increase the productivity of scarce irrigation water, it is desirable to understand the complex and integrated relationships among crops, soils, and irrigation systems. These factors are described in some depth by McNeal et al. (1979). We will first focus on the items affecting irrigation efficiency: runoff and deep percolation. Typical rooting depths for common irrigated crops range from 2 ft for peas, beans, and potatoes to 4 ft or more for irrigated alfalfa. Deep percolation is defined as that quantity of water which moves vertically through the soil below the crop root zone. Coarse-textured soils will store less water per unit of soil and will allow more rapid penetration of water than fine-textured soils. In a similar fashion, the amount of water that leaves the field without penetrating the soil is defined as runoff.

The period during which water is continuously applied to the field with any irrigation technology (excluding center pivot systems) is the set time, and the length of time between irrigations will determine the frequency of irrigation. Both will optimally vary with the crop, soils, water availability, and irrigation technology. However, in practice, the length of set is more commonly determined by convenience rather than by exact field needs. Farmers will tend to substitute water for irrigation labor when possible, allowing for some irrigation inefficiency. Also, under normal conditions farmers will irrigate to minimize the field portion subjected to water stress; that is, to reduce risk by maintaining a high level of moisture adequacy for the crop. These irrigation tactics are quite rational but they also create opportunities for water conservation when drought occurs.

A factor which interacts with irrigation adequacy to affect irrigation efficiency is irrigation uniformity. Most systems of irrigation result in non-uniform application of water over the field surface. Methods of gravity flow or furrow irrigation are generally the worst but even the best sprinkler systems will result in some level of non-uniformity. While irrigation uniformity is strongly affected by the irrigation technology, it can also be influenced by irrigation management, particularly on furrow irrigated crops. The effect of non-uniform water application is that some portions of the field receive more than others. Uniformity has various technical definitions (Heermann and Kohl 1983) but is generally measured as the ratio of mean field application to the sum of absolute deviations from the mean. The lower the coefficient of uniformity, the greater the portion of a field that is over- or underirrigated. When combined with the level of adequacy, uniformity is a major determinant of irrigation efficiency. For example, a field irrigated to a level of 90% adequacy will have 90% of the field with excess water and 10% with insufficient water. The coefficient of uniformity will then determine the amount of over- and under-irrigation, hence the level of irrigation efficiency. Together, these factors will largely determine the opportunities for short-term water conservation when drought occurs. Patterson et al. (1982) describe short-term drought management strategies for irrigated agriculture based on a survey of Idaho farmers after the drought of 1976–1977. Their observations are incorporated into the following comments.

For a given level of uniformity, improved water use efficiency is accomplished by relying on more frequent irrigation of shorter set time, a tactic to reduce the level of irrigation adequacy and water losses to percolation or runoff without proportionately reducing crop yield. With any irrigation system, this managerial technique may improve irrigation efficiency. Irrigation uniformity can also be influenced by irrigation management. On furrow irrigated crops, it is possible to increase the furrow advance rate with a large stream flow and then reduce the stream flow (cut back) after water is through the field. Runoff water can be captured and reused on the same farm. Reducing length of field run will also help to improve irrigation uniformity.

With wheel move and hand line sprinkler systems, it is possible to improve uniformity by paying more attention to wind conditions during irrigation. These techniques for improving the efficiency of water use allow the substitution of labor for water as the latter becomes more scarce.

Another short-term tactic for managing water shortage is to reduce the amount of water consumed by the crop. As adequacy levels are reduced to improve water use efficiency, some parts of the field are under-irrigated and will not meet full crop water requirements. As this tactic is consciously applied (deficit irrigation), the reduction in crop yield will generally be less than proportional to the reduction in water consumption.

If the seasonal water supply is known prior to the spring planting period, it is possible to adjust the acreage of annual crops. For example, the acreage of small grains can be increased while reducing the acreage of crops requiring more water. This tactic will help to maintain the total acreage of irrigated crops while requiring less total seasonal water use. However, since maximization of farm income is the driving force in irrigation management, it may be most profitable to idle cropland to provide sufficient water for remaining crops. For example, perennial crops (forages, orchards, vineyards, etc.) need sufficient water to ensure long-run survival even if short-term productivity is reduced. The acreage or productivity of annual crops will normally be sacrificed to protect the investment in perennial crops.

12.2.2 Long-Term

Long-term adjustment to drought is almost a contradiction in terms. Nevertheless, while drought is generally perceived as a temporary phenomenon, it can persist for more than one production period. Moreover, the risk of loss to drought can be more severe in some areas or for some farmers than others. While short-term is defined as a period insufficient to change investments in irrigation technology, water supply, or perennial crops, the long-term would allow such changes to occur. If drought is expected to be persistent or frequent, some changes in the capital structure of the farm may be in order. These changes are specifically a response to the risks of drought, though they may affect the efficiency of water use on a permanent basis.

Some irrigation technologies are more efficient in water use than others. If a permanent shift in water use efficiency will aid farmers in times of drought, it is possible to invest in more efficient irrigation systems. In general, the order of irrigation systems by level of efficiency would be flood (without field leveling), furrow, wheel move and hand move sprinkler, center-pivot, solid set, flood (with laser leveling), and drip. The potential for moving from one technology to another will depend upon the soils, crops, field size and configuration, slope, and water quality, among other factors. The profitability of such a change will depend upon the added cost, increased crop productivity, and water saving in time of drought.

Permanent changes in the mix of crops, including perennial crops, can be another method of reducing the risk of losses during drought. However, since drought is not permanent by definition, one has to be careful not to reduce long-term farm income just to reduce the risk of losses during drought.

New irrigation wells can be installed as an alternative source of water during drought. This strategy was used by Washington Yakima Valley farmers in the drought of 1977. Large investments in new wells were made to avoid the risk of losing orchards to the drought. In the end most farmers could have avoided the well cost with small loss to the orchard investment.

12.3 Policy Considerations

The ability of farmers and competing water users to reach socially beneficial adjustments to drought conditions is influenced by the legal and institutional framework within which they operate. This section discusses some of the major factors affecting farmers in this regard.

12.3.1 Competing Uses of Water

The uses of water that compete with agriculture will vary from one setting to another. At certain periods and locations it is possible to have complementary but varied uses of the same water. The water can serve fisheries, recreation, hydropower, or other uses and still meet the needs of irrigated agriculture. However, it is also common for the alternative uses of water to compete with agricultural diversions. In the Pacific Northwest the diversion of water for irrigation directly competes with the instream uses of hydropower, fisheries, and navigation (Whittlesey and Hamilton 1986). In the southwestern US, municipal and industrial uses of water compete with agriculture. The relative scarcity of water and the level of competing water needs will generally determine the degree of competition for water used by irrigated agriculture in any setting. The extent to which these competing uses will be satisfied in time of drought will depend upon the value of the alternative uses, the property rights structure for water uses, and the legal constraints on water allocation.

12.3.2 Value of Water in Alternative Uses

While the literature of water value is extensive, there is insufficient space here to provide even a brief review of this literature. Instead the reader is referred to Colby (1989) for a short but excellent review of water values in alternative uses. In general, the values of water as published in research and market information are shown to be lower for agriculture than for hydropower, municipal or industrial uses, but comparable to uses for fisheries, recreation, navigation, or wetlands uses. However, even these generalizations are misleading because it is difficult to know the conditions for estimating the values in each case. The published prices of water commonly refer to average values or, at least, values for large increments of water. All manner of life and human activity requires some water, but the willingness to pay for water in any use will depend upon the need for an additional unit of water. It is the marginal value of water that is important, particularly in a drought period. Under normal weather conditions most irrigated agriculture would be willing to pay very little for another acre foot (1230 m^3) of water. However, in a drought period the same agriculture may be willing to pay several thousand dollars for another acre foot of water to preserve an orchard or vineyard. These alternative values of water for annual crops are described in a discussion by Whittlesey (1984). He shows in an example that the short-term value of water for an annual crop could be as high as $104/ac-ft ($84/10^3 m^3) when payment to all other fixed factors of production (land and capital) is zero, conditions that might be observed in a drought. The value drops to $38/ac-ft ($31/10^3 m^3) when paying all capital costs and the dryland value of land. If all capital costs are paid and land receives its irrigated worth the water value drops to $1/ac-ft ($0.8/10^3 m^3), an appropriate condition for a normal year with abundant water supply. Permanent, long-term exchanges in water rights would probably take place at prices somewhere between the latter two values. Similar scenarios could be constructed for the value of water in other uses. The value of water will thus depend upon the relative scarcity of water for the period of exchange.

12.3.3 Legal and Institutional Constraints

An array of legal and institutional constraints limits water management during drought. In the West, many of these constraints relate to the "appropriation doctrine", or "first in time, first in right" basis of water law. Under the appropriation doctrine, water rights specify amount, location, use and a priority date when the water was first used. During drought, water rights are generally filled in order of priority until the supply is exhausted. Senior right holders get their full water supply; junior holders get none. This applies whether water rights are held by individual farmers or in common by water delivery organizations. Figure 12.1 illustrates how the appropriation doctrine allocates water among individual farmers during a drought. The figure shows three farms, each of which began diverting irrigation water from a single stream in the 1880s.

The first farm began diversion part way downstream in 1884. The next year another diversion was developed upstream to serve farm number 2.

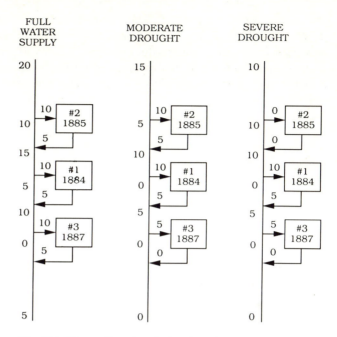

Fig. 12.1. Water allocation during drought

Two years later another farm was developed downstream. Each farm has a right to divert 10 cfs (0.28 m³/s), which it typically uses with 50% efficiency and returns 5 cfs (0.14 m³/s) to the stream as wastewater.

The normal stream flow of 20 cfs (0.56 m³/s) is enough to give all three farms their full water supply. If drought cuts the flow to 15 cfs (0.42 m³/s), there is no longer enough for everyone and other priority dates determine water allocation. Farm number 2 can take its full 10 cfs and still leave 10 cfs at the diversion point of the senior right to serve farm number 1. However, not enough water remains to meet the most junior rights of farm number 3. All of the impacts of moderate drought fall on the most junior water user, and the more senior diverters are not required to make any adjustments. As drought becomes more severe, farm number 2 eventually has its water supply cut off. Note the anomaly that number 2 is cut off before number 3 because the latter has access to wastewater from the most senior right. In addition to priority dates, the actual layout of streams, diversions, and return flows determines who gets water and who does not.

Note that this institutional arrangement gives little incentive to conserve water during drought. The most senior rights are secure, with no incentive to conserve. The most junior users do not get any water to use or conserve. Only the farms at the margin are faced with a partial supply which they must try to use as effectively as possible.

Instead of three individual farms, Fig. 12.1 could also apply to three water delivery organizations (WDOs), each serving a set of individual farms. The appropriation doctrine would treat WDOs in the same way it treats individuals. However, a WDO might have greater scope for creative adjustment to drought. Because WDOs are more likely to have multiple water rights with different priority dates, they can find themselves with a partial water supply, and they may have more latitude in managing that limited supply under the bylaws of the WDO.

The rigidity of the appropriation doctrine is beneficial in some ways. It assures the security of the senior water rights, providing the stable environment necessary to justify investment in irrigated farming. The rigidity also reduces the likelihood that one farmer acting in his own self-interest will damage the interest of his neighbors in unexpected ways.

12.3.4 Options for Reallocation

In addition to the managerial tactics discussed in Sect. 12.2, farmers or irrigation districts may react to water shortages in one of the following ways: (1) rotation – rather than assert their right to a reduced continuous flow, users can agree to take turns receiving a greater flow for a shorter period. The rotation often reduces delivery system losses, and the larger delivery flow may be easier to use efficiently at the farm; (2) prorationing – when water is scarce, farmers can agree to be altruistic and share the shortage, rather than following a strict priority. WDO's often have bylaws which mandate prorationing of available water supplies within the organization; and (3) transfers – the quantity of most water rights under normal conditions is more than ample to grow a crop. Senior appropriators often have water which they could transfer to less fortunate junior appropriators at a small cost to themselves. A user growing lower-valued grain or pasture could transfer water to another growing higher-valued vegetables or perennial fruits. However, western water law recognizes that water right transfers are neither as simple nor as clearly beneficial as they seem. The transfer of water rights is rigidly restricted, unless it can be shown that the transfer: (1) does not injure any other water right; and (2) is not an enlargement of use. These criteria are very difficult to satisfy. Water markets generally require special authorizing legislation.

12.3.5 Risk Tradeoffs in Allocating Stored Water

Some farmers rely upon surface stored water for irrigation where reservoir storage capacity exceeds normal annual requirements. There is an element of risk in depleting storage supplies in a year of drought to maintain irrigation as usual. The drought may continue for another year or storage space may not refill, forcing even more severe irrigation cutbacks the next year. However, not using stored water the first year may incur an unnecessary risk if the drought breaks before the following season. Farmers will adopt strategies based on their risk perceptions and preferences. The following sections report on research that has focused specifically on agricultural water use under conditions of water storage.

12.4 Modeling Farmer Response to Water Shortage

To evaluate irrigator response under limited water supplies, a two-stage, farm-level irrigation management model was developed and applied in Washington State's Columbia River Basin (Bernardo et al. 1987). In the first stage of the analysis the SPAW-IRRIG water budget-crop yield model was employed to evaluate yield response to specified irrigation schedules. SPAW-IRRIG uses two components to estimate water requirements and the associated yield for each irrigation activity. First, daily soil-plant moisture calculations are made to distribute available water among various uses (e.g., soil evaporation, transpiration, runoff, etc.). Next, yield estimates are calculated from measures of accumulated water stress derived from daily predictions of evapotranspiration (ET). The water stress measures express relative yield as a function of the ratio of actual to potential evapotranspiration as described in Doorenbos and Kassam (1979).

One-acre (0.4 ha) irrigation activities were developed by running SPAW-IRRIG for a number of irrigation scheduling criteria available to area irrigators. Each activity represented the output of a single run of the model and consisted of the derived irrigation schedule (dates and depths of applications) and the resulting yield. Irrigation scheduling alternatives employed included criteria based upon fixed time intervals, soil moisture percentage, soil tension, and accumulated ET since the previous irrigation.

The generated irrigation activities provided the physical component of a farm-level optimization model used in the second stage of modeling to represent producer response to water supply restrictions. The specific objective of the mathematical programming model was to allocate available land, water and other limiting resources among the various irrigation activities to maximize farm-level net returns.

The analysis was formulated to consider several short-run management options available to producers responding to water supply limits. Irrigation scheduling alternatives were incorporated through the availability of the numerous irrigation activities included in the model. These activities represented various degrees of water deficit as well as several levels of scheduling sophistication. The adoption of labor-intensive irrigation practices to increase application efficiency was also represented. For example, under surface irrigation, application efficiency can be increased by adopting such practices as reducing set-time, cutting back stream size, and monitoring runoff. Several application practices (represented by labor rates and application efficiencies) were evaluated within the optimization model. Crop selection, reallocation of water among crops, and idling land were also represented as responses to water supply limits.

The model was applied to a representative 520-ac (211 ha), furrow or rill irrigated farm. Four crops – grain corn, wheat, dry beans, and alfalfa – were included in the analysis and restricted to a maximum area of 130 acres (53 ha). Efficient seasonal irrigation plans are presented for four alternative annual water allotments in Table 12.1. These scenarios would be representative of a case where a producer is given a seasonal water allotment which must be optimally allocated both spatially (among fields) and temporally (over the irrigation season). Since irrigation systems are assumed fixed, the analysis focuses on short-run responses to water supply reductions.

Currently, irrigators in the study area pay a fixed per-acre delivery charge which entitles them to divert as much water as they deem necessary. These conditions of unlimited water supply are represented in column 1 of Table 12.1. A total of 26 884/ac-in (27 690 ha-cm) was applied, resulting in a return to land, management, and fixed costs of irrigation of $84,462. Irrigation schedules selected were high in water use with yields approaching the maximum attainable.

Table 12.1. Optimal seasonal irrigation plans for alternative annual water allotments on a 520-acre surface irrigated farm, Washington State's Columbia River Basin

	Annual water allotment (acre-inches)			
	26 884	22 000	16 000	12 000
Net returns	84 462	82 352	76 749	61 336
Land (A)	520	520	520	475
Irrigation labor (hrs)	4550	3990	3840	3910
Application efficiency (%)	45	48	55	65
Crop information:				
Dry beans				
Average water applied (AI/A)	39.6	37.1	26.9	19.5
Average yield (cwt/A)	23.9	23.9	22.6	20.9
Grain corn				
Average water applied (AI/A)	54.5	44.1	37.4	32.3
Average yield (T/A)	4.77	4.49	4.39	4.30
Wheat				
Average water applied (AI/A)	45.3	42.6	24.9	16.9
Average yield (bu/A)	89.8	89.8	80.6	76.8
Alfalfa				
Average water applied (AI/A)	67.5	45.2	40.4	35.0
Average yield (T/A)	6.46	6.16	6.16	6.10

Note: All crop fields are 130 acres, except corn which is 85 acres in the 12 000 AI solution.

Incremental reductions in seasonal water supply were met through adoption of several water conserving strategies, including irrigation scheduling, improving application practices, reallocating water among crops, and removing land from production. Large reductions in water applications to each crop were attained with only marginal yield losses. Through improved scheduling and application techniques, water applications were reduced without significantly affecting crop consumptive use. For example, in attaining a 40% reduction in water applied to corn, crop consumptive use was reduced by only 19%, and yield by 12%. Water allocation among crops was based upon each crop's relative value, as well as its drought tolerance. For example, wheat, the lowest valued and relatively drought tolerant crop, incurred the highest percentage reduction in water application and crop yield.

Irrigation scheduling criteria changed considerably as water became more constraining. Under unrestricted water supplies, optimal schedules were generated using fixed time intervals and application amounts. As water availability was limited, the sophistication of irrigation scheduling required to maximize the efficiency of the water supply increased. Under the most limiting water supply, irrigation schedules were generated based upon soil moisture percentages and accumulated ET since the previous irrigation. Because irrigations were directly related to water consumed since the last irrigation, these criteria tended to maximize the efficiency of water use. In addition, irrigations were discontinued in growth stages where yields were least susceptible to water stress. Application efficiencies appearing in row 4 indicate modifications in irrigation practices as water supplies become more constraining. Initially, the average application efficiency was 45% and 0.7 h/ac (0.28 h/ha) were required for each application. As water supply was reduced, and its scarcity value increased, application efficiency was increased through the adoption of labor-intensive application practices. In meeting the first reduction, a 3% increase in efficiency was attained through runoff monitoring. Average application efficiency was improved in the 16 000 ac-in (16 480 ha-cm) solution by additional monitoring of runoff and adjusting set time. In the final solution, efficiency increased to 65% by adopting cutback methods. As a result, the labor requirement increased by 0.4 h/ac/irrig (0.16 hr/ha/irrig) over the baseline level.

In addition to changes in the quantity of water applied, efficient response to limited water supply requires several other resource adjustments. In meeting the most restrictive water allotment, 45 ac (18 ha) of land were idled. Economic returns were maximized by spreading the available water supply over fewer acres rather than imposing additional water stress to the crops. Also, despite an increase in the labor intensity of individual irrigations, total labor use was reduced by 640 h.

The results presented in Table 12.1 demonstrate a large potential for water conservation by surface irrigators in the study area. For example, returns to land, management, and fixed costs of irrigation declined by only

10% as a result of a 40% reduction in water supply. However, nominal income losses associated with water supply reductions did not prevail over the entire range of water restrictions. To achieve an additional 15% reduction in water supply (from 16 000 to 12 000 ac-in, 16 480 to 12 360 ha-cm), net returns decreased an additional 14%.

12.5 Contingent Water Markets for Drought Management: The Idaho Case

Agriculture, along with municipal, industrial, fisheries, recreation, and hydro-power, suffers during drought. When water is short, many of these uses have a higher value for the water than irrigation. This section suggests how irrigated agriculture's flexibility in responding to water shortage can be used to help other users obtain needed water during drought.

Contingent water markets would create long-term option contracts on water presently owned and used by farmers for irrigation. Having sold such an option, the farmer could continue using the water for irrigation in most years. In years of specified water shortage, the purchaser would exercise the option and take delivery of the water, for municipal, industrial power, fishing or recreational use. Both the irrigator and the purchaser expect to be better off in the long run through market participation.

12.5.1 The Idaho Setting and Study Results

A recent Idaho study looked at the feasibility of contingent water markets to move water from irrigation to hydropower generation (Hamilton et al. 1989). Idaho irrigated farms are located high in the Snake-Columbia river basin, so water purchased from irrigation could be used to generate electricity at several downstream dams.

Average annual Snake River flow in south central Idaho (1928–78 corrected to 1980 levels of irrigation diversions) varied from 7178 to 16 701 cfs (4220 to 9820 m^3/s). Since Idaho depends heavily on hydropower, its electricity supply closely parallels river flow. Utilities limit firm power sales, the power of highest value, to that available in the lowest stream flow years. Uncertain supplies produced with stream flows above the minimum historical level bring lower prices as nonfirm or surplus power. Contingent water markets would allow utilities more water for power generation in drought years, increasing firm power supplies in all years and raising the average value of power sales over time.

The probability and magnitude of market intervention needed to maintain selected flows in the upper Snake River is determined from the historic flow record. A contingent water market would, for example, require that

112 425 ac-ft (3830 ha-m) of marketable water be available in the driest years to boost flow from the 7178 cfs (4220 m³/s) lowest flow of record, up to an assured average flow of 7333 cfs (4310 m³/s), or that 625 715 ac-ft (76 960 ha-m) be available to assure a minimum annual average flow of 8042 cfs (4730 m³/s). A contingent water market to assure the 8042 cfs (4730 m³/s) flow level would interrupt irrigation 19.6% of the time. However, not all the 625 715 ac-ft (76 960 ha-m) under contract is needed in every interruption year because of variations in severity of water shortage. All contract water is required once in 51 years, 82% is needed in another year, 70.5% in a third, with lesser deliveries needed in seven other interruption years. While some deliveries from agriculture to hydropower are needed in 10 years out of 51, in the long-term only 8.3% of contract water would actually be delivered.

Hydropower value of flow augmentation has two parts. First, power generated with market water in firm power. The long-term avoided cost value for firm power set by the Idaho Public Utilities Commission is 5.65 ¢ per kWh. Second, this assured dry year generation enables utilities to increase firm power sales in all other years. This increases the value from the 2.77 ¢ per kWh of surplus power to the 5.65 ¢ per kWh value of firm power, an increase of 2.88 ¢ for each kWh raised from surplus to firm category.

Electric utilities could benefit form a contingent water market if the price for option contracts were less than their benefits. However, the price also depends on what it costs farmers to give up the water. The Idaho study used a version of Bernardo's model to estimate farmer costs of participating in a contingent water market. This model allowed farmers to change irrigated acreage, crop mix, water application rates, and irrigation efficiency in response to drought year effects of the contingent market. Because the farm water supply is interrupted only in occasional years, the model assumes farmers would not change irrigation systems in response to market participation. The model also assumes that farmers would know the seasonal water supply prior to the spring planting period, allowing all possible annual adjustments to the restrained water supply. Politically, the study constrains irrigators to market no more than half their normal consumptive use of water.

The study results show that average net returns for a center pivot irrigated farm would be $169/ac ($417/ha) with no market. With market participation, annual net returns range from $124 to $169/ac ($306 to $417/ha) and average $166/ac ($410/ha) for this farm, reducing average income by $2.59/ac ($6.40/ha) through market participation. For the seven representative farms used in the study, average annual farm income losses from market participation range from $2.06 to $2.70/ac ($5.09 to $6.67/ha). To determine market feasibility, these farm income losses must be compared to the hydropower benefits of the contingent water market.

The contingent water market increases hydropower value even in normal years when agriculture continues to irrigate with an undiminished supply of water, because it allows the utility to sell more power at firm rates rather than surplus rates. The increased value of this firmed up power in normal water supply years is $22 to $30 for each irrigated acre ($54 to $74/ha) participating in the market. In interruption years, the power generated with market water is all firm, ranging in value from $43 to $58 per participating acre ($106 to $143/ha). For all years the average annual power benefits from the market averaged $23 to $32/ac ($57 to $79/ha). Hence, for all representative farms, power benefits were at least eight times the farm income loss from market participation. Utilities should be able to pay farmers enough to make a contingent water market feasible.

12.5.2 Legal, Institutional and Physical Considerations

In spite of their potential for drought year management, there are a number of reasons why contingent water markets do not now exist. The legal constraint of the appropriation doctrine pressures Idaho water users to "use it or lose it". Portions of a water right not used for the originally intended purpose are in danger of appropriation by someone else. If farmers market water in this manner, does this mean the exchanged water is in excess of their needs? If so, farmers could lose that portion of their permanent water rights. Idaho law does not yet provide for the long term transactions implicit in a contingent water market without endangering farmer-owned water rights.

Normally, with willing buyers and sellers, a market will arise naturally. This is difficult to achieve for contingent water markets, however. The potential sellers are a diverse group. Some rights are held by individual farmers, others by irrigation districts. There are only a few potential buyers, mainly power utilities, M&I users, and agencies concerned with instream flows. Institutions that allow these diverse groups to bargain for water do not now exist. A contingent water market also faces physical barriers. The irrigation system is extremely complex and difficult to manage precisely. The Snake River and the Snake Plain Aquifer are linked in complex ways, and in many cases one irrigator's waste water is another's water supply. Irrigation diversions are rarely measured with precision. Irrigation structures are often designed to work with a full water supply and will not function properly with less. And water supply forecasting does not achieve the precision or timeliness that would allow market participants to anticipate or adjust fully to drought.

12.5.3 Other Applications

The Idaho study focused on a contingent water market to move water from irrigation to hydropower, but this is only one possible application. In the

Pacific Northwest there is potentially an even more important alternative use of water; that is, to flush juvenile salmon down the Snake and Columbia Rivers. Several species of salmon existing in the Snake River basin are now considered endangered. One reason for their endangered status is insufficient flows during their spring downstream migration. Contingent water markets might be part of a recovery plan to boost flows during critical periods in drought years.

In the West, there is growing concern over water supplies for urban use. Adequate water supplies can usually meet normal year demands. However, water rights to supplies sufficient for municipal needs in dry years can be very expensive, and some of this water would be unused in normal years. A contingent water market would allow agriculture to use the water in most years, while only allocating water for urban use in drought years.

12.6 Global Climate Change and Western Irrigation

Global climate change arising from increases in atmospheric CO_2 and trace gas concentrations is an international concern. Adams and Whittlesey (1991) recently estimated the potential effect of such climate changes on the productivity and net farm income in Pacific Northwest (PNW) irrigated agriculture. Three global climate change models (GCM) were used to generate estimates of the regional climatic effects that would be expected by the year 2030. These were models provided by the Goddard Institute for Space Studies (GISS), Geofluid Dynamics Laboratory (GFDL), and United Kingdom Meteorological Office (UKMD). Using these GCMs, Scott of Battelle Northwest Laboratories provided hydrological data on streamflow and water supply. Rosenzweig and Iglecias at the Goddard Institute for Space Studies estimated crop yield and water consumptive use changes that would result from the climate change.

Only two of the GCMs will be discussed here to illustrate the diversity of impacts that may be expected from the global climate change. Using corn as an example, the GISS model projected a 68% yield increase for the Upper Columbia Basin (UCB) and an 18% yield decrease for the Snake River Basin (SRB). Corn yield projections by the GFDL model were similar. Water consumptive use requirements for irrigated corn were projected by the GISS model to increase by 18% in the UCB and by 7% in the SRB. However, water requirements for corn were estimated by the GFDL model to increase by 31% in the UCB and by 41% in the SRB. Other crops such as wheat and potatoes had different responses to the climate change. Crop yields were increased in all cases, but for wheat, water requirements were decreased under the GISS model and increased under the GFDL model. Water supplies for irrigation were estimated by Scott to decrease by 50% in the UCB and by 10% in the SRB, a permanent drought condition.

Results of these changes in global climate were evaluated by applying a modification of the Bernardo model to representative farms throughout the PNW. The combined effects of yield, water requirement and water supply changes on PNW irrigated agriculture for the GISS model showed a 1% decrease in cropped acres and an 11% increase in farm net returns, the latter due to the substantial expected gains in crop yield. By comparison, the aggregate effects of the GFDL model resulted in no change in cropped acreage and a 24% increase in net farm income.

This study indicates that global climate change could have substantial impacts on the availability and use of water in agriculture. The results are short-term in that they give no consideration to potential changes in irrigation technology, nor do they consider crop market impacts of global climate change. Moreover, caution should be exercised in interpreting these results, given the preliminary nature of data used in the analysis.

12.7 Conclusion

This chapter has described some strategies for dealing with drought as it affects irrigated agriculture. While the focus has been on irrigation in the western US, the implications should have a global reference. Managerial, technological, and institutional adjustments to drought conditions will always be site-specific, but there will also be similarities in the problems created by drought and the strategies for combating it.

References

Adams R, Whittlesey NK (1991) The effects of climate change on water demand: a case study of irrigated agriculture in the Pacific northwest. Rep USEPA, Off Policy, Planning and Evaluation, Washington, DC

Bernardo DJ, Whittlesey NK (1987) Optimal irrigation management under conditions of limited water supply. Res Bull, Coll Agric Home Econ, Washington State Univ, 27 pp

Bernardo DJ, Whittlesey NK, Saxton KE, Day L Bassett (1987) An irrigation model for management of limited water supplies. W J Agric Econ 2(2)

Colby BG (1989) Estimating the value of water in alternative uses. Nat Resour J 29:87–96

Doorenbos J, Kassam AJ (1979) Yield response to water. FAO, Rome Pap 33

Felch RE (1978) Drought: characteristics and assessment. In: Rosenberg NJ (ed) North American droughts. AAAS Select Symp 15

Gardner RL (1985) The potential for water markets in Idaho. Id Econ Forecast 7, 1:27–34

Haas JE (1978) Strategies in the event of drought. In: Rosenberg NJ (ed) North American Droughts. AAAS Select Symp 15

Hamilton JR, Whittlesey NK, Halverson P (1989) Interruptible water markets in the Pacific northwest. Am J Agric Econ 71, 1

Heermann DF, Kohl RA (1983) Fluid dynamics of sprinkler systems. In: Jensen ME (ed) Design and operation of farm irrigation systems. ASAE Monogr 3:583–618

Huschke RA (1959) Glossary of American meteorology. Am Meteorol Soc, Boston

McNeal BL, Whittlesey NK, Obersinner VF (1979) Control of fertilizer nutrient and sediment losses in irrigated portions of the Pacific northwest. EPA, ORD, Ada, OK, Proj no R805037010, 245 pp

Patterson PE, Walker DJ, Hamilton JR (1982) Drought impact on three selected irrigated agricultural areas in southern Idaho. AE Extension Ser 382, Dep Agric Econ, Univ Id

Riefler RF (1978) Drought: an economic perspective. In: Rosenberg NJ (ed) North American droughts. AAAS Select Symp 15

Rosenzweig C, Iglecias A (1991) Crop yield effects associated with alternative climate forecasts for the Pacific northwest. Rep USEPA, Off Policy, Planning and Evaluation, Washington, DC

Saliba BC, Bush DB (1987) Water markets in theory and practice: market transfers, water values, and public policy. Westview, Boulder

Schwinden T (1984) One state's strategy for putting water to beneficial use. In: Englebert EA, Schuering AF (eds) Water scarcity: impacts on western agriculture. Univ Cal Press, Berkeley, pp 347–444

Scott M (1991) Hydrologic implications of global climate change on the Pacific northwest. Rep USEPA, Off Policy, Planning and Evaluation, Washington

Tannehill IR (1947) Drought: its causes and effects. Univ Press, Princeton NJ

Wade JC (1986) Efficiency and optimization in irrigation analysis. In:Whittlesey NK (ed) Energy and water management in western irrigated agriculture. Studies in water policy and management. Westview, Boulder, pp 73–100

Whittlesey NK (1984) Local and regional economic impacts: discussion. In: Englebert EA, Schuering AF (eds) Water scarcity: impacts on western agriculture. Univ Cal Press, Berkeley, pp 269–271

Whittlesey NK, Hamilton JR (1986) Energy and the limited water resource: competition and conservation. In: Whittlesey NK (ed) Energy and water management in western irrigated agriculture. Studies in water policy and management. Westview, Boulder, pp 307–327

Yevjevich V, Hall WA, Salas JD (eds) (1978) Drought research needs. Water Resour Publ, Fort Collins, CO

Young R, Gray SL (1972) Economic value of water: concepts and empirical estimates. Tech Rep US Nat Water Commiss, NTIS, PB210356

13 Evaluation of Management and Policy Issues Related to Irrigation of Agricultural Crops

A. Dinar and J. Letey

13.1 Introduction

As agricultural, urban and environmental interests increasingly compete for scarce global resources, many parts of the world are suffering from a rapid deterioration of water quality. Water is expected to severely limit further development unless conservation and efficiency measures are implemented (McCalla and Learn 1985). Agriculture, the most voracious consumer of water, is also a major polluter of soil and water resources with its by-products of pesticides, nitrates, and other hazardous materials (Batie et al. 1985).

Efficient water use in irrigated agriculture can achieve several goals, such as conserving fresh water, maintaining high standards of public safety, and improving environmental quality. Management of irrigated agriculture at the field, farm, and regional levels involves making decisions on input levels for a given crop, allocation of input among competing crops, investment in appropriate technologies needed for the production process, and substitution among several complementary inputs. Consideration must be given to the effects of these decisions on yields and accumulation of pollutants in soil and water bodies. Both short- and long-term effects on profitability and pollution must be analyzed.

All these decisions necessitate the full understanding of the physical relationships among input, output and pollution levels. The literature documents several approaches for capturing these relationships. For a detailed review of these approaches see Vaux and Pruitt 1983; Letey et al. 1990; Dinar and Zilberman 1991; Just 1991; Letey 1991.

The economic analysis of agricultural production, and specifically the allocation of irrigation water for food and fiber production, has focused mainly on maximizing production or profits for the farmer. Externalities, third party effects, and social costs have usually been ignored, although it is apparent that these factors are significant in agricultural production. Agricultural production can be modeled to include pollution by considering the relationships among input, output and pollution (Just 1991).

Crop production functions are central to the evaluation of management and policy issues. The purpose of this chapter is to evaluate a variety of

Adv. Series in Agricultural Sciences, Vol. 22
K.K. Tanji/B. Yaron (Eds.)
© Springer-Verlag Berlin Heidelberg 1994

management and policy issues using a particular crop-water production function that includes the salinity as well as the quantity of applied water. This function considers the output from the process that includes both crop yields and the amount of water which percolates below the root zone. The following section provides a description of production functions with greater emphasis on the one chosen for this chapter. Later sections are devoted to analysis and evaluation of a variety of management and policy issues using the production function. The chapter concludes with a discussion of future needs for a better and more comprehensive analysis of management and policy issues relevant to irrigated agriculture.

13.2 A Crop-Water Production Function Model

Two approaches are available to estimate crop-water production functions. The first synthesizes production functions from theoretical and empirical models of individual components of the crop-water process. The second approach estimates production functions by statistical inference from observations on alternate levels of crop yield, water application, soil salinity and other relevant variables. The large number of variables requires extensive field testing by the second approach. Thus, the discussion in this chapter will be limited to the use of synthesized production functions.

The crop-water production function is the relationship between input (applied water) and output (crop yield and deep percolation) of a crop production process. Although deep percolation (DP) is not commonly considered to be an output of a crop-water production function, it becomes important as a source of externalities in the economics of agricultural water management, the focus of this chapter. In our discussion, all DP ends as drainage volume (V).

Water is but one of several inputs to the crop production process. This chapter assumes that other inputs have been applied at a sufficient level for water to be a limiting factor in crop production. Furthermore, yields will generally be reported on a relative rather than an absolute basis. In other words, a relative yield of 1.0 indicates that water was not a limiting factor in production and that the absolute yields would be determined by the level of other input factors. Irrigation water supplies have differing levels of dissolved salts (salinity). Since water salinity influences crop yields, the crop-water production functions described and used in this chapter account for salinity.

Models to compute production functions can be broadly classified as either transient or seasonal. Seasonal models compute yield from total applied water of a given salinity during the season. Different yields may be obtained with the same amount of water depending on how water is allocated during the season. A unique relation between crop yield and seasonal water application can be obtained by defining the production function as the maximum yield that can be obtained from a given level of seasonal water applications. For the economic analysis of a season, this relation is useful since growers are assumed to maximize profits.

Transient models use basic water flow and salt transport equations with initial soil conditions to compute salt and water distributions in the soil at various times. These models allow consideration for the time and amount of water applied for each irrigation. A water uptake (root extraction) term is added to the flow equation to account for water removal by transpiration. The root extraction term provides linkage between the soil-water-salinity status and crop yield.

Cardon (1990) developed a transient state model by combining routines of R.J. Hanks and M.Th. van Genuchten with appropriate modifications to simulate crop yield and deep percolation for a given irrigation schedule. The transient model has the advantage of simulating production over multi-season, multi-crop and dynamic irrigation and precipitation events. Furthermore, it has utility under conditions with high water table as well as free drainage conditions. The major shortcoming is that considerable input data and computer time are required to simulate a growing season, the time being dependent upon the complexity of irrigation events. The transient model is rather cumbersome to use in deriving a production function curve. The computed seasonal yield is related to the programmed irrigation events. A series of simulations with different irrigation applications is required to produce a production function curve. Thus the transient model is extremely useful in simulating the effects of a proposed management scheme but is not convenient for an optimization analysis.

Letey et al. (1985) and Letey and Dinar (1986) described the development of a seasonal production function model. The relationships of yield versus evapotranspiration, yield versus average root zone salinity, and average root zone salinity versus leaching fraction were combined to develop an equation that relates yield to the amount of seasonal applied water of a given salinity. Irrigation water salinity levels are quantified by the electrical conductivity of the water (C). Letey and Dinar (1986) used the model to compute production functions for several crops, and included these functions in that report. A problem with the seasonal model occurs when the soil salinity at the beginning of the crop season differs greatly from the salinity of irrigation water. Since time is required to establish steady-state conditions, the model is less reliable under these conditions. Knapp (1991) replaced the steady-state soil salinity relations by a dynamic equation for soil salinity which assumed piston-flow conditions. This modification makes it more useful for a dynamic analysis.

The seasonal model allowed computation of relative yield or deep percolation to applied water (W) where W is expressed as an equivalent depth of water. The value of W in the field can be ambiguous. It can represent water discharged to the field including runoff or it can represent the amount of water that infiltrated the soil, which is the difference between discharge and runoff. In this chapter, W denotes infiltrated water because runoff does not contribute to crop production or deep percolation. Applied water is computed by dividing the volume of discharged water that does not run off by the area of the field. Applied water computed in this way represents the average value for the field. The actual W at any given location in the field may be considerably different from the average value because irrigation water is never applied uniformly. Letey et al. (1984) presented numerically computed production functions for non-uniform irrigation. The seasonal production function model which accounts for irrigation water salinity (Letey et al. 1985) can be combined with one to account for non-uniform irrigation (Letey et al. 1984) to compute crop-water production functions which account for both irrigation water salinity and uniformity of irrigation (Dinar et al. 1985).

Production functions for corn (*Zea mays*) as computed by the seasonal model which accounts for irrigation water salinity and irrigation uniformity are illustrated in Figs. 13.1–13.5. The relationships between relative yield, W and C are presented in Fig. 13.1 for the case where irrigation is uniform. The maximum evapotranspiration used in the model computations was 68 cm. Note that the range of applied water depicted in the figures ranges from considerably less than to more than twice the maximum evapotranspiration. The highest depicted applied water values are not likely to be used in a farm operation.

For a given amount of W, the relative yield decreases with increasing C (Fig. 13.1). The relative yield increases with increased W for each C. For the lower C, the relative yield plateaus at a value of 1.0 at the higher levels of W. However, at the three highest C values, maximum yields are never achieved regardless of the quantity of W.

The relationships among deep percolation (V), W, and C under uniform water applications are illustrated in Fig. 13.2. The deep percolation increases

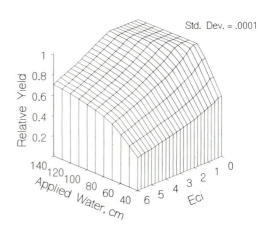

Fig. 13.1. Relative corn yield as affected by applied water quantity and quality, for application uniformity of SD = 0.0001

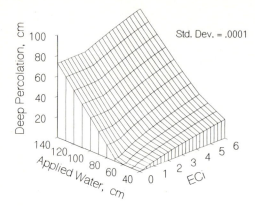

Fig. 13.2. Deep percolation of irrigation water for corn as affected by applied water quantity and quality, for application uniformity of SD = 0.0001

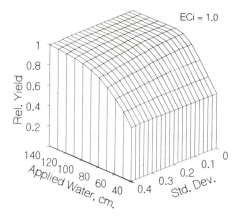

Fig. 13.3. Relative corn yield as affected by applied water quantity and uniformity, for water quality of EC = 1.0

with increasing W for each C. For a given value of W the deep percolation increases with increasing level of C. Of practical significance is the fact that for saline irrigation waters, some deep percolation occurs even under very low water application rates. This phenomenon is the result of decreasing plant growth from salinity, which decreases evapotranspiration, thus allowing some of the irrigation water to cause deep percolation.

The effects of irrigation uniformity are illustrated in Fig. 13.3 for the case where the water has C equal to 1 dS/m. The standard deviation identifies the degree of non-uniformity and a standard deviation of 0 represents perfectly uniform irrigation. Except at very high or very low W levels, the yield decreases with decreasing irrigation uniformity for a given level of W. More water must be applied to achieve the high yields for non-uniform irrigation as compared to those of the more uniform irrigation.

Relationships between V, W and irrigation uniformity are depicted in Figs. 13.4 and 13.5 for irrigation water salinities of 0.1 and 1.0 dS/m, respectively. Considering the non-saline case (Fig. 13.4), essentially no deep

Fig. 13.4. Deep percolation of irrigation water for corn as affected by applied water quantity and uniformity, for water quality of EC = 0.1

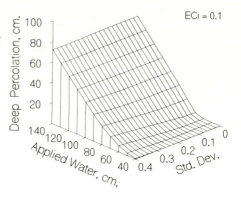

Fig. 13.5. Deep percolation of irrigation water for corn as affected by applied water quantity and uniformity, for water quality of EC = 1.0

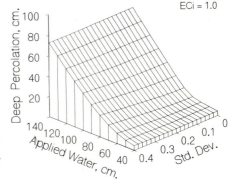

percolation occurs under the lowest W levels for uniform irrigation. However, under non-uniform irrigation, some deep percolation occurs even at relatively low average water application rates. For a given level of W, V increases with decreasing irrigation uniformity. The main effect of increasing C (Fig. 13.5) is to increase the amount of V at the lower W levels. The significant point is that when saline waters are used for irrigation, significant amounts of V can result even at very low water application rates.

Clearly, there are significant interactions among irrigation water salinity, uniformity, applied water levels, and crop yield and deep percolation. These crop-water production functions will serve to evaluate optimal irrigation methods under various economic and environmental constraints, which will be reviewed in the following section.

13.3 Evaluation of Management and Policy Issues

The problems associated with management of water in agriculture can be handled at several levels. First, field level management issues include decisions on crop, type of irrigation technology, water application level, drainage facilities, and mix of different water qualities. Second, farm level management decisions include crop mix and farm-level drainage treatment facilities. Third, regional level management decisions include investment in regional drainage and water resource development. To achieve a given social or private goal, various policies such as quality standards on agricultural contamination, or quotas, taxes, and subsidies on inputs and outputs, can be adopted at farm and/or regional levels. Several management and policy issues are evaluated in the following sections using the seasonal crop-water production function model presented in the previous section.

It should be mentioned that there are several additional extensions and modifications to and applications of the crop-water production function model that will not be described here. These include modifications to account for transient root zone salinity [Vinten et al. 1991); for sweet corn and cotton in Israel, Knapp (1991); and for cotton and tomatoes in California, Knapp et al. (1991)], and extensions to account for management of regional water quantity and quality problems in California (Weinberg and Wiley 1991; Weinberg et al. 1992).

13.3.1 Optimal Combination of Fresh and Saline Waters

Semi-arid regions face increasing competition for a limited fresh water supply. In these regions substantial amounts of saline water are often available from many sources, such as ground water aquifers, agricultural drainage, municipal wastewater, etc. If both saline and non-saline irrigation water at the same locale are available for irrigation, it can offset the demands for fresh water, as well as reduce environmental contamination from agricultural drainage water.

The field-level decision maker, having full knowledge of the relevant physical relationships between yield and water quantity and quality, needs to optimize the mix of saline and non-saline water for irrigation of a given crop and should know the prices for crop yield and water of all qualities in the field.

Various strategies have been proposed to use saline waters. One approach blends the saline water with good quality water to an acceptable salinity for use in irrigating crops. A second method uses cyclic irrigation strategy to initially irrigate a crop with high-quality water when the crop is more sensitive to salinity, and then uses a lower quality water for later irrigation when the crop is more tolerant to salinity (Rhoades 1984, 1987). A third strategy cycles water of varying salinity in a crop rotation scheme as a function of the crop's salt tolerance. This strategy allows osmotic stress to be applied to the crop which is the most tolerant to this stress (Rhoades et al. 1989).

The cyclic strategy does not lend itself well to analysis by the seasonal model. Bradford (1991) used the transient state model to simulate the effects of the cyclic strategy on crop production. The analysis was conducted assuming that alfalfa (*Medicago sativa* L.) was grown continuously for several years and also for the case where corn and cotton (*Gossypium hirsutum* L.) were grown in a crop rotation during alternate years. Corn is a salt-sensitive crop and cotton is a salt-tolerant crop. The simulations were run to compare the cyclic versus blending strategy. The same amounts of salt and water were applied to the crop for both blending and cycling strategies for direct comparison. For the continuous crop of alfalfa, the average yields with equal salt and water application were the same regardless of blending or cycling strategy. However, in the crop rotation, the cotton yields were the same whether the cyclic or blending strategy was used; however, the corn yields were higher under the cyclic rather than the blending strategy.

When one crop is grown continuously or the crop rotation includes species of about the same salt tolerance, it makes very little difference whether the waters are blended or cycled. The seasonal model allows an optimization analysis for the blending of waters of two salinities. The following describes that analysis.

Assuming that the decision maker is a profit maximizer, then his objective function is:

$$\pi = Y[(W/e)_x, C_x] \cdot P^y - (W/e)_1 \cdot P^{w_1} - (W/e)_2 \cdot P^{w_2} , \qquad (13.1)$$

where π is returns to land and management, Y is yield affected by water quality, C_x and quantity $(W/e)_x$ of the water mix, e is pan evaporation, P^y is yield price net of harvesting cost, $(W/e)_1$ is applied fresh water, P^{w_1} is cost of unit of fresh water, $(W/e)_2$ is applied saline water, and P^{w_2} is cost of unit of saline water.

Given water and crop prices, the decision maker optimizes the mix ratio of the water sources over the decision variables $(W/e)_1$ and $(W/e)_2$. To demonstrate the use of the model, isoquants were derived (Dinar et al. 1986a) for several crops (Table 13.1, and Fig. 13.6) for relative yields of 1, 0.9, and 0.8, and for mixing non-saline waters (C = 0) with saline water of C = 4, 6, or 11 dS/m.

The values for the mixed water quality and quantity are given by:

$$C_x = \frac{(W/e)_1 \cdot C_1 + (W/e)_2 \cdot C_2}{(W/e)_1 + (W/e)_2} , \qquad (13.2)$$

and

$$(W/e)_x = (W/e)_1 + (W/e)_2 .$$

The isoquants (Fig. 13.6) characteristically summarize the physical and technical conditions of production to obtain relative yield of 1. The slope of the isoquant describes the extent to which substitution between the two inputs can occur. A negative slope indicates production conditions in which inputs can be substituted for each other. A positive slope indicates non-efficiency substitution between the inputs since an additional unit of one

Table 13.1. Various parameters derived from isoquants for various crops irrigated to achieve maximum yield

Crop	EC (dS/m)	Price ratio[a] mix	Price ratio[a] allowing max. saline in mix	Max mix ratio
Alfalfa	4	0.30	0.03	1.00
Cauliflower	6	0.15	0.00	1.00
	4	0.45	0.30	3.00
Celery	4	0.45	0.25	0.35
Corn	4	0.14	0.00	0.33
Cotton	11	0.75	0.22	2.60
	6	0.80	0.43	∞
	4	0.85	0.67	∞
Cowpea	11	0.30	0.00	0.45
	6	0.60	0.22	2.00
	4	0.75	0.55	∞
Lettuce	------------------------------ No mix ------------------------------			
Oats	6	0.60	0.00	0.20
	4	0.85	0.12	1.00
Sugarbeet	6	0.35	0.15	2.00
	4	0.65	0.38	∞
Tomato	6	0.60	0.00	0.37
	4	0.90	0.40	1.00
Wheat	11	0.60	0.00	2.00
	6	0.75	0.20	∞
	4	0.95	0.62	∞

[a] Price ratio of saline to non-saline water.

input can be accommodated only by adding more units of the other input. The dashed line "a" represents the upper limit of a slope that can be tangent to a given isoquant, and therefore represents the upper limit of the ratio of prices of saline to non-saline water for which the use of saline water is efficient. The dashed line "b" represents the lowest slope that is tangent to a given isoquant, and this slope specifies the price ratio at which the largest amount of saline water would be used.

Scrutiny of Fig. 13.6 shows that saline water of C = 4, 6, and 11 cannot be substituted with fresh water in the case of lettuce, and saline water of C = 11 cannot be substituted with fresh water in the case of oats. Note that for saline water of C = 6 there is a limited range over which substitution can be obtained in the case of oats to produce a relative yield of 1.

Key parameters for several crops (Dinar et al. 1986a) appear in Table 13.1. The price ratio mix is the price ratio of saline to non-saline water for which mixing becomes feasible. For a higher price ratio only non-saline waters will be optimal. The price ratio allowing maximum saline water in the mix is equivalent to the slope of line "b" in Fig. 13.6. The maximum mix ratio is the maximum amount of saline water that can be efficiently used. If the value in

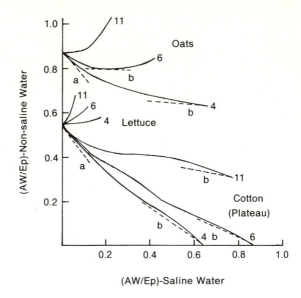

Fig. 6. Isoquants of RY = 1 for cotton, lettuce, and oats when irrigated with combinations of nonsaline and saline waters with EC = 4, 6, or 11 dS/m. *a* and *b* represent tangents with the highest and lowest slope to the isoquant, respectively

Table 13.1 is ∞, then only saline water would be used for the specified C and price ratio values.

With information on specific water and crop prices, the optimal management decision could be made using the information in Table 13.1. Farm-level decisions of allocating fresh and saline water for on-farm irrigation of several crops and finding optimal mix ratios of these waters is also presented in Knapp et al. (1986); Dinar and Knapp (1988); Dinar et al. (1990).

13.3.2 Evaluation and Adoption of Irrigation Technologies

On-farm source control is one means to conserve water and reduce agricultural drainage. Production functions which relate crop yield and quantity and quality of drainage water to: (1) quantity and quality of applied irrigation water; (2) irrigation technology; and (3) environmental conditions, may assist in comprehensive decision making regarding the determination of applied water and choice of irrigation and drainage-reduction technologies. In this section we present three field-level applications for the crop-water production function model, used for evaluation of irrigation technologies.

Dinar et al. (1985) evaluated optimal applied water at various levels of irrigation uniformities (used as surrogates for irrigation technologies) and conditions of water quality and drainage. The results serve to determine the profitability of improving the uniformity of the irrigation system. (In that paper the economic value associated with improving crop tolerance to salinity was also estimated. These results are not presented and the interested reader is referred to that paper.)

The objective is to maximize profits from the field. The following relationships apply:

$$\pi = P^y \cdot Y - P^w \cdot W - P^p \cdot V - P^{cm} \cdot Z - C^{op} - T - F , \qquad (13.3)$$

$$Z = V/e^p , \qquad (13.4)$$

$$C^{op} = (P^y \cdot Y - P^w \cdot W - P^p \cdot V - T - F) \cdot Z , \qquad (13.5)$$

where F is non-water production costs, T is cost associated with drainage tiles, Z is ratio of drainage water to evaporation, used to calculate fraction of field area to be devoted to evaporation pond (it can be viewed also as area), and e^p is the annual evaporation rate from the pond. P^{cm} is annual capital and maintenance costs for the pond, C^{op} is the opportunity cost of land being taken out of production to construct evaporation ponds, and P^d is drainage disposal costs (including pumping). A field with adequate natural drainage is simulated by allowing, P^p, P^{cm} and C^{op} and T to equal zero. The model calculates optimal values π^*, W^*, Y^*, and V^*. Irrigation uniformity is given as the Cristiansen Uniformity Coefficient (CUC), expressed in values ranging from 0 to 100, with 100 representing perfect uniformity.

Results in Table 13.2 are for selected scenarios where prices of surface water and pumping drainage water were fixed at $1 and $0.08/ha-cm, respectively. Here we will discuss only the effect of irrigation uniformity, since other aspects were discussed in the previous section.

Irrigation uniformity affects values of returns to land and management, applied water, yield and drainage volume (not shown). In all cases decreasing CUC leads to an increase or no change in W^*, a decrease in π^*, a decrease in Y^*, and an increase in V^*. The effect of CUC on Y^* and π^* is least when no drainage system is required and is greatest when an evaporation pond must be constructed on productive land. The opposite effect is noted for W^* and V^*. When the cost of drainage water is particularly high, proportionally greater benefits are achieved from increasing irrigation uniformity. In principle, irrigation uniformity improvement can be accomplished by changing irrigation management and/or irrigation systems. The difference $\pi^*(cuc = i) - \pi^*(cuc = i - 1)$, where all other variables are held constant, provides a measure for the economic value for irrigation uniformity improvements. From Table 13.2 one concludes that the difference in profits associated with low and high uniformities increases as management strategy becomes more expensive, and decreases as water quality deteriorates. This information may assist policy makers to identify conditions under which the farmers are likely to invest in improving their irrigation uniformity.

Table 13.2. Optimal values of returns to land and management, water application rates and yields for corn and cotton, under various drainage conditions, water quality, and irrigation uniformity

CUC	C_i	No drainage requirements			Off-farm facility			On-farm evaporation pond on productive land		
		Optimal values of								
		Returns ($/ha)	Water (ha cm)	Yield (ton/ha)	Returns ($/ha)	Water (ha cm)	Yield (ton/ha)	Returns ($/ha)	Water (ha cm)	Yield (ton/ha)
Corn										
72	0	215	140	8.66	200	130	8.58	71	90	7.90
	3	111	190	8.24	83	180	8.16	0	0	0
	5	0	0	0	0	0	0	0	0	0
100	0	320	80	9.00	317	80	9.00	266	80	9.00
	3	158	160	8.37	137	150	8.29	0	0	0
	5	0	0	0	0	0	0	0	0	0
Cotton										
72	0	1256	140	1.53	1240	130	1.52	970	70	1.43
	5	1225	160	1.53	1205	150	1.52	775	80	1.39
	11	1115	200	1.48	1104	200	1.48	412	90	1.19
100	0	1359	80	1.56	1358	80	1.56	1334	70	1.54
	5	1329	110	1.56	1322	100	1.55	1077	80	1.34
	11	1186	200	1.51	1177	170	1.51	580	90	1.32

In a study by Letey et al. (1990), specific CUC values were assigned to a set of irrigation technologies under conditions prevailing in the San Joaquin Valley of California. The irrigation systems that were evaluated based on their performances and costs in relation to cotton production and drainage volumes were: (1) furrow (0.9 km long) siphon tubes; (2) furrow (0.45 km long) surge-gated pipe; (3) subsurface drip; (4) hand move sprinkler; (5) linear move sprinkler; and (6) low energy precise application (LEPA).

Profit (π_i) associated with technology i, defined as returns to land and management, was computed using the following equation:

$$\pi_i = (P^y) \cdot Y_i - (P^w + P^{e_i}) \cdot W_i - (P^p + P^n) \cdot V_i - K_i , \qquad (13.6)$$

where P^{e_i} is energy cost associated with the technology, P^n is cost of nitrogen leached with drainage water, and K_i is technology fixed cost. All other variables are as defined above.

CUC values, non-water production fixed costs, and capital costs for the evaluated technologies are presented in Table 13.3. Profits were calculated for a variety of conditions and compared among the technologies. Here we will discuss only the relationship between cost of drainage disposal and returns to land and management. The range of values for drainage disposal cost was $0 to $240/ha-cm of drainage water, with 0 simulating the case with no drainage problems. Results are depicted in Fig. 13.7.

Table 13.3. Summary of irrigation technology characteristics and costs, and fixed non-water cotton production costs associated with these technologies

Technology	CUC	Annual capital cost[a]	Fixed non-water production cost	Total cost
		---------------($/ha/yr)-----------------		
Furrow 0.90 km	70	49.00	1029.15	1078.15
Furrow 0.45 km	75	96.00	1055.35	1151.35
Hand move sprinkler	80	129.00	1087.85	1216.85
Linear move sprinkler	90	242.00	990.25	1232.25
LEPA	85 (90)[b]	242.00	998.85	1240.85
Subsurface drip	90	476	813.25	1289.25

[a] Including maintenance and insurance.
[b] Two CUC values were used in the analysis.

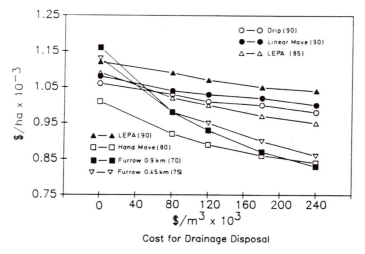

Fig. 7. Drainage disposal cost effect on returns to land and management for several irrigation systems. Numbers within parentheses for each irrigation system are the assumed Christiansen uniformity coefficient

With no costs or restrictions on drainage water disposal (or with no drainage problems), maximum profits were obtained with furrow irrigation technologies (the 0.90 km is the most profitable). Profitability decreased as the cost of drainage water disposal increased. The rate of profit decrease is strongly dependent on the CUC assigned for each technology. The 0.90 long run furrow with the lowest CUC (70) has the rapid drop in profitability with increasing drainage disposal costs. Whereas furrow is currently the most profitable irrigation system, it would become economically justifiable to switch to another system if significant costs for drainage water disposal were imposed.

The differences in profitability for the various systems are related to both yield differences (not presented) and drainage volumes. Of the two, the effect on drainage volume is the predominating one.

Dinar and Zilberman (1991) suggested further modification of the crop-water production function model to include the effects of environmental conditions, such as weather and land quality, on selection of irrigation technologies. To save space, model development and numeric results will not be presented. However, a policy-oriented discussion is provided based on the analysis in Dinar and Zilberman (1991). Irrigation technologies associated with higher CUC values (pressurized "modern" technologies) substantially reduce impacts of input quality and weather conditions on profitability, and in essence change differences in production conditions because of locational variability. Thus, the introduction of these technologies is beneficial to certain locations, especially when water price and drainage costs are high.

Pressurized modern technologies (which are more capital intensive) are more likely to be most profitable for new rather than existing operations. The discount rates used to derive annual irrigation capital costs have a significant impact on technology profitability. These technologies may provide incentives to conserve resources and reduce pollution if coupled with appropriate policies of input and output prices and regulations on pollution. It can also be realized that the economic advantages of a given irrigation technology depend on an array of factors – under or beyond the control of the grower. Therefore, the reader should be cautious in drawing general conclusions regarding a given irrigation technology.

13.3.3 On-Farm Management Practices for Drainage and Salinity Problems

Agricultural water pollution has externality effects caused by an upstream producer on either a downstream producer or the environment. Agriculture is considered a non-point source of pollution; therefore, the usually preferred control strategy is adoption of measures to reduce or eliminate the source of pollution. In this section we will discuss the use of the production function model for evaluation of several on-farm management options for the case where the farm has a high water table which must be controlled by a drainage system. The analysis assumes a profit maximizing decision maker, and treats social and environmental considerations exogenously to the model.

Several options are available for reducing and disposing of drainage water produced on a farm in the absence of a natural outlet or an external drainage facility. These include reducing applied water quantities per unit area, changing cropping patterns, reusing (either directly or mixing with fresh water) drainage water for irrigation of more salt-tolerant crops, constructing evaporation ponds, and improving irrigation application efficiency through changes in irrigation technology and management practices.

The farm has a limited land area which can be used for growing crops and constructing evaporation ponds, and it has a limited fresh water allotment that can be amended by reused drainage water. X_i is area of croping activity i which is a specific combination of crop type and quantity and quality of irrigation water. This definition allows a linear formulation of the problem.

$$w_i X_i = W_i + \sum_{j \in J_i} Q_{ji} \quad i = 1, \ldots, n \,. \tag{13.7}$$

Here, w_i is quantity of irrigation water per unit area applied to the ith cropping activity, W_i is total quantity of fresh water applied to ith cropping activity, Q_{ji} is quantity of drainage water from jth cropping activity which is reused on ith cropping activity, and J_i is the set of cropping activities that can supply drainage water (timing and quality) to the ith cropping activity.

The fresh water provided to the farm is of a given quality $\bar{C}$. Each crop can be irrigated by either fresh water only, reused drainage water of salt concentration d_j, or a mix of both sources that has a weighted concentration of C_i.

$$C_i w_i X_i = \bar{C} W_i + \sum_{j \in J_i} d_j Q_{ji} \quad i = 1, \ldots, n \,. \tag{13.8}$$

Quantity of drainage water produced by cropping activity i ($q_i X_i$) can either be reused on cropping activity j, or sent to on-farm (Q_i^p) and off-farm (Q_i^e) facilities. The quantity of lateral inflow from upstream farms that ends on that farm can be sent either to the on-farm evaporation pond (D^p) or to an off-farm facility (D^e). The area of the on-farm evaporation pond is constrained by engineering requirements and determined endogenously by the optimal solution to contain all drainage water sent to it. The drainage water sent to an off-farm facility is constrained by an annual quota.

Several additional mass balance and cropping pattern constraints are also implemented to keep the model result reasonable with known agrotechnical practices.

The problem addressed here is a dynamic one. The use of a seasonal production function was justified by the empirical fact that in a dynamic environment, optimal steady-state solutions were achieved after a relatively short period (Dinar and Knapp 1986) often which were maintained for the entire planning horizon.

Annual returns to land and management (π) of the farm are:

$$\pi = \sum_i (p_i^y y_i - F_i - T) X_i - p^w \sum_i W_i - \sum_i \gamma (Q_i^p + Q_i^e)$$

$$- \gamma (D^p + D^e) - \sum_j \sum_i \tau_{ij} Q_{ij}$$

$$- M - P^{cm} \cdot Z - a \cdot \sum_i d_i Q_i^p - a \cdot d' D^p, \tag{13.9}$$

where g is the variable cost associated with shipping drainage water to the on-farm pond and/or off-farm facility, and t_{ij} is the variable cost of drainage water reused from cropping activity i on cropping activity j. Capital cost associated with the reuse system is noted by M, a is the annualized cost of salt removal per unit concentration of drainage water, and d' is the salt concentration of lateral inflows to the farm.

The objective is to find values for X_i, W_i, Q_i^p, Q_i^e, D^p, D^e, Q_{ij}, and Z which maximize (9) subject to the constraints explained above. For more details the reader is referred to Dinar and Knapp (1988). The model provides economic and agronomic parameters for decision makers so that they can compare the effects of various potential regulations to be implemented on-farm.

Results for several on-farm strategies that were evaluated in Dinar and Knapp (1988) are presented in Table 13.4. The first management strategy simulates a situation where farmers do not change their current behavior, unless they are considering installation of an evaporation pond. In the second strategy, they consider changing water application rates, but all other practices remain as in strategy 1. Strategies 3 and 4 allow changes in water application rates and cropping patterns. However, strategy 3 does not consider reuse of drainage water, but strategy 4 does. Use of fresh water is 15 to 20% larger for strategy 1 than for the other strategies. Average relative yield is also the highest for strategy 1, but the difference in yields is not great. Returns to land and management are highest for strategies 2, 3 and 4 and lowest for strategy 1.

For comparison purposes, the model was also run assuming unlimited natural drainage and other conditions as in strategy 3. In this case cotton and sugar beets were grown to their limits and alfalfa took up the remaining area. Average optimal yields exceeded 99% of maximum. Total fresh water use on the farm was 268 ha-cm/yr, drainage volume was 87 ha-cm/yr and returns to land and management were $1,124 ha/yr. These values are all significantly larger than those for the farm with no natural drainage.

13.3.4 Taxes and Subsidies for Water Conservation and Pollution Reduction

Costs of pollution resulting from irrigated agriculture drainage are usually not apparent to those who create them. Drainage water disposal cost is not a determinant in the private decision making process, although society bears increasing environmental costs associated with agricultural pollution.

It is recognized that point-source pollution problems can be addressed by imposing direct taxes on the pollutant. However, agricultural drainage pollution is a non-point source problem, and as such it is difficult to identify and impose direct taxes. Dinar et al. (1989) explored, using the seasonal crop-water production function model, the consequences of various irrigation water pricing policies on optimal water use, generation of drainage volumes, and private and social benefits. Dinar and Letey (1989) extended

Table 13.4. Results for a farm with no natural drainage and no access to off-farm facilities under alternative management strategies

Variable	Management strategies			
	Fixed application rates	Variable application rates	Optimal management	
	Current crops No reuse	Current crops No reuse	No reuse	Reuse
Quantity of fresh water used (ha cm)	206.08	172.95	168.40	166.47
Size of evaporation pond (% of farmed area)	19	7	5	3
Cropping pattern (% of cropped area)				
Alfalfa	13	13	0	0
Wheat	3	3	10	0
Sugarbeet	6	6	20	20
Cotton	70	70	70	70
Barley	8	8	0	10
Weighted average yield (% of maximum yield)	98	94	95	97
Drainage water produced (ha cm/year)	66.69	24.01	17.91	35.47
Drainage water reused (ha cm/year)	–	–	–	24.82
Drainage water evaporated (ha cm/yr)	66.69	24.01	17.91	10.65
Returns to land and management ($/ha)	494	595	756	756

Table 13.5. Management practice incentives (irrigation technology subsidies) when the price of irrigation water is $2/ha cm and drainage disposal costs are $1/ha cm

Applied water ha-cm	Technology			
	Furrow 0.45 km	Furrow 0.9 km	Linear move/drip	LEPA
	---------------------($/ha/year)---------------------			
70	7.2	9.0	2.3	4.0
75	10.0	11.8	4.8	6.6
80	13.1	14.9	8.3	9.8
85	16.1	18.4	12.5	13.7
90	20.4	22.1	17.1	17.9
95	24.5	26.0	22.0	22.4
100	28.8	30.1	26.9	27.1
105	33.2	34.3	31.9	31.9
110	37.7	38.7	36.9	36.8

the study to include effects on private and social benefits of subsidies for irrigation technology equipment. Knapp et al. (1990) developed a general model that considered taxes and subsidies for both continuous (water, fertilizers, pesticides) and discrete (irrigation technology) inputs that may be associated with the non-point pollution problem. They employed the seasonal crop-water production function.

Their economic model was applied to cotton growth under conditions prevailing in the San Joaquin Valley of California. Several policies were analyzed in the three studies cited above. Results for only one application are presented here: suggested charges on various irrigation technologies for given irrigation and drainage water costs and applied irrigation water (Table 13.5). The charges per ha are simply the predicted drainage times drainage disposal costs. Furrow 0.9 km long is the most charged technology and linear move sprinklers and drip are the less charged technologies for each level of applied water. In general, these charges increase as applied water increases. Note that the technology specification is the same as in the analysis in Section 13.3.3.

Pricing policies have comparative advantages and disadvantages. In the case of a non-point source control, the advantage of using input-intensive policies is contingent upon the existence of a reliable crop-water production function system that provides the necessary information about physical production-pollution relationships.

13.3.5 Institutions for Water Conservation and Drainage Reduction

It has been proposed that institutions developed to manage water resources also affect the use efficiency of this resource (Frederick and Gibbons 1986).

Water marketing has been suggested, among other institutional changes, to
facilitate water transfers between agriculture and the urban sector (Wahl
1989) to increase water use efficiency. In addition to economic efficiency in
utilizing scarce water, water markets may be expected to lead to reduction in
environmental pollution created by a non-efficient water use.

A modification to the crop-water production function model was incorporated into an economic decision-making framework aimed at evaluating
the effects of a water market on water conservation and environmental
pollution from a private and social perspective (Dinar and Letey 1991).

The model considers the case of a profit-maximizing farmer having one
field and one crop over one growing season. The field has drainage problems
in the form of a high water table; and to maintain agricultural production
the farmer must collect and dispose of excessive drainage water. This is
associated with additional private and social costs. The farmer utilizes a
given irrigation technology. Crop yield and the amount of drainage water
produced are functions of the quantity of applied water. The farmer has a
fixed quantity of water which he purchases from a given source, and water
not utilized by the farmer can be sold in a water market. Assume that the
demand for water in that market is given and that actions by individual
farmers do not affect its price.

The farmer's profit function is:

$$\max \pi = y(w) \cdot P^y - w \cdot P^w - F - V(w) \cdot P^d$$
$$+ m \cdot P^m \text{ s.t. } m + w \leq B , \qquad (13.10)$$

where $V(w)$ is the drainage function and m is the surplus quantity of water
sold in the free market. B is the farmer's annual water allotment; P^d is the
cost associated with drainage disposal (in this case it may also include off-
farm cost), and P^m is the effective water market price ($P^m = P^a - P^w$) net of
purchasing price, where P^a is the actual water market price net of transaction costs. Water market transaction costs might be quite high. However,
for a transaction to be executed, the net price a farmer receives should at
least be greater than the price he paid for this water. For the purpose of the
current analysis, consider only the case where $P^a > P^w$.

For the purposes of our analysis assume that y and V are smooth
functions, twice differentiable having:

$$\frac{\partial y}{\partial w} \geq 0, \quad \frac{\partial^2 y}{\partial w^2} < 0; \quad \frac{\partial V}{\partial w} > 0, \quad \frac{\partial^2 V}{\partial w^2} > 0 . \qquad (13.11)$$

If $P^m > 0$, then $m = B - w$, meaning that the farmer will sell his remaining quota, and (13.11) becomes:

$$\pi(w) = y(w) \cdot P^y - w \cdot P^w - F - V(w) \cdot P^d + (B - w) \cdot P^m. \qquad (13.12)$$

Maximizing (13.12) with regard to water allocation provides:

$$\frac{\partial y}{\partial w}P^y - \frac{\partial V}{\partial w}P^d = P^w + P^m. \tag{13.13}$$

At the optimum, the net value of marginal product (which accounts for drainage) expressed by the left side of Eq. (13.13) equals the marginal revenue expressed by the right side of this equation.

In the case with no water market (simulated by $P^a = 0$), profits are determined only by P^y, P^d, and P^w. Increasing the level of P^a or P^w or both (such that $P^a > P^w$ still holds) has the same effect, resulting in decreased water used for irrigation, and increased water sold in the market. The extent to which this occurs is also affected by the level of P^d. Higher drainage water disposal costs lead to a decreased rate of water applied, and an increased quantity of water sold in the market. In contrast, under these conditions a water quota determines only the level of profit through $(B - w) \cdot P^m$. No change in optimal water application occurs.

The model in Dinar and Letey (1991) also analyzed the case (not presented here) where the farmer has the capability of selecting both the irrigation technology and the level of applied irrigation water.

The model was applied to conditions prevailing in the San Joaquin Valley of California that include range of values for water quota, water market price (value of zero indicates a non-market case), and cost of drainage disposal. Under essentially all conditions, increasing the water market price induced a reduction in water application and thus an increase in water available for transfer from agriculture to urban uses. Increasing water market price induced a reduction in drainage water compared to the no-market level with the consequence effect of reducing environmental degradation caused by drainage water. The water market provided a source of income to be used by the farmer to mitigate the costs of drainage disposal. Thus, the water market provided funding to invest in upgrading irrigation technology as well as to mitigate the environmental degradation associated with drainage water pollution.

The effect of a water market on social costs when drainage disposal costs are both internalized (borne by the farmer) and not internalized are presented in Table 13.6 for different values of water market price. The social benefits are always higher when drainage disposal costs are internalized, but this difference is reduced as the water market price increases.

13.3.6 Regional Drainage Quality and Quantity Control Policies

Policy analysis is usually performed on a regional scale rather than at the field or even farm levels. Regional approaches to water quantity/quality control are needed for two reasons: (1) agricultural pollution problems are,

Table 13.6. Private returns, social costs and net benefits for different water market prices when environmental costs are either internalized or not internalized

Actual mkt price	Drainage costs not internalized				Drainage costs internalized
	Profit	Drainage	Social cost	Social net benefits	Social net benefits
($/ha cm)	$/ha	cm	$/ha	$/ha	$/ha
0	1158	24	288	870	1052
3	1158	24	288	870	1071
5	1158	18	216	942	1111
8	1201	3	36	1165	1178
12	1242	2	24	1218	1226

at least in part, a result of regional water delivery and institutional systems; and (2) both surface and subsurface drainage create externality effects. Therefore, one should also include third party effects in the analysis; since irrigated agricultural pollution is a non-point source, the individual polluter is less likely to be identified. A regional approach may capture the total pollution more accurately.

The seasonal crop-water production function was incorporated into a given regional model aimed at evaluating resource use, pollution, and income effects of drainage-control policies. The model was applied to an agricultural region suffering drainage problems. The region is represented here as a water body (river) and a group of several farms served by a water district and subject to a regulatory agency. Either the district or the agency may impose drainage control policies. The water body that may serve to absorb the regional drainage effluent has a limited assimilative capacity.

A number of alternative crops can be grown on each farm, which can be irrigated with different combinations of water quantity and quality. Sub-surface drain tiles have already been installed in farms where shallow ground water and drainage problems affect farming. The district collects drain water from sumps on each farm for treatment and disposal. The disposal outlet may be constrained in both quality (salinity) and total volume allowed. The district may use part of its surface water allocation to dilute drainage if quality constraints are imposed.

Several on-farm and district-wide management options for reducing the agricultural drain water quantity and/or quality will be evaluated here. Most of these options have been considered and described at field and farm levels (Dinar and Knapp 1988). Individual farmers and the district can select one or a combination of these options in response to policy measures.

The regional model maximizes regional net income subject to the following constraints:

$$\sum_i X_{ij} \le (1 - I_j) \cdot L_j \quad \text{for each J} , \tag{13.14}$$

where i stands for crop and j for farm, L_j is available land, and I_i is a portion of land to be idle $(0 \leq I \leq 1)$. Idle land can be increased to represent a conversion of land to non-irrigated uses.

The seasonal crop-water production function model has been used to estimate continuous functions for yield, drainage volume, and quality (salinity). Relative yield (f_{ij}), deep percolation volume (v_{ij}), and salt concentration in the deep percolation water (d_{ij}) are functions of the quantity (a_{ij}) and salt concentration (c_{ij}) of applied water, the application uniformity (u_{ij}) measured by Christiansen uniformity coefficient (CUC), and environmental conditions expressed by pan evaporation during the growing season (e_{ij}).

$$y_{ij} = y_{ij}(a_{ij}, c_{ij}, u_{ij} \,|\, e_{ij}) \cdot Y_{ij} \,, \tag{13.15}$$

$$v_{ij} = v_{ij}(a_{ij}, c_{ij}, u_{ij} \,|\, e_{ij}) \,, \tag{13.16}$$

$$d_{ij} = d_{ij}(a_{ij}, c_{ij}, u_{ij} \,|\, e_{ij}) \,. \tag{13.17}$$

The variable y_{ij} is used here to express absolute crop yields. Y_{ij} represents the maximum potential yield that a given farm can achieve under optimal conditions, and reflects differences in management other than those considered in the production function. The pan evaporation variable allows the model to be transferred to any location (Letey and Dinar 1986).

The total amount of drainage water produced on farm j is:

$$V_j = q_j \sum_i v_{ij} X_{ij} - \sum_{n \neq j} \beta_{jn} q_j \sum_i v_{ij} X_{ij} + \sum_{k \neq j} \mu_{jk} q_k \sum_i v_{ik} X_{ik} \,, \tag{13.18}$$

where q_j represents the severity of the drainage problem on farm j, and $(0 \leq q_j \leq 1)$; $q_j = 1$ means that all deep percolation results in drainage. It is assumed that each farm has homogeneous soil properties, so q_j represents on-farm drainage conditions. β_{jn} $(0 \leq \beta_{jn} \leq 1)$ is the fraction of drainage produced on farm j that ends beneath farm n $(n \neq j; \sum \beta_{jn} = 1)$, and μ_{jk} $(0 \leq \mu_{jk} \leq 1)$ is the fraction of drainage from farm k that arrives to farm j $(k \neq j; \sum \mu_{jk} = 1)$. Subsurface lateral flow is one source of externality effects within the region. Where there are no lateral drainage flows, $\beta_{jn} = 0$ and $\mu_{jk} = 0$ for each n and k.

Annual quantity of drainage water that the district must handle is:

$$\sum_j V_j \leq V^d, \tag{13.19}$$

where V^d is the maximum allowed discharge by the district.

A quality (salinity) standard (C^d) may be imposed on discharged drainage. If the salinity exceeds that standard, the district must dilute the drainage with fresh water of a better quality. The quality constraint is:

$$\left[\sum_j \sum_i d_{ij} V_j + S^d \cdot C^s \right] \Big/ \left[\sum_j V_j + S^d \right] \leq C^d, \tag{13.20}$$

where S^d is the amount of surface water with a given quality C^s (C^s is a better quality than d_{ij}) used by the district for dilution. Both the quantity and the quality constraints (as a matter of fact, the product D^dC^d is the pollution load) reflect the assimilative capacity value that society assigns to the water body.

Each farm has an annual quota (S^{fj}) of fresh water (also of quality C_s) provided by the district. Farms can supplement their surface supply by pumping groundwater. Additional technical mass balance equations related to surface and ground water quantity and quality are not presented here.

Uniformity of applied water is used as a surrogate for irrigation technology and irrigation management activities, with a more advanced technology being associated with a higher CUC value. Higher CUC values are associated with greater costs in irrigation hardware and/or management. The total irrigation cost (except for the cost of the water) is:

$$K_{ij} = r_{ij}(u_{ij}) \cdot X_{ij} , \tag{13.21}$$

where K_{ij} is the annual irrigation cost. It is assumed that $\partial r/\partial u > 0$ and $\partial^2 r/\partial u^2 \geq 0$, meaning that the cost of achieving a better irrigation uniformity application is increasing. It is assumed also that $\partial K/\partial X = \text{Const}$; that is, no economies of scale exist with regard to the size of the irrigated field.

The district purchases surface water for a given price per unit volume (P^{ds}) and then provides it to the farmers for a given price of P^{sij} per unit volume. The model allows the district to discriminate among farms and crops. The district may try to control water consumption by increasing and decreasing this price. In addition, the model allows the district to charge either a flat or a tiered rate for water.

$$Pw_{ij} = \begin{array}{ll} P & \text{for } S_{ij}/X_{ij} \leq H_{ij} \\ (h(S_{ij}/X_{ij}) & \text{for } S_{ij}/X_{ij} > H_{ij} \end{array}, \tag{13.22}$$

where Pw_{ij} is the price per unit volume of surface irrigation water applied and H_{ij} is a parameter determining the maximum amount of water per unit area that will be charged the basic rate ($P \geq Pds$ and includes only the overhead of the district). The function v has a positive first derivative with regard to the per unit area water volume.

The district can also impose (or relay) a tax (H_j) on volume of drainage created by each farm. This is done assuming that the district monitors each farm's outlet and that the monitoring costs are either zero or are already included in the district services charged to the farms.

The objective function of the region can be formulated as follows:

$$\text{Max } P = \sum_i P^{yi} \sum_j y_{ij} X_{ij} - \sum_j \sum_i F_{ij} X_{ij} - P_p \sum_j V_j - \sum_j \sum_i K_{ij}$$

$$- \sum_j P^{gj} \sum_i G_{ij} - \sum_j \sum_i Pw_{ij} S_{ij} - P^{ds} S^d - H_j \sum_j V_j . \tag{13.23}$$

Here P is the regional net income, P^{gj} is the cost of pumping ground water, and P^{ds} is the cost of diluting the drainage water to be discharged (assuming no additional dilution cost except fresh water price to the district). All other variables are as defined earlier.

Several policies aimed at reducing drainage quantity and salinity resulting from the agricultural activity may be considered. A case with no regulation is used as a base for comparison. The policies include restrictions on pollution in the form of: (1) volume of drainage water; (2) concentration of pollutants in the drainage water; and (3) load of pollutant to be disposed of to the environment. Policy instruments include quotas, standards, and taxes on input and output levels. These policies have been evaluated by Dinar et al. (1990, 1991).

Only a few examples are presented in Table 13.7. Regional income, acres farmed, surface and ground water used (for irrigation and dilution), drainage volume and quality disposed of, and regional taxes collected from drainage and/or irrigation water fees are determined by the policy instruments. Some policy instruments such as drainage tax are superior to others in reducing drainage volumes. Regional income was most sensitive to drainage and water taxes (including redistribution of collected taxes that must stay in the region). Regional income is least sensitive to load constraint.

Farmers respond to the policy instruments by changing crop mix and by increasing the share of ground water used in the irrigation water for the case where surface water supply is constrained and no restriction on drainage quality is imposed. In such cases, this policy may result in increased pollution. It was found (not presented) that farmers do not prefer to switch technologies, particularly because of the relatively high cost of improved technologies. The farmers shifted to more efficient technologies only when the policy instrument values were very constraining.

13.4 Discussion and the Need for Further Development

The production process of agricultural commodities involves desicions which determine levels of multiple inputs. In certain cases, the output may have several components, some of which have negative external effects. Production functions have been developed and used in the literature as a means to analyze many management and policy issues. This chapter has demonstrated the application of a particular crop-water production function model for a variety of management and policy issues associated with irrigated crops. These issues include field, farm, and regional-level decisions, covering a spectrum of variables such as irrigation rate, mix of water qualities, cropping patterns, investment in irrigation technologies and drainage facilities, water markets, regional arrangements, standards, taxes on agricultural pollution, and constraints on input levels.

Table 13.7. Regional income, acres farmed, applied surface and ground water, drainage quantity and quality, and collected taxes, as affected by policy measures

Policy and policy var. value	Regional income[a] (10^6)	Acres farmed (ha)			Applied Water (10^3 ha cm)		Drainage		Collected taxes (10^6)
		Alf.	Cot.	Tom.	Surface	Ground	Quantity	Water Quality (EC)	
No regulation	1148.4	200	2174	1150	355.5	53.7	63.0	8.3	0
Water price (flat)									
$3/ha cm	970.7	200	2174	1150	355.5	53.2	62.8	8.6	177.7
$6/ha cm	300.4	200	1425	1150	101.8	198.9	33.2	14.7	356.2
Drainage fee (flat)									
$10/ha cm	763.1	200	1425	1150	255.3	23.1	24.6	21.5	246.8
$40/ha cm	417.1	200	1106	937	204.7	2.5	10.4	56.7	417.1
Surface water quota									
300 000 ha cm	1108.0	200	1713	1150	300.0	53.1	50.7	9.15	0
200 000 ha cm	892.0	200	1124	950	200.0	57.9	29.2	11.3	0
Drainage constraint									
60 000 ha cm	1146.0	200	2177	1150	357.6	46.8	60.0	9.4	0
30 000 ha cm	902.3	400	1842	891	340.8	8.5	40.0	11.2	0

[a] Not including redistribution of taxes.

It was demonstrated that a policy maker may benefit from employing production functions, such that the one in this chapter, for a variety of analyses. However, there is still a need for additional research to address remaining problems. First, there is a vector of residuals resulting from irrigated agriculture. Although it is not simple to model the various "output" components, it is desirable to include them in any management or policy analysis. Currently, this kind of development is not available for the policy maker; those production models that do include some of the components are often cumbersome and not compatible with management. Second, production processes are dynamic in nature. This is true both within a particular season and between seasons. Some changes in the biological systems associated with the production process are irreversible. In the case of salinity, the approach we employ can be justified under certain conditions, but this is not the general case. Some agricultural residuals accumulate overtime in the soil and water to potentially harmful levels for human beings, plants, or animals. Therefore, future research should emphasize the dynamic aspects of the production and pollution processes associated with management and policy. Third, the literature has less strength, to some extent, in modeling production of perennial crops. Fruit crops consume roughly twice the amount of water needed for field crops on the same land unit. Therefore, a greater potential exists for improvements in efficiency and pollution reduction for perenial crops, which produce higher levels of revenue and are more vulnerable to environmental risks. An effort to develop production functions for perennial crops would perhaps be a timely one. And finally, the approaches to model production functions are usually deterministic ones. It is argued that in agricultural production, uncertainty matters (in all respects–environmental conditions, prices, information). This element, if brought under control, will probably improve our ability to address the production process.

References

Batie SS, Shabman LA, Kramer RA (1985) US agriculture and natural resource policy. In: Price KA (ed) The dilemmas of choice. Resources for the Future, Washington, DC, pp 127–146

Bradford S (1991) Transient state modeling of various irrigation management strategies. MS Thesis, Dep Soil Environ Sci, Univ Cal, Riverside

Cardon GE (1990) A transient-state model of water and solute movement and root water uptake for multi-seasonal crop production simulation. PhD Diss, Dep Soil Environ Sci, Univ Cal, Riverside

Dinar A, Knapp KC (1986) A dynamic analysis of optimal water use under saline conditions. W J Agric Econ 11(1):58–66

Dinar A, Knapp KC (1988) Economic analysis of on-farm solutions to drainage problems in irrigated agriculture. Aust J Agric Econ 32(1):1–14

Dinar A, Letey J (1989) Economic analysis of charges and subsidies to reduce agricultural drainage pollution. In: 2nd Pan-Am Reg Conf Toxic substances in agricultural water supply and drainage – an international environmental perspective, Ottawa, Can, June 8–9, 1989

Dinar A, Letey J (1991) Agricultural water marketing, allocative efficiency and drainage reduction. J Environ Econ Manag 20:210–223

Dinar A, Zilberman D (1991) The economics of resource-conservation, pollution-reduction technology selection: the case of irrigation water. Resour Energ 13:323–348

Dinar A, Letey J, Knapp KC (1985) Economic evaluation of salinity drainage and non-uniformity of infiltrated irrigation water. Agric Water Manag 10:221–33

Dinar A, Letey J, Vaux HJ Jr (1986a) Optimal ratios of saline and non-saline irrigation waters for crop production. Soil Sci Soc Am J 50(2):440–43

Dinar A, Knapp KC, Rhoades JD (1986b) Production function for cotton with dated irrigation quantities and qualities. Water Resour Res 22(11):1519–25

Dinar A, Knapp KC, Letey J (1989) Irrigation water pricing to reduce and finance subsurface drainage disposal. Agric Water Manag 16:155–171

Dinar A, Hatchett SA, Loehman ET (1990) A model of regional drainage quality and quantity control. In: 3rd Interdisciplinary Conf Natural resource modeling & analysis, Cornell Univ, Ithaca, NY, Oct 11–13, 1990

Dinar A, Rhoades JD, Nash P, Waggoner BL (1991) Production functions relating crop yield, water quality and quantity, soil salinity and drainage volume. Agric Water Manag 19(1):51–66

Frederick KD, Gibbons DC (eds) (1986) Scarce water and institutional change. Resources for the Future, Washington, DC

Johanson Per-Olov (1987) The economic theory and measurement of environmental benefits. Univ Press, Cambridge NY

Just RE (1991) Estimation of production systems with emphasis on water productivity. In: Dinar A, Zilberman D (eds) The economics and management of water and drainage in agriculture. Kluwer, Boston, pp 251–274

Knapp KC (1991) Optimal intertemporal irrigation management under saline, limited drainage conditions. In: Dinar A, Zilberman D (eds) The economics and management of water and drainage in agriculture. Kluwer, Boston, pp 599–616

Knapp KC, Dinar A (1988) Production with optimum irrigation management under saline conditions. Eng Costs Prod Econ 14(1):41–46

Knapp KC, Dinar A, Letey J (1986) On-farm management of agricultural drainage problems: an economic analysis. Hilgardia 54(4):1–31

Knapp KC, Dinar A, Nash P (1990) Economic policies for regulating agricultural drainage water. Water Resour Bull 26(2):289–298

Knapp KC, Stevens BK, Letey J, Oster JD (1991) A dynamic optimization model for irrigation investment and management under limited drainage conditions. Water Resour Res 26(7):1335–1343

Kneese AV (1984) Measuring the benefits of clean air and water. Resources for the Future, Washington, DC

Letey J (1991) Crop-water production function and the problems of drainage and salinity. In: Dinar Z, Zilberman D (eds) The economics and management of water and drainage in agriculture. Kluwer, Boston, pp 209–227

Letey J, Dinar A (1986) Simulated crop-water production functions for several crops when irrigated with saline waters. Hilgardia 54(1):1–32

Letey J, Vaux HJ Jr, Feinerman E (1984) Optimum crop water application as affected by uniformity of water infiltration. Agron J 76:435–441

Letey J, Dinar A, Knapp KC (1985) Crop-water production function model for saline irrigation waters. Soil Sci Soc Am J 49(4):1005–9

Letey J, Dinar A, Woodring C, Oster J (1990) An economic analysis of irrigation systems. Irrig Sci 11:37–43

McCalla AF, Learn EW (1985) Public policies for food, agriculture and resources: retrospect and prospect. In: Price KA (ed) The dilemmas of choice. Resources for the Future, Washington, DC, pp 1–22

Portnoy PR (ed) (1991) Public policies for environmental protection. Resources for the Future. Washington, DC

Rhoades JD (1984) Use of saline water for irrigation. Cal Agric 38:42–43

Rhoades JD (1987) Use of saline water for irrigation. Water Qual Bull 12:14–20

Rhoades JD, Bingham FT, Letey J, Hoffman GI, Dedrick AR, Pinter PJ, Replogle JA (1987) Use of saline drainage water in irrigation: Imperial Valley study. Agric Water Manag 16:25–36

Smith CK, Desvousges WH (1986) Measuring water quality benefits. Kluwer, Boston

Steiner JL, Williams JR, Jones OR (1987) Evaluation of the EPIC simulation model using a dryland wheat-sorghum-fallow crop rotation. Agron J 79(4):732–38

Vaux HJ, Pruitt WO (1983) Crop-water production functions. In: Advances in irrigation, vol 2. Academic Press, New York, pp 61–97

Vinten AJA, Frenekl H, Shalhevet J, Elston Da (1991) Calibration and validation of a modified steady-state model of crop response to saline water irrigation under conditions of transient root zone salinity. J Contamin Hydrol 7:123–144

Wahl RW (1989) Markets for federal water. Resources for the Future, Washington, DC

Weinberg M, Wiley Z (1991) Creating economic solutions to environmental problems of irrigation and drainage. In: Dinar A, Zilberman D (eds) The economics and management of water and drainage in agriculture. Kluwer, Boston, pp 531–556

Weinberg M, Kling C, Wilen J (1992) Analysis of policy options for the control of agricultural pollution in California's San Joaquin River Basin. In: Russell C, Shogren J (eds) Theory, modelling and experiencing in the management of non-point source pollution. Kluwer, Boston (in press)

Yaron D, Dinar A (1982) Optimal allocation of irrigation water on a farm during peak season. Am J Agric Econ 64:681–89

Subject Index

Printing: Saladruck, Berlin
Binding: Buchbinderei Lüderitz & Bauer, Berlin